Green Energy and Technology

Climate change, environmental impact and the limited natural resources urge scientific research and novel technical solutions. The monograph series Green Energy and Technology serves as a publishing platform for scientific and technological approaches to "green"—i.e. environmentally friendly and sustainable—technologies. While a focus lies on energy and power supply, it also covers "green" solutions in industrial engineering and engineering design. Green Energy and Technology addresses researchers, advanced students, technical consultants as well as decision makers in industries and politics. Hence, the level of presentation spans from instructional to highly technical.

Indexed in Scopus.

Indexed in Ei Compendex.

Alois Peter Schaffarczyk

Introduction to Wind Turbine Aerodynamics

Third Edition

 Springer

Alois Peter Schaffarczyk
Department of Mechanical Engineering
Kiel University of Applied Sciences
Kiel, Schleswig-Holstein, Germany

ISSN 1865-3529   ISSN 1865-3537 (electronic)
Green Energy and Technology
ISBN 978-3-031-56926-5   ISBN 978-3-031-56924-1 (eBook)
https://doi.org/10.1007/978-3-031-56924-1

This Springer imprint is published by the registered company Springer Nature Switzerland AG
The registered company address is: Gewerbestrasse 11, 6330 Cham, Switzerland

Paper in this product is recyclable.

This book is dedicated to my wife Ulrike and our children Claudia, Carola and Kai.

Foreword

The modern development of wind power is a remarkable story of the combined effort of enthusiastic entrepreneurs and skilled engineers and scientists. Today, wind power forms the most rapidly advancing renewable energy resource with an annual growth rate of about 30. The aerodynamics of wind turbines concerns modeling and prediction of the aerodynamic forces on the solid structures of a wind turbine, and in particular on the rotor blades of the turbine. Aerodynamics is the most central discipline for predicting performance and loadings on wind turbines, and it is a prerequisite for the design, development and optimization of wind turbines. From an outsider's point of view, the aerodynamics of wind turbines may seem simple as compared to the aerodynamics of, e.g. fixed-wing aircraft or helicopters. However, there are several added complexities. Most prominently, the aerodynamic stall is always avoided for aircraft, whereas it is an intrinsic part of the wind turbine operational envelope. Furthermore, wind turbines are subjected to atmospheric turbulence, wind shear from the atmospheric boundary layer, wind directions that change both in time and in space and effects from the wake of neighboring wind turbines. *Introduction to Wind Turbine Aerodynamics*, written by an experienced teacher and researcher in the field, provides a comprehensive introduction to the subject of wind turbine aerodynamics. It is divided into ten chapters, each giving a self-contained introduction to specific subjects within the field of wind turbine aerodynamics. The book begins with an overview of different types of wind turbines, followed by a comprehensive introduction to fluid mechanics, including different forms of the basic fluid mechanical equations and topics such as potential flow theory, boundary layer theory and turbulence. The momentum theory, which still forms the backbone in the rotor design of wind turbines, is presented in a separate chapter, which, in a compressed form, describes the basic momentum theory as well as the important blade element momentum theory. State-of-the-art advanced aerodynamic models are also presented. This includes vortex models as well as computational fluid dynamics (CFD) techniques and an introduction to the most common turbulence models. The last part of the book gives an overview of experiments in wind turbine aerodynamics and the impact of aerodynamics on blade design. Each chapter contains some homework problems and concludes with a bibliography. *Introduction to Wind Turbine Aerodynamics* is a highly topical and timely textbook of great value for undergraduate students as well as for trained engineers who wish to get a fast introduction to the many aspects of the amazing world of wind turbine aerodynamics.

April 2014

J. N. Sørensen
Technical University of Denmark
Kongens Lyngby/Copenhagen
Denmark

Preface to the Third Edition

Thanks to Springer (especially A. Doyle, Springer, UK), this third edition was prepared by reflecting the progress made since the appearance the second edition.

During the last four years obviously, a lot of events have heavily influenced energy policy which resulted in, for example the so-called Esbjerg declaration, in which a huge trans-European project to promote off-shore wind energy in the North Sea was set forward.

In my opinion progress in wind turbine aerodynamics was most notable for

- a 16 MW prototype with 246 meter rotor diameter was erected,
- on-shore machines of 3 MW are now state of the art,
- much more investigation on transition on wind turbine blades,
- a new IEAwind Annex 41 *turbinia* was established,
- a rather extensive handbook of wind energy aerodynamics [1] appeared.

Statistics were up-dated whenever reasonable data was available.
The following new sections were added:

- An aerodynamics analysis and design code is described in much detail,
- An vortex ring code for a constantly loaded actuator disk is presented,
- A section of Basics Wind Farm Aerodynamics is added and,
- Section 8.1.2 was enlarged for a more detailed discussion on recent developments on very thick airfoils.

Kiel, Germany

Alois Peter Schaffarczyk

December 2023

Reference

1. B. Stoevesandt, et al. *Handbook of wind energy aerodynamics*. Springer, Cham, Switzerland, 2022.

Preface to the Second Edition

About five years have passed since the first edition appeared. Thanks to Springer (esp. Dr. Baumann) a 2nd edition was prepared reflecting the progress made during this time. As predicted by Ben Backwell [1] the market sees some manufacturers to disappear and a lot of merging. Nevertheless Renewables will account for about 200 GW **new** installed capacity in 2019 with 100 GW from Photovoltaics, 55 GW from onshore and about 5 GW from off-shore wind energy.

To my opinion progress in wind turbine aerodynamics was most notable for

- a 12 MW machine with 220 meter rotor diamater is now in operation,
- on-shore machines of 3 MW are now state-of-the-art,
- much more investigation on transition on wind turbine blades,
- IEAwind Annex 29 was extended to *Aerodynamics or Analysis of Aerodynamic Measurements*.

Statistics were up-dated whenever reasonable data was available. Values in Table 5.1 were corrected (My thanks go to Emmanuel Brandlard, NREL). A new Section 4.6 about artificial turbulence esp. for Large Eddy Simulation was prepared by B. Lobo.

Kiel, Germany
December 2019

Alois Peter Schaffarczyk

Reference

1. B. Backwell. *Wind Power—The struggle for control of a new global industry*. earthscan, Oxon, UK, 2nd Ed., 2018.

Preface to the First Edition

Wind energy has flourished over the last 20 years. At the time of this writing, approximately 40 GW of wind power production capacity, equivalent to 80 Billion US Dollars in investment, is added to the worldwide energy portfolio annually.

About 240000 (IEAwind 2012 annual report) engineers are working in this field, and many of them are confronted with challenges related to aerodynamics. Even special academic programs (one of them being the MSc Program in Wind Engineering at the consortium of universities in Schleswig-Holstein, the northernmost province of Germany) now are educating young people in this field. From personal experience, the author knows how difficult it is to navigate through the jungle of scientific literature without being injured before having learned all the basics. This textbook intends to help young and old interested people to gain easier access to the basics of the fluid mechanics of wind turbines.

The scope of this book is to take the reader on a interesting journey from the basics of fluid mechanics to the design of wind turbine blades. The reader's basic knowledge of mathematics (Vector calculus) seems to be necessary for understanding, whereas fluid mechanics will be explained in what amounts to a crash course. But there should be no mistake: fluid mechanics being *non-linear*, even as an applied science, is a complicated affair and can only be learned and understood by one's **own** problem solving. This is especially true when investigating *turbulent* flow which is of utmost importance when investigating mechanical loads on wind turbine parts. Therefore, the estimated amount of effort necessary for solving our selected problems at the end of each chapter ranges from 10 minutes to 10 months.

The author will try to give the beginning reader a glimpse of established knowledge about the aerodynamics of wind turbines. For the more familiar reader, this book will facilitate the discovery of (and hopefully understanding of) relevant advanced papers. In any event, all readers will glean a deeper understanding which will allow them to perform analyses either by hand or by using computer codes.

The structure is as follows: an introduction is given to inform the reader about the specific relevance of wind turbine aerodynamics, while touching on some interesting historical examples. Then a chapter on installed windmills and wind turbines is added to describe the basics of mainstream and unconventional technology. After that we are ready to tackle our general framework: the laws of fluid mechanics. I will try to limit the contents to those topics which are closely related to wind, understood as turbulent flow in the lower (boundary layer) part of the atmosphere. Because the rotor swept area now easily reaches several *hectares* ($= 10^4$ m^2) something must be said about real-world inflow conditions, which are not constant from a spatial or temporal point of view. Chapter 5 is the core of the book and explains the various versions of *Momentum Theory* as well as its limitations, followed by a description of applications of its counterpart: *Vortex Theory*. After exhausting these classical methods, we then have to present the modern approach of solving the differential equations called *Computational Fluid Dynamics* in some detail. Many efforts have been expended over the last 15 years to bridge the gap from older, sometimes called simpler, theories to actual measurements using these computational methods. This leads directly to a transition in the next chapter to a discussion of free-field and wind tunnel measurements. After that we will try to get even closer to practical, real things. This means that we try to give examples of actual blade shapes from industry. We conclude with summarizing remarks and an overview of possible future developments.

As this text grew out of my lectures on wind turbine aerodynamics in the first year of our international Master of Science Program, it is targeted toward this student audience. Nevertheless I have tried as much as possible to make it valuable and informative for a much broader group of readers.

Kiel, Germany   Alois Peter Schaffarczyk
April 2014

Acknowledgments

I want to thank several people for introducing me to wind turbine aerodynamics: First of all Sönke Siegfriedsen from aerodyn Energiesysteme GmbH, Rendsburg and Gerard Schepers, ECN, the energy research center of the Netherlands, who gave me a first chance in 1997. My very first contact with wind energy was actually somewhat earlier (1993) during supervising a Diploma-Thesis which is now summarized as the *strange farmer* problem (see Problem 5.1). Participation in several IEAwind tasks (11, 20 and 29) gave me the opportunity to interact with leading international experts. State of Schleswig-Holstein, Federal German and European funding helped contribute to important projects and finally an International M.Sc. program in *Wind Engineering* was established in 2008 in which now more than 160 young people from all around the world (hopefully) learned about wind energy technology.

Mr. Lippert prepared most of the graphs and Mr. John Thayer brought the text into more correct English. Any errors, however, are the responsibility of the author only.

Contents

Nomenclature

$<\bullet>$	Averaging
u_τ	Friction velocity
α	Angle-of-attack
α, β	Component counting tensor indices
β	Camber of Joukovski airfoil
$\bullet^+$	Dimensionless turbulent quantities
$\delta(\mathbf{u}r)$	Structure function
$\delta^\star$	Displacement thickness of boundary layer
$\dot{m}$	Mass flow, unit kg/sec
ℓ	Arbitrary extension of profile
ℓ	Pitch in
ℓmix	Prandtl's mixing length
ϵ	Dissipation
η	Dynamic viscosity
η	Kolmogorov's length scale
η	Similarity parameter: normalized distance from wall
Γ	Circulation
$\Gamma(x)$	Gamma function
$\gamma(x)$	Linear distributed vorticity
κ	Adiabatic exponent
κ	Vortex strength of a vortex sheet
Λ	Pressure parameter in wedge-flow
λ	Parameter of Joukovski's transformation
λ	Tip-speed-ratio
λ_T^2	Taylor's microscale
$\mathcal{F}$	Complex (2D) force
μ	Aspect ratio
∇	3D vector of Cartesian partial derivatives
ν	Kinematic viscosity
ν_t	Turbulent viscosity
Ω	Angular velocity of turbine

Ω	Rotational velocity of turbine (rad/sec)
ω, ω_1	Angular wake velocity at disk and far downstream
Φ_{ij}	Spectral tensor
ψ	Stream function
$\psi_m(z/L)$	Function specifying the state of the ABL
ρ	Density of air, in most cases about $1.2 \ \mathrm{kg/m^3}$
$\rho(s)$	Autocorrelation
σ	Solidity of blade
σ_v	Variance of wind-speed measurement
τ	Shear stress
τ_w	Wall shear stress
$\tau_{\alpha\beta}$	Reynold's stress tensor
θ	Momentum thickness of boundary layer
Φ	Velocity potential
φ	Flow angle
ζ	Complex quantity
A	Constant in Hill's spherical vortex
a	Axial induction factor
a'	Tangential induction factor
A, k	Scaling factors in Weibull distribution
A_n	Fourier coefficients in thin airfoil theory
A_r	The reference area for capturing wind
c	Chord of Joukovski airfoil
c	Velocity of sound
C^∞	Hilbert space of infinitely differentiable functions
c_D	Drag coefficient
C_f	Friction coefficient
C_K	Kolmogorov's constant
c_L	Lift coefficient
c_N	Normal force coefficient
c_P	Fraction of the power of the wind converted to useful power
c_p	Specific heat at constant pressure
c_{D+}	Drag coefficient in wind direction
c_{D-}	Drag coefficient opposite to wind direction
c_{tan}	Tangential force coefficient
$D = 2R_{tip}$	Rotor diameter
D	Drag force in wind direction
d	Spacing of surrogate model, see Eq. (12.66)
$E(k)$	Elliptic integral
F	Prandtl's tip (loss) factor
$f(r)$	Function in Eq. (3.107)
$F(z)$	Complex 2D velocity potential
G	Goldstein function
g	Earth's acceleration
$g(r)$	Function in Eq. (3.107)

H	Shape parameter
H	Turbulent flux
I	Turbulence intensity
J	Advance ratio
K	Drag coefficient for wind vehicle
k	Kinetic turbulent energy
k	Von Karman's constant
$K(k)$	Elliptic integral
L	Integral correlation length
L	Lift force perpendicular to inflow velocity
L	Monin-Obukhov length-scale
M	Torque (or moment)
M	Torque of rotor
N	Amplification exponent
N	Number of disks
P	Probability
p_0	Ambient pressure
P_{Wind}	The power of wind
Q	Torque on low-speed shaft
$Q_{\alpha\beta}(\mathbf{r})$	Correlation tensor
r_2, r_3	Distance of regarded streamline form center at disk and far downstream
R_i	Ideal gas constant
Re	Reynolds number: local ratio between viscous and inertia forces
Re_x	Reynolds number with respect to a running length
$S(\mathbf{r})$	Source function for pressure from Poisson's equation, Eq. (3.60)
S_n	Structure functions
T	Absolute temperature in K(elvin)
t	Thickness of Joukovski airfoil
T_R	Return time of maximum wind speed
Ti	Turbulence intensity
u	Speed of moving vehicle
u'	Fluctuating part of velocity
$U = <u(t)>$	Averaged velocity
u_1	Axial velocity far up-stream
u_2	Axial velocity at disk
u_3	Axial velocity far down-stream
u_e	Velocity at the edge of the boundary layer
U_S	Normalized gust wind-speed
v	Wind velocity
z	Vertical coordinate, positive upwards
z_0	Roughness heigh, location where wind is zero
κ	Wave vector
ω	Local vorticity
σ	Cauchy stress tensor

A	Vector potential
a	Acceleration in an Eulerian (Earth fixed) frame of reference
b	Unit bi-normal vector of a space curve
D	Deviation (deformation) tensor
F	Force in Newton's second law
H	Bernoulli's constant or specific enthalpy
n	Unit normal vector of a space curve
T	Thrust—force in wind direction
t	Unit tangent vector of a space curve

List of Figures

Chapter 1
Introduction

1.1 The Meaning of Wind Turbine Aerodynamics

Wind energy has been used for a long time. In this chapter of [1] gives a short account of this history and describes the first scientific approaches. It is immediately seen that even this applied branch of fluid mechanics suffers from a general malady: Prior to the twentieth century, mathematical descriptions and practical needs were far removed from one another. Impressive reports about the history of fluid mechanics can be found in [3, 4, 6]. There it is shown how this dilemma was resolved at least partly. We will follow this approach and will try to combine applied theory with practical needs as far as possible.

How much power is in the wind ? Imagine an area—not necessarily of circular shape—of A_r. The temporally and spatially constant wind may have velocity v perpendicular to this area; see Fig. 1.1. Within a short time dt this area moves by $ds = v \cdot dt$. This then gives a volume $dV = A_r \cdot ds$. Assuming constant density $\rho = dm/dV$ we have a mass of $dm = \rho \cdot dV = \rho A_r v \cdot dt$ in this volume. All velocities are equal to v, so the kinetic energy content $dE_{kin} = \frac{1}{2} dm \cdot v^2$ increases as the volume linearly with time. Defining power as usual by $P := dE/dt$ we summarize:

$$P_{wind} = \frac{dE}{dt} = \frac{1}{2} \cdot \rho \cdot A_r \cdot v^3 \; . \tag{1.1}$$

We may transfer this to an energy density (W/m^2) by dividing by the reference area A_r. Using an estimated air-density of 1.2 kg/m^3 we see that for $v = 12.5$ m/s this power-density matches the so-called solar constant of about 1.36 kW/m^2. This energy-flux ($J/m^2 \cdot s$) is supplied by the sun at a distance of one astronomical unit (the distance from the sun to the earth) and may be regarded as the natural reference unit of all renewable power resources.

As $p \cdot \dot{V}$ (static pressure times volume flow) also gives power, it is sometimes argued (in Sect. 5.2) that by decreasing ambient pressure (around 1013 hPa) up to

A. P. Schaffarczyk, *Introduction to Wind Turbine Aerodynamics*, Green Energy and Technology, https://doi.org/10.1007/978-3-031-56924-1_1

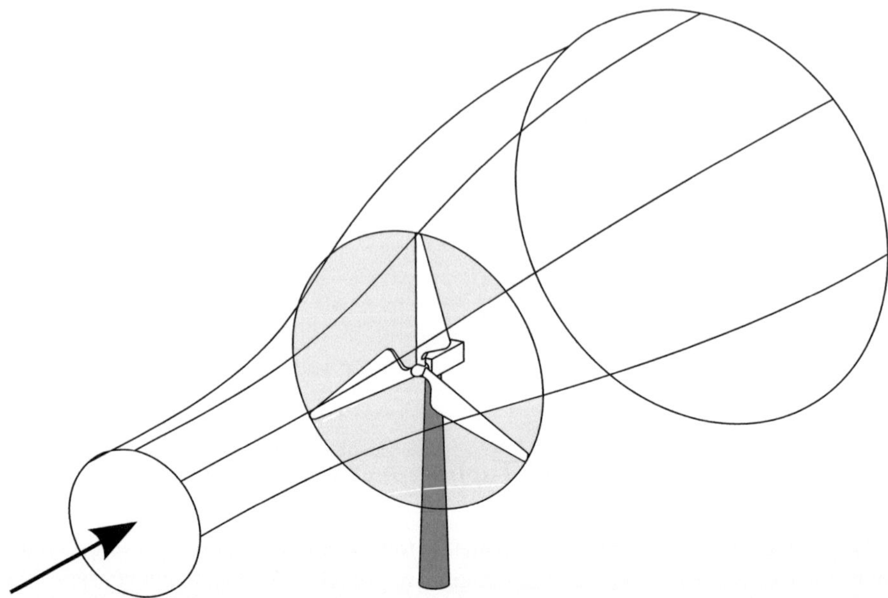

Fig. 1.1 Wind turbine with slip stream

far downstream even more power may be extracted. We will come back to this idea later (See Sect. 5.3).

Everybody knows that wind is NOT constant in time, and therefore statistics must be involved. In the simplest form, a family of two-parameter functions is used:

$$P(v; A, k) := \frac{k}{A} \left(\frac{v}{A}\right)^{k-1} \exp\left(-\frac{v}{A}\right)^{k}. \tag{1.2}$$

is a commonly used probability distribution attributed to Weibull.[1] A is connected with the annual averaged mean velocity $\bar{v}$ and k is the shape factor with values between 1 and 4. For most practical applications:

$$\bar{v} := \int_0^\infty P(v; A, k)v \cdot dv \approx A \left(0.568 + \frac{0.434}{k}\right)^{\frac{1}{k}}. \tag{1.3}$$

Figure 1.2 shows some examples of Weibull-Distributions with shape parameters varying from 1.25 to 3.00. In Figs. 1.3 and 1.4 we compare fitted (by the least square algorithm) distributions with a measured distribution. We see that the 2008 data fits rather well whereas (note the logarithmic y-scale !) 2011 is much worse. Thus, unfortunately we must conclude that the distributions derived from annual data

---

[1] A Weibull distribution with k = 2 is called a *Rayleigh-distribution*.

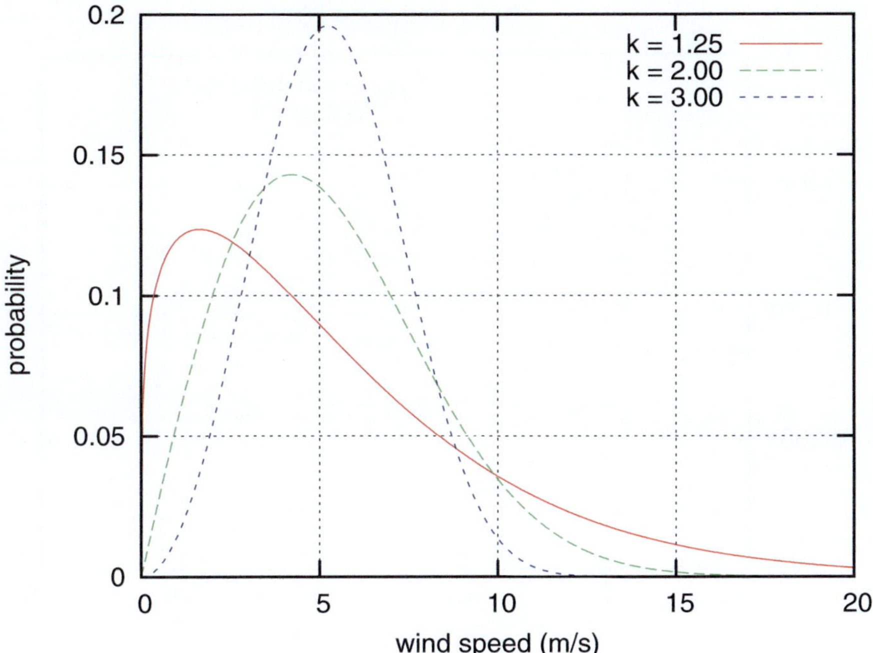

Fig. 1.2 Weibull distributions

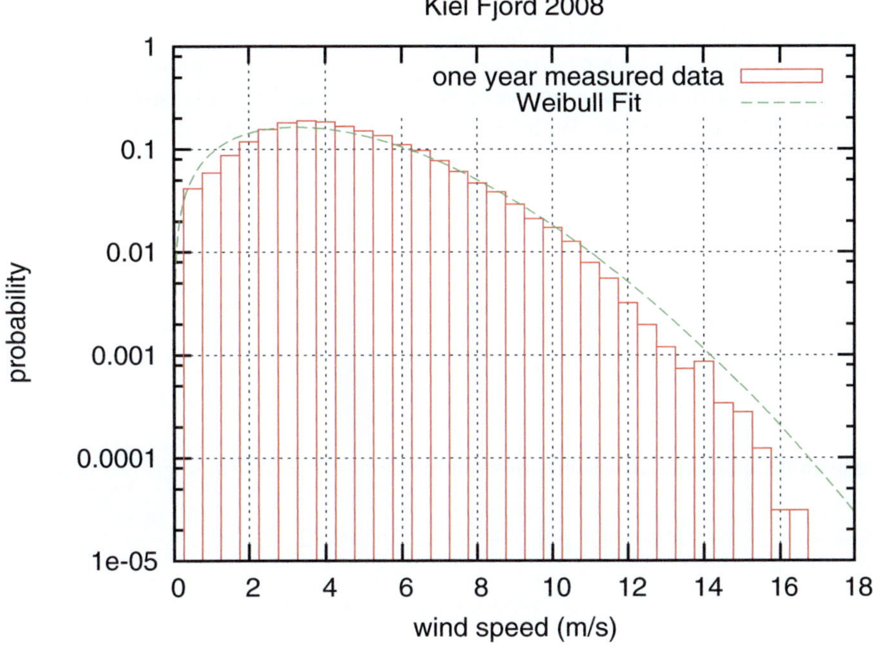

Fig. 1.3 Comparison with measured data from Kiel, Germany, 2008

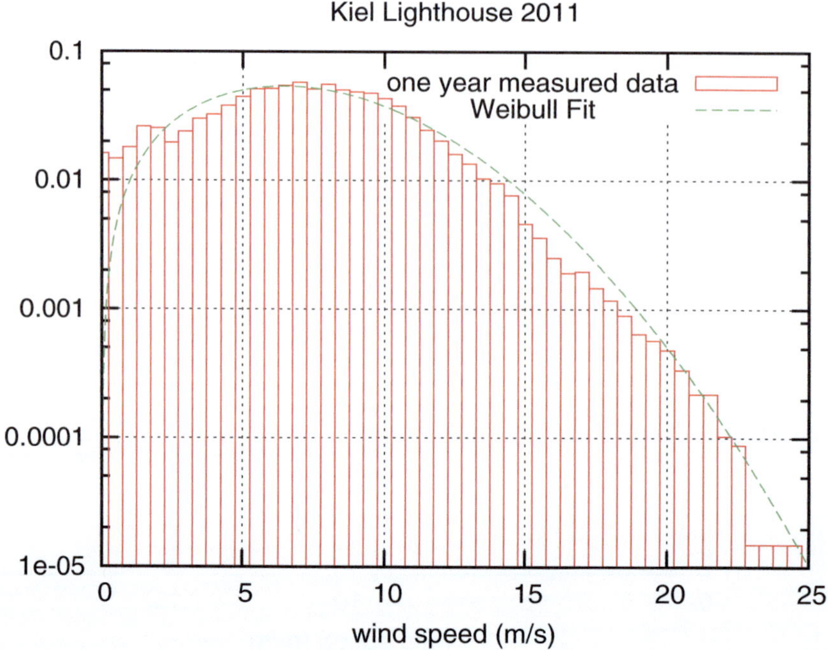

Fig. 1.4 Comparison with measured data from Kiel, Germany, 2011

are not constant in time. This has long been known in wind energy as the *Wind Index Problem*. It means that a long-term (e.g. 50 year) average may not be constant over time. Figure 1.5 shows that even over a period of about 10 years, large fluctuations of about 20% are possible. With Eq. (1.2) some useful limits of annual yield (= useful mechanical work = power × time = P × 24 × 365.25) can be given, under the assumption that only a fraction $0 \leq c_P \leq 1$[2] of Eq. (1.1) can be given. Using

$$m_n := \int_0^\infty P(v; A, k) v^n \cdot dv = A^n \cdot \Gamma\left(1 + \frac{n}{k}\right) \tag{1.4}$$

and

$$P := c_P \cdot \frac{\rho}{2} A_r \cdot v^3 \tag{1.5}$$

we have for an annual averaged power in the case of $k = 2$ (Rayleigh-Distribution):

$$\bar{P} := c_P \cdot \frac{\rho}{2} A_r \cdot R^3 \cdot \Gamma(2.5) = 0.92 \cdot c_P \cdot P . \tag{1.6}$$

---

[2] c_P is the efficiency of a wind turbine.

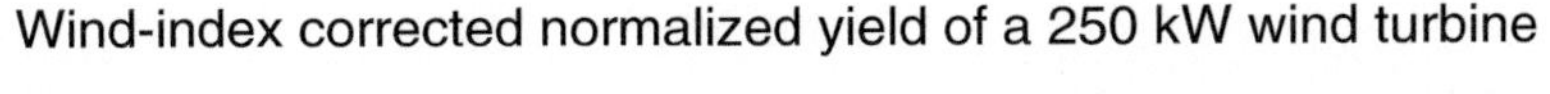

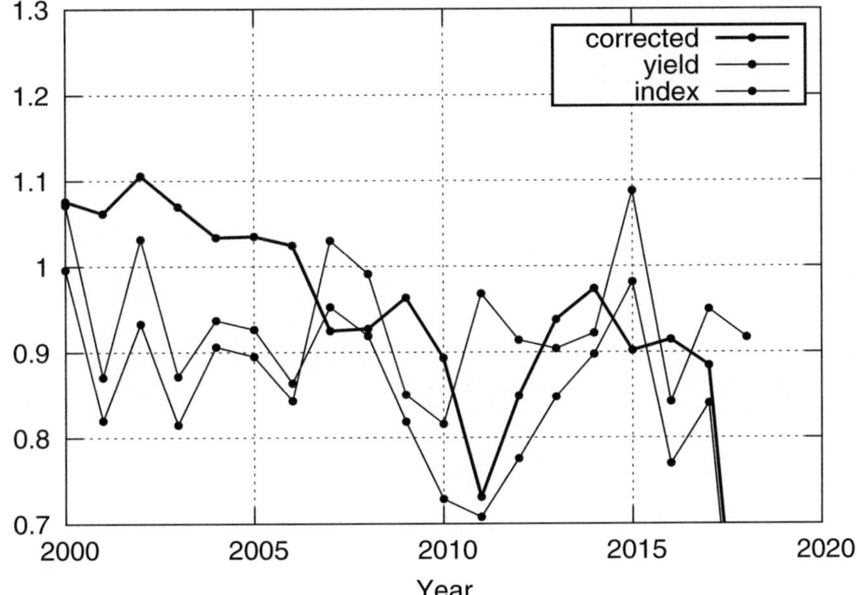

Fig. 1.5 Wind Index of Schleswig-Holstein applied to the energy yield of a research wind turbine. 1 = nominal output, Note that there was an exchange of blades in 2011. Rated power was partly throttled in 2018 because of extension of operating permit

Sometimes this efficiency number is translated into *full-load-hours* which must be less than 8760 h per year, or into capacity factors. In a later chapter we will come back and see how state-of-the-art wind turbines fit into this picture.

Of course the question arises how we may justify or even derive special types of wind occurrence probability distributions. A very simple rationale starts from the *central limit theorem* of probability theory together with the basic rules for transformation of stochastic variables. van Kampen [7] describes it in a somewhat exaggerated manner:

The entire theory of probability is nothing but transforming variables.

We will come back to this interesting question in the context of the statistical theory of Turbulence; Sect. 3.7.

There is one concluding remark about comparing the economical impact of wind turbine blade aerodynamics to helicopters [5] and ship propellers [2]. Assuming a worldwide annual addition of 50 GW of installed wind power, the resulting revenue for just the blades is about 12.5 Billion Euros. About one thousand helicopters (80% of them for military purposes) [5] may produce approximately 50 Billion Euros with an unknown smaller amount for just the rotor—estimated less than 10%. Assuming

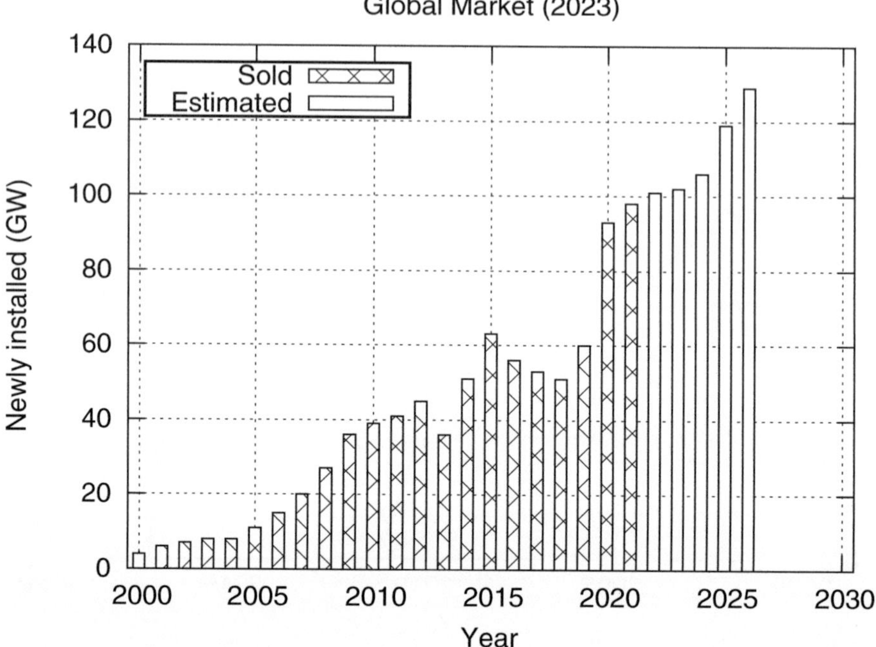

Fig. 1.6 Wind Energy Market 2000–2023, partly estimated. *Sources* BTM consult, wind power monthly and global wind energy council

also 1000 new ships are built every year, each of them having a propeller of about 100 (metrics) tons of mass, the resulting revenue is 1.5 Billion Euros.

All in all, we see that in less than 25 years, wind energy rotor aerodynamics has exceeded the economic significance of other longer-established industries involving aerodynamics or hydrodynamics (Fig. 1.6).

1.2 Problems

Problem 1.1 Calculate the instantaneous power-density for v-wind $= 6, 8, 10$ and 12 m/s. Compare with the annual averaged values (Rayleigh-distributed, $k = 2$, then $A = \bar{v}$) of 4, 5, 6 and 7 m/s. Take density of air to $\rho = 1.26$ kg/m^3.

Problem 1.2 Assume $c_P = 0.58$, $c_P = 0.5$ and $c_P = 0.4$. Calculate the corresponding annual averaged $\bar{c}_P$.

Problem 1.3 Assume that a 2D (velocity) vector has a Gaussian 2-parameter distribution

$$G(x; m, \sigma) := \frac{1}{\sigma\sqrt{2\pi}} \exp -\frac{1}{2}\left(\frac{x-m}{\sigma}\right)^2 \tag{1.7}$$

for both components $\mathbf{v} = (u, v)$. Calculate the distribution of length $y = \sqrt{u^2 + v^2}$.

Problem 1.4 A German reference site is defined as a location with vertical variation of wind speed by

$$u(z) = u_r \cdot \frac{log(z/z_0)}{log(z_r/z_0)} \tag{1.8}$$

$$u_r = 5.5 \text{ m/s at } z_r = 30 \text{ m and } z_0 = 0,1 \text{ m} . \tag{1.9}$$

A historical Dutch wind-mill with $D = 21$ m, $z_{hub} = 17.5$ m is quoted to have a (5-year's) reference yield of 248 MWh. Calculate an efficiency by relating the average annual power with that the same rotor would have with $c_P = 16/27$ and a Rayleigh-distribution of wind.

References

1. Beurskens J (2023) History of wind energy, Ch. 1. In: Understanding wind power technology, Schaffarczyk AP (ed), 3rd ed. SpringerNature
2. Breslin A, Andersen P (1996) Hydrodynamics of ship propellers. Cambridge University Press
3. Darrigol D (2005) Worlds of flow. Oxford University Press, Oxford, UK
4. Eckert M (2006) The dawn of fluid mechanics. Wiley-VCH, Weinheim, Germany
5. Leishman J (2006) Principles of helicopter aerodynamics, 2nd edn. Cambridge University Press, Cambridge, UK
6. McLean D (2013) Understanding aerodynamics. Wiley, Boeing, Chichester, UK
7. van Kampen N (2007) Stochastic processes in physics and chemistry, 3rd edn. Elsevier, Amsterdam, The Netherlands

Chapter 2
Types of Wind Turbines

Ich halte dafür, daß das einzige Ziel der Wissenschaft darin besteht, die Mühseligkeit der menschlichen Existenz zu erleichtern (B. Brecht, Life of Galileo, 1941) [8].[1]

Equation (1.5) from Chap. 1 may also be used to define an efficiency or *power coefficient* $0 \leq c_P \leq 1$:

$$c_P = \frac{P}{\frac{\rho}{2} A_r \cdot v^3} .$$ (2.1)

Wind turbine aerodynamic analysis frequently involves the derivation of useful equations and numbers for this quantity. Most (in fact: almost all) wind turbines use *rotors* which produce *torque or* moment of force

$$M = P/\omega.$$ (2.2)

with $\omega = RPM \cdot \pi/30$ the angular velocity. Comparing tip speed $v_{Tip} = \omega \cdot R_{Tip}$ and wind speed we have

$$\lambda = v_{Tip}/v_{wind} = \frac{\omega \cdot R_{Tip}}{v_{wind}}$$ (2.3)

the (TSR). Figure 2.1, sometimes called the *map of wind turbines*, gives an overview of efficiencies as a function of non-dimensional RPM. The numbers are estimated efficiencies only. A few remarks are germane to the discussion. From theory (Betz, Glauert) there are clear efficiency limits, but no theoretical maximum TSR. In contrast, the semi-empirical curves for each type of wind turbine have a clearly defined maximum efficiency value.

---

[1] Presumably for the principle that science's sole aim must be to lighten the burden of human existence.

A. P. Schaffarczyk, *Introduction to Wind Turbine Aerodynamics*, Green Energy and Technology, https://doi.org/10.1007/978-3-031-56924-1_2

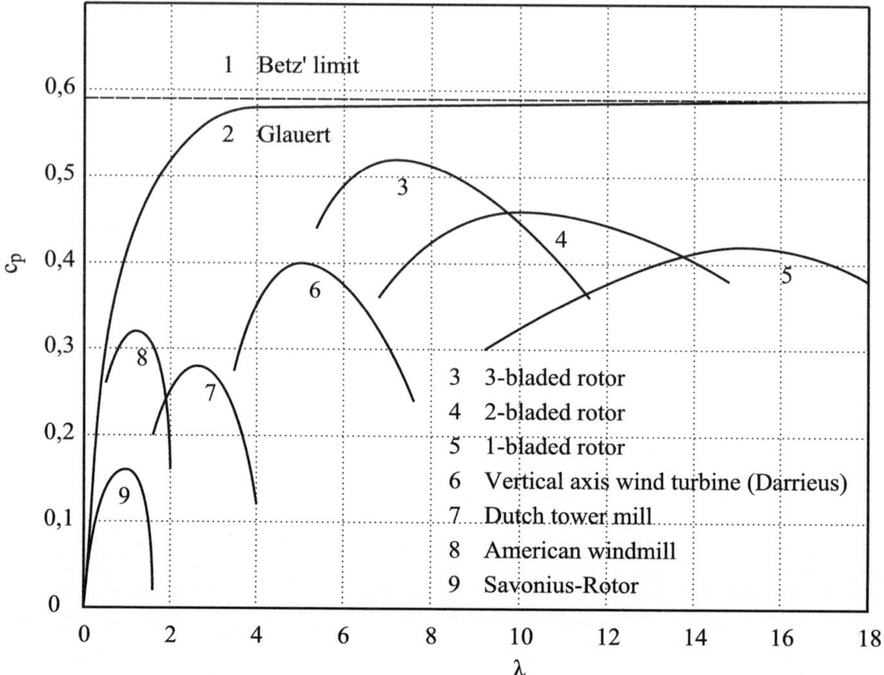

Fig. 2.1 Map of wind turbines

2.1 Historical and State-of-the-Art Horizontal-Axis Wind Turbines (HAWT)

The wind energy community is very proud of its long history. Some aspects of this history are presented in [7, 31]. Apparently the oldest one [31] is the so-called *Persian* windmill (Fig. 2.2). It was first described around 900 AD and is viewed from our system of classification (see Sect. 2.6) as a drag-driven windmill with a vertical axis of rotation.

Somewhat later the *Dutch* windmill appeared as the famous *Windmill Psalter* of 1279 [31]; see Fig. 2.3. This represented a milestone in technological development: the axis of rotation changed from vertical to horizontal. But also from the point of view of aerodynamics, the Dutch concept began the slow movement toward another technological development: lift replacing drag. These two types of forces simply refer to forces perpendicular and in line with the direction of flow. The reason this is not trivial stems from *D'Alembert's* Paradox.

Theorem 2.1 *There are no forces on a solid body in an ideal flow regime.*

We will continue this discussion further in Chap. 3. These very early concepts survived for quite a few centuries. Only with the emergence of electrical generators and

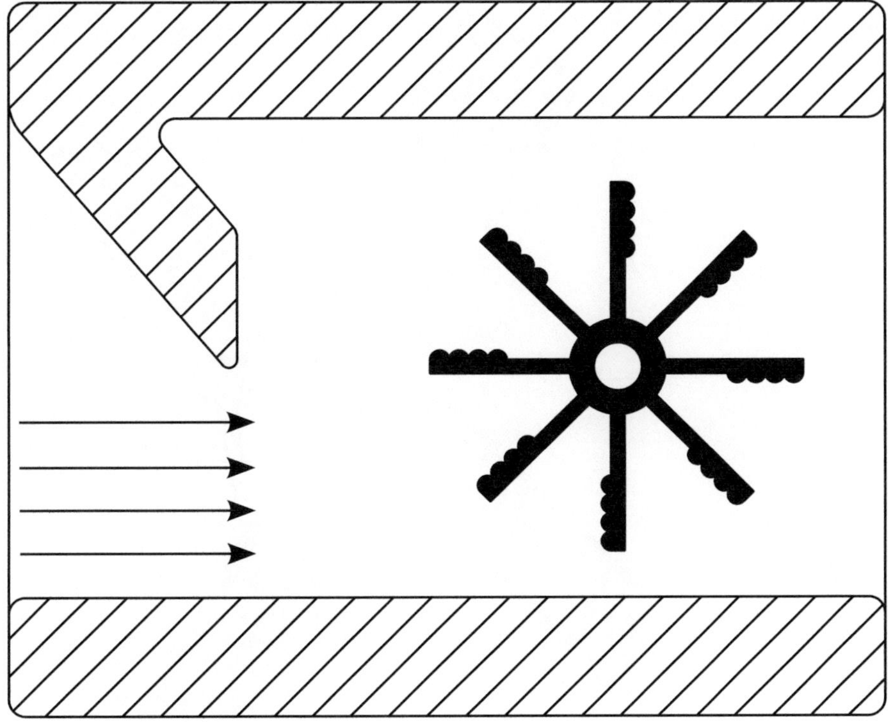

Fig. 2.2 Persian Windmill

airplanes were these new technologies adapted, in the course of a few decades, to what is now called the standard horizontal-axis wind turbine:

- horizontal axis of rotation,
- three bladed,
- driving forces mainly from lift,
- upwind arrangement of rotor; tower downwind,
- variable speed/constant TSR operation and
- pitch control after rated power is reached.

2.2 Non-Standard HAWTs

With this glimpse of what a *standard* wind turbine should be, everything else is *non-standard*:

- **no** horizontal axis of rotation,
- number of blades other than three (one, two or more than three),

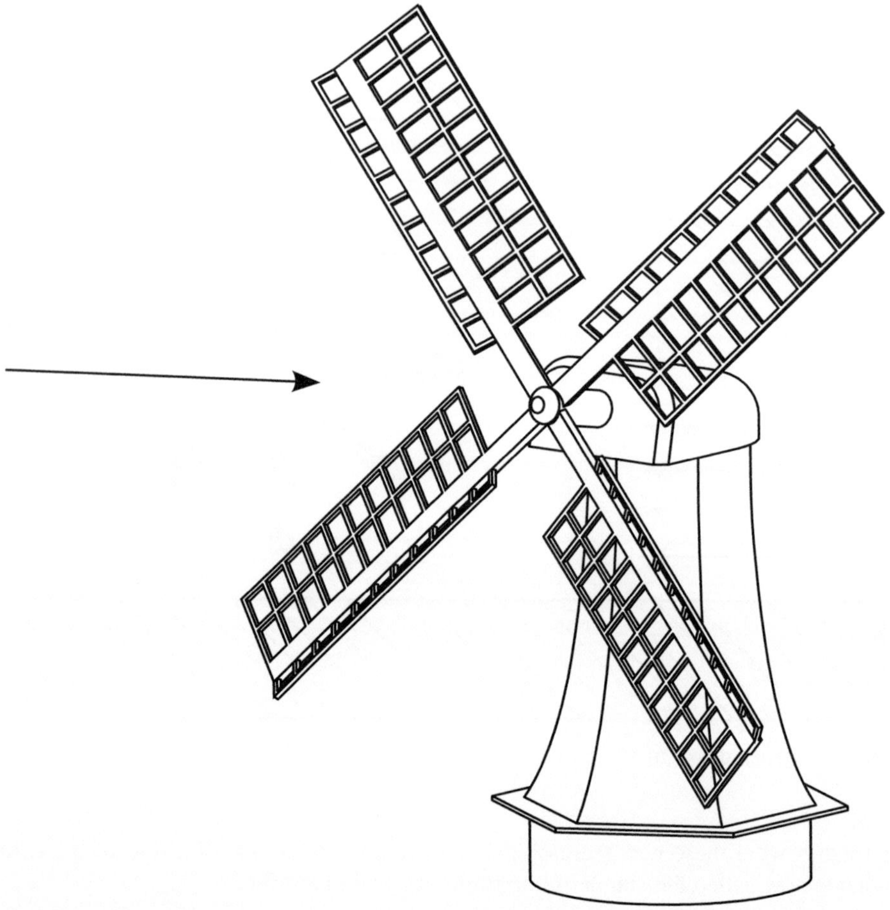

Fig. 2.3 Dutch Windmill

Fig. 2.4 Lift and Drag
forces on an aerodynamic
section

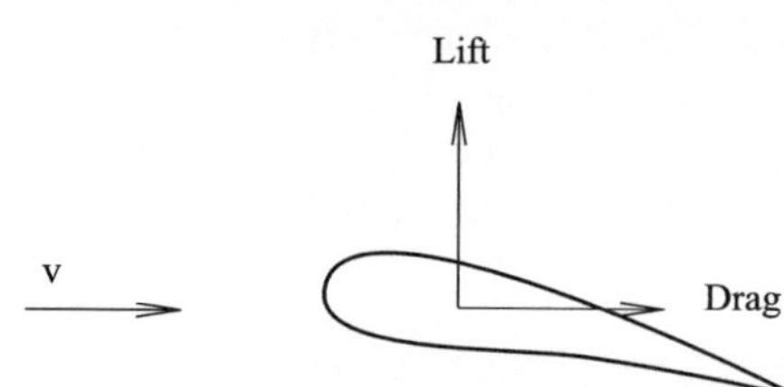

- drag forces play important role,
- downwind arrangement of rotor; tower upwind,
- constant speed operation and
- so-called **stall** control after rated power is reached.

From these characteristics, we may derive a large number of different designs. Only a few of them became popular enough to acquire their own names.

The *American* or *Western type* turbine [31], Chap. 1, uses a very high (10 to 50) number of blades which in most cases are flat plates with a small angle between plane of rotation and chord. These turbines were used mainly in the second half of the nineteenth century; see Fig. 2.5.

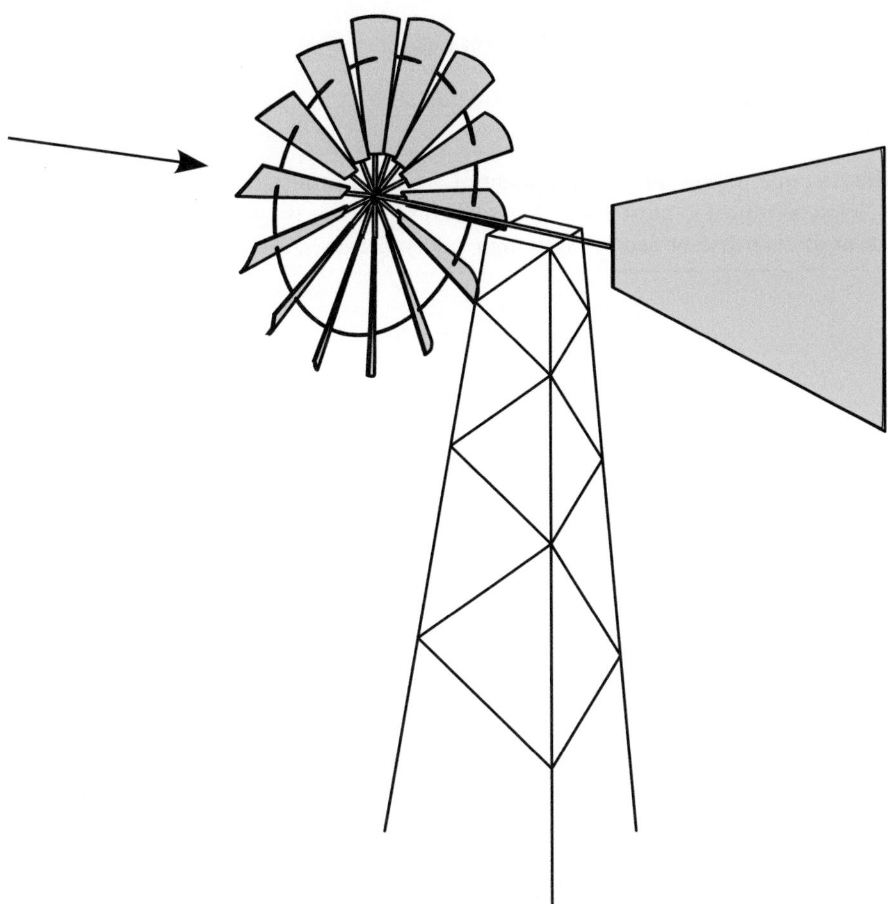

Fig. 2.5 American or Western Windmill

The *Danish Way* of extracting wind energy [24] used most of the now classical properties with a fixed-pitch blade arrangement and a constant-RPM operation mode. The development of this design philosophy started in the 1940s and died off slowly in the 1990s.

2.3   Small Wind Turbines

Small wind turbines are defined by IEC [10] as a wind turbine with a rotor swept area no greater than 200 m^2. Therefore the diameter is limited to 16 m. However, most of them have much smaller diameters starting at about 1m. More can be found in (9) [35]. Figure 2.6 gives an account of scaling. The main problem with safety approval of [10] is that it offers two very different methods:

- the usual aero-elastic simulation modeling and
- a simplified load model.

The first one implies the same amount of work as for a state-of-the-art turbine and is not economical in most cases. The second procedure is much easier (see (9) [35]) but at the expense of exceedingly high safety factors. As an example the required

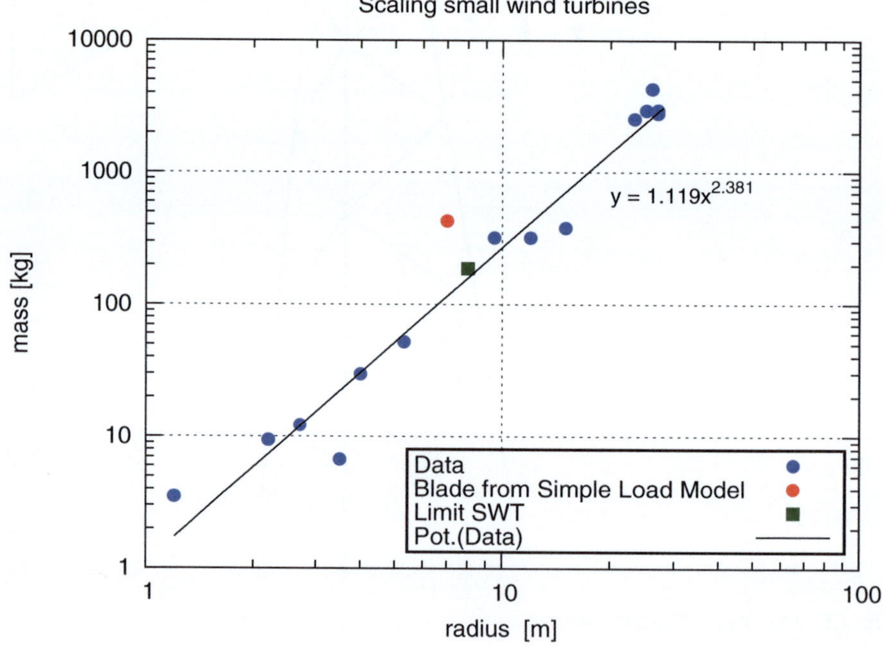

Fig. 2.6   Scaling of blade masses of smaller wind turbines

Savonius-Rotor   Darrieus-Rotor   H-Darrieus-Rotor

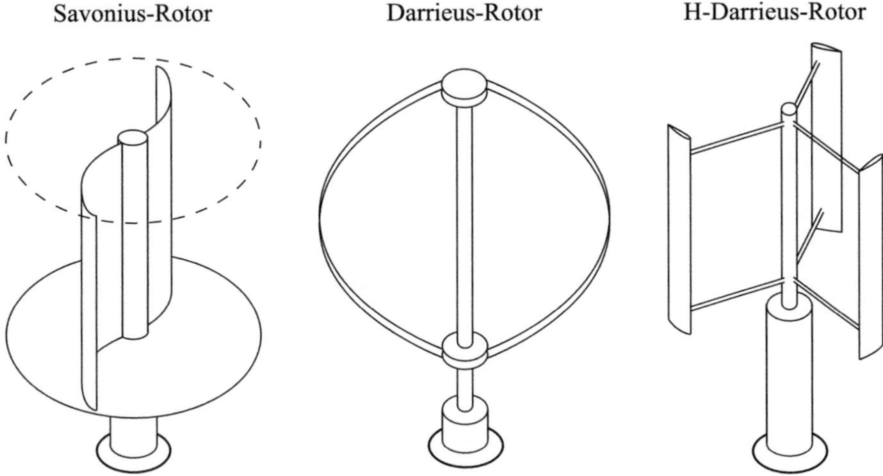

Fig. 2.7 Types of Vertical-Axis Wind Turbines

blade mass for the simple load model is shown in Fig. 2.6. The mass required by the simple load model has to be more than 300 kg, compared to only 120 kg for a blade designed without these high safety factors (Fig. 2.7).

2.4 Vertical-Axis Wind Turbines

As was explained earlier, vertical-axis windmills and the subsequent vertical axis wind turbines seem to be older than those with an horizontal axis of rotation. Mainly due to the inventions of Darrieus [11] and Savonius, [28][2] interest in these vertical turbines was renewed in the early twentieth century. Then, after the first so-called oil crisis at the beginning of the 1970s, many US [5, 6, 17, 25] and German [2–4, 15, 23, 32, 33] vertical turbine development efforts were undertaken which lasted until the 1990s, while HAWTs also progressed. A summary may be found in [26]. Now, after some 20 years of dormancy, interest in VAWTs seems to be returning slowly; see for example [13].

One of the big advantages is the independence of directional change in wind. See Fig. 2.8 for a typical distribution of wind direction at a typical site. Also heavy components may be installed close to the ground, as illustrated in Fig. 2.9. The largest VAWT manufactured so far was the so-called Éole-C made in Canada. Its height was about 100 m, the rotating mass was 880 metric tons and the rated power was supposed to be 4 MW. Unfortunately due to severe vibration problems, the rotational speed was limited to such low values that only 2 MW was reached [26].

---

[2] A more detailed discussion can be found in Sect. 2.6.

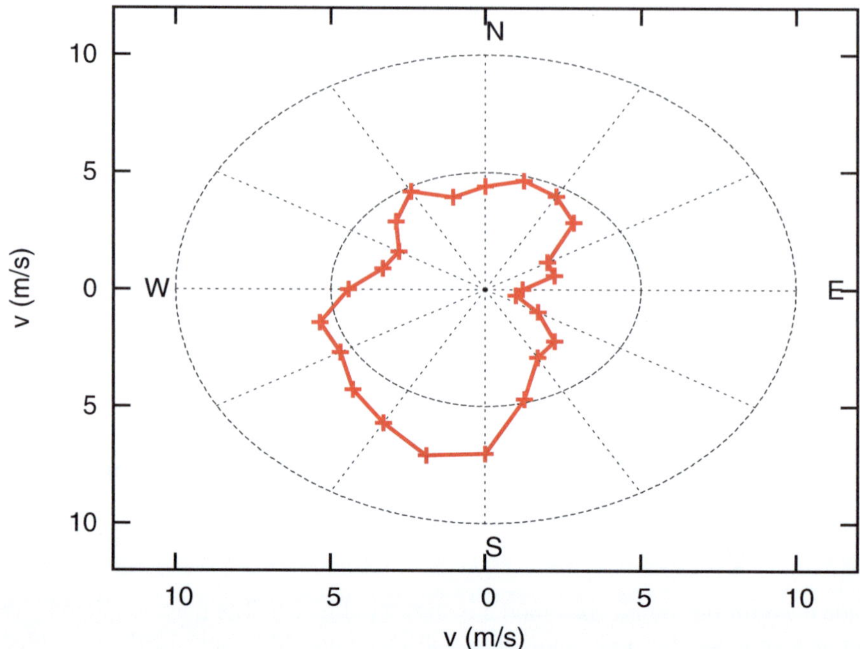

Fig. 2.8 Typical wind direction distribution near Hamburg at 10 m height

It has been noted that the flow mechanics behind these designs (see Problems 6.1 and 6.2 in Chap. 6) are much more interesting—but also more complicated—due to the fact that the sphere of the influence of the rotor is modeled as a real volume and not a 2D disk as is assumed for the HAWT actuator disk model. Therefore at least for each half-revolution, the blades are operating in a wake, meaning that load fluctuations have a much greater influence on the blades. The resulting so-called aerodynamic fatigue loads for a VAWT rotor blade are much higher and are one of the reasons that VAWTs are much more prone to earlier failure of components—mostly at joints—than are HAWTs.

One way out of this difficulty is the so-called *Gyro-Mill* or articulated VAWT [9], where the pitch angle (the angle of the chord line in relation to the circumference) is changed periodically so that—in an ideal case—the driving force remains almost constant (Figs. 2.10 and 2.12).

2.5   Diffuser Augmented Wind Turbines

As we have seen, the power contained in the wind is proportional to the swept area. An obvious extension of this concept is to look for *wind-concentrating* devices resembling a cone or funnel; see Fig. 2.11. Such devices are very common in wind

Fig. 2.9 Dornier's Eole-D, height $= 20\,\text{m}$, $P_{rated} = 50$ kW, Reproduced with permission of EnBW Windkraftprojekte GmbH, Stuttgart, Germany

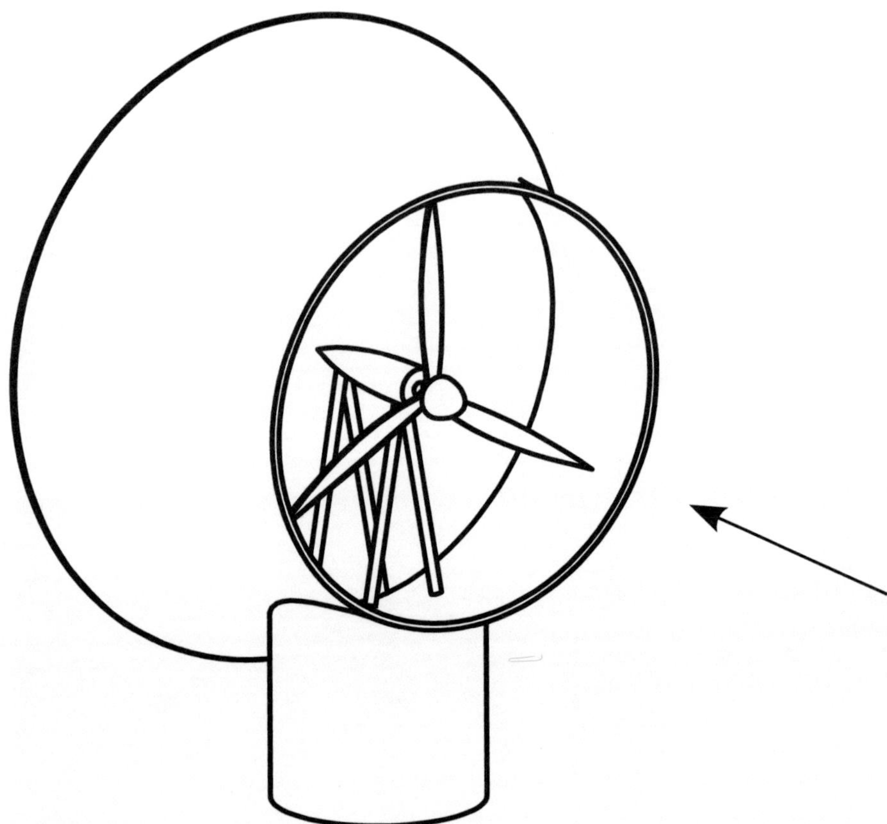

Fig. 2.10 Example of a DAWT, Wind from right

turbines and are called draft tubes or suction tubes. From first principles of fluid mechanics, the **exit** area of such a device has to be larger than the inflow area. At a glance this is clearly counterintuitive. Then, by closer inspection of the basic laws of conservation of mass and energy—called *Bernoulli's* law—it follows that an increase of mass-flow proportional to the area ratio A_{exit}/A_{inflow} is possible if the flow follows the contour of the cone.

Unfortunately nature is not that generous. The increase of diameter has to be very moderate. To be more precise: opening-angles less than $10\,°C$ have to be used to avoid what is called flow separation. It immediately follows that, to have reasonable area-ratios, we have to use very long diffusers with very large weights. The engineering task then is to find a reasonable compromise—if possible at all. Serious work started in 1956 by Liley and Rainbird [18] and efforts up to 2007 are summarized by van Bussel [34]. Some applications to small wind turbines may be found at [29].

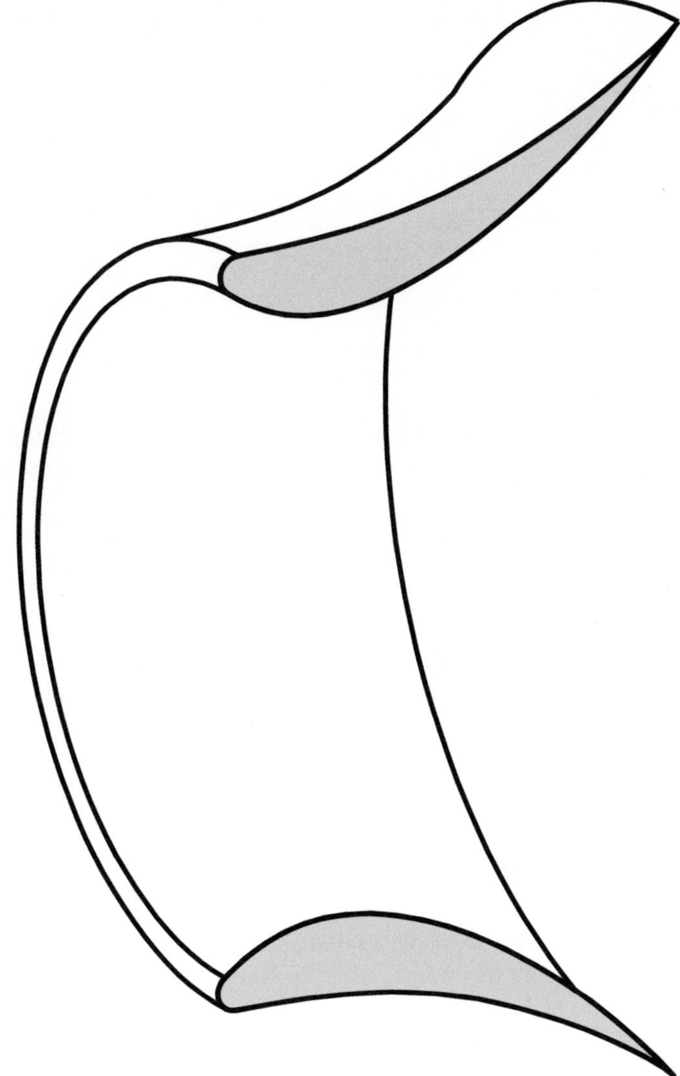

Fig. 2.11 Cross section of a diffuser or ring-wing

2.6 Drag-Driven Turbines

Drag was defined earlier as a force on a structure subjected to a stream of air **in line** with the flow; see Fig. 2.4. We therefore define a simple number, the *drag coefficient*:

$$c_D := \frac{D}{\frac{\rho}{2} \cdot A_r \cdot v^2} \, .$$

(2.4)

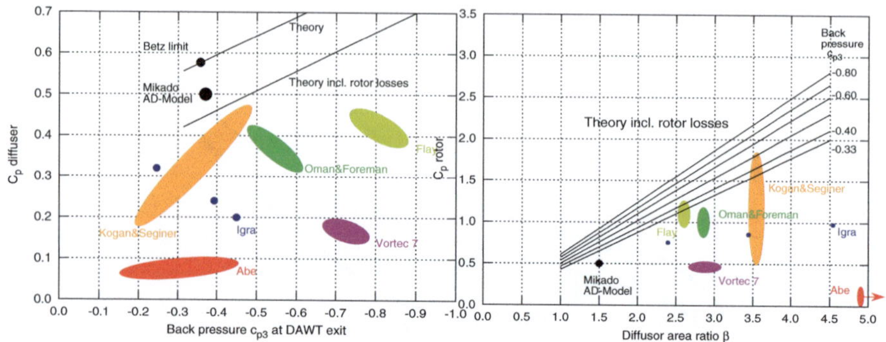

Fig. 2.12 Collected Performance data for diffusers, from van Bussel (2007)

To imagine this, we consider at first a simply translating sail of velocity u and Area $A = c \cdot \ell$ and v the wind velocity as usual. The power $P = D \cdot u$ is using the expressions above:

$$P = \frac{\rho}{2}(v - u)^2 \cdot c_D \cdot c \cdot \ell \cdot u \,. \tag{2.5}$$

Now using ($P_{wind} = \frac{\rho}{2}v^3$) we have

$$c_P = c_D \left(1 - u/v\right)^2 (u/v) \,. \tag{2.6}$$

Setting $a := u/v$ and solving for a by making $\frac{dc_P}{da} = 0$, we see that the maximum power of such a *drag-driven vehicle* may not exceed

$$c_P^{max,D} = \frac{4}{27} \cdot c_D \quad \text{at} \quad a = \frac{1}{3} \tag{2.7}$$

$c_P^{max,D} \approx 0.3$ if **u = 1/3 v**. For example a sailing boat at wind-force 7 (Beaufort $\approx$ 30 knots $\approx$ 16.2 m/s) may not travel faster than 10 Knots. If it has 30 m^2 sail area the maximum power will be P = 24 kW.

The next step is to discuss the *Persian windmill* or the closely related anemometer; see Fig. 2.13. The idea is to use a specially shaped body (semi-sphere) which has different c_D when blown from one side or the other. A common pair of values for c_D is $c_{D+} = 1.33$ and $c_{D-} = 0.33$. We then arrive at

$$F_+ = c_{D+} \cdot \frac{\rho}{2} A_r \left(u - \Omega r\right) \tag{2.8}$$

$$F_- = c_{D-} \cdot \frac{\rho}{2} A_r \left(u + \Omega r\right) \text{ and finally} \tag{2.9}$$

$$c_P = (F_+ - F_-) \cdot v/A_r = \lambda \left(c_1 - c_2 \cdot \lambda + c_1 \cdot \lambda^2\right) \tag{2.10}$$

with $c_1 = c_{D+} - c_{D-}$ and $c_2 = 2 \cdot (c_{D+} + c_{D-})$. Figure 2.14 shows then a very small efficiency at very low tip speed ratios ($c_{D+} = 1.33$ and $c_{D-} = 0.33$).

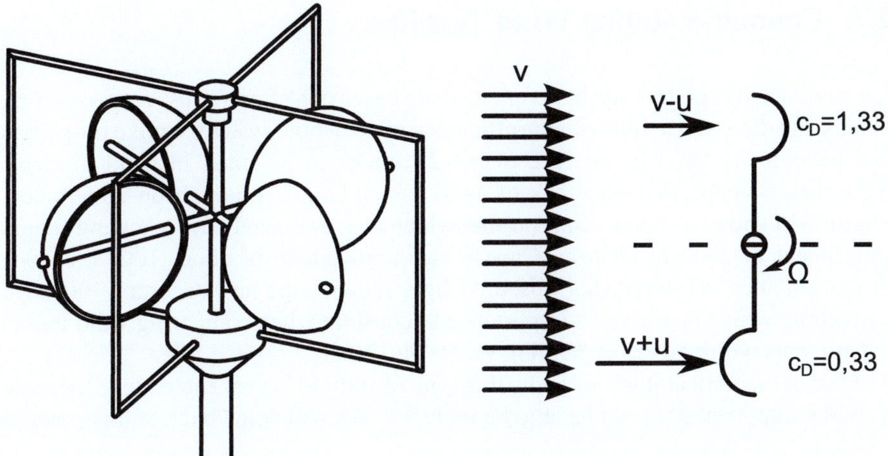

Fig. 2.13 Drag-driven turbines

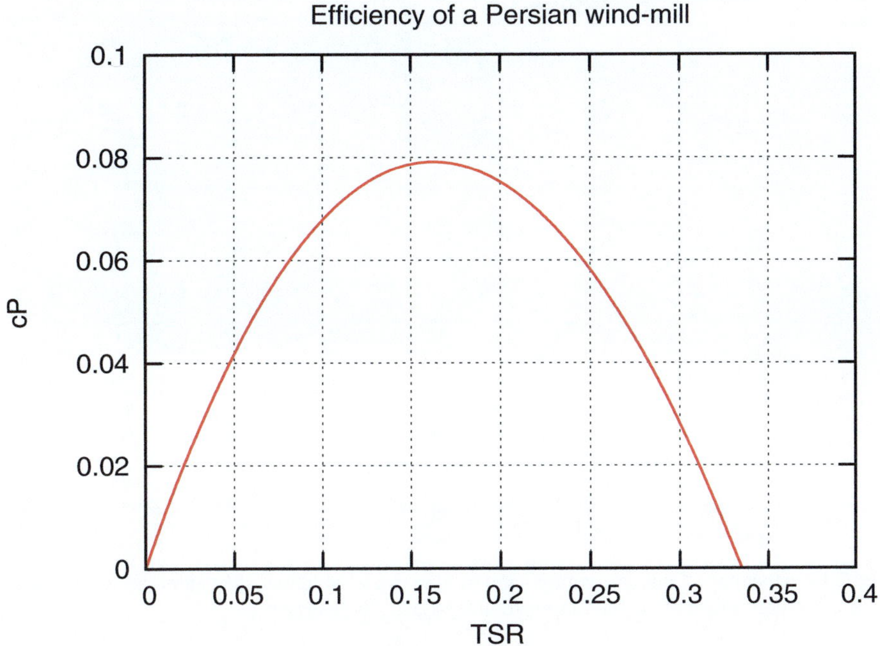

Fig. 2.14 Performance curve of an anemometer

2.7   Counter-Rotating Wind Turbines

For some ship propellers and helicopters, there have been efforts to use rotors rotating in opposite direction to improve the efficiency of the whole system. Unlike propellers and helicopters, swirl losses (see Sect. 5.2) appear to be small in wind turbines. Nevertheless many investigations of swirl losses [16, 30] have been undertaken. Figure 2.15 shows a model wind turbine which was evaluated during a wind tunnel experiment in 2009 in Geneva. Clearly an improvement of about 10% was seen in performance but somewhat obscured by a rather large and uncertain blockage correction, which is of utmost importance to consider when comparing wind tunnel experiments with a freely expanding wake (Fig. 2.16).

From a theoretical point of view, this type of turbine is very interesting, because a swirl component does not have to be included. We will come back to this point in Chap. 6.

Fig. 2.15 Model of a Counter-rotating wind turbine in a wind tunnel, diameter = 800 mm. Reproduced with permission of HEIG-VD, Iverdon-les-Bains, Switzerland

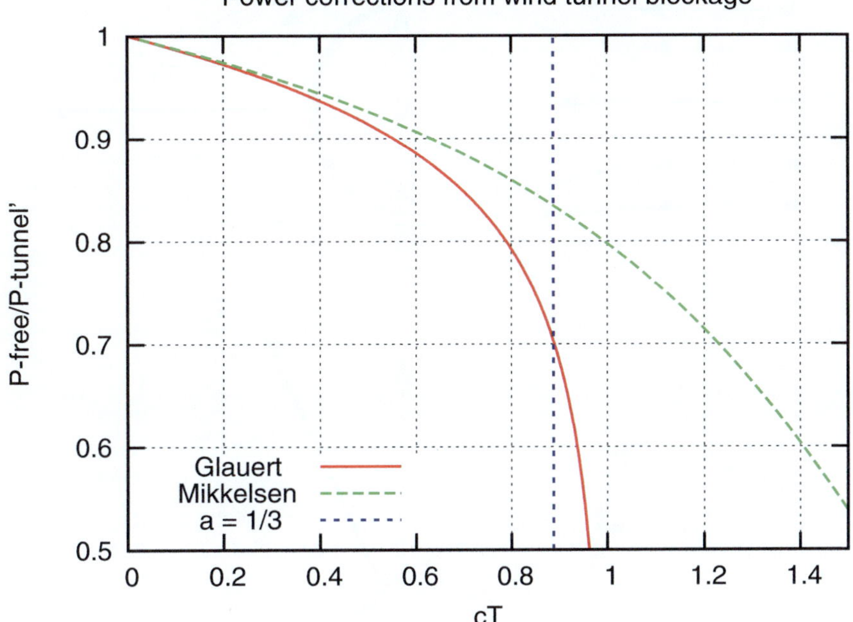

Fig. 2.16 Blockage Corrections from Glauert (3) [14] and Mikkelsen [20]

2.8 Airborne Wind Turbines

Ahlers and Loyd [1, 19]. The mechanism for driving airborne devices is somewhat similar to a simple translating device driven by an lifting airfoil (compare to Fig. 2.17).

Geometry is 2D and let's assume that wind and driving direction are perpendicular.

Power is $P = N \cdot v$. with N normal force also perpendicular to wind and parallel to v. U and v may be completed for an triangle, N consists of

$$N = N_L + N_D = L \cdot cos(\varphi) - D \cdot sin(\varphi). \tag{2.11}$$

Some simple algebra gives

$$c_P = a \left(\sqrt{1 + a^2} \right) \cdot (C_L - C_D \cdot a). \tag{2.12}$$

Maximum power coefficient (see Fig. 2.18) now is

$$c_P^{max,L} = \frac{2}{9} C_L \left(\frac{C_L}{C_D} \right) \sqrt{1 + \frac{4}{9} \left(\frac{C_L}{C_D} \right)^2}. \tag{2.13}$$

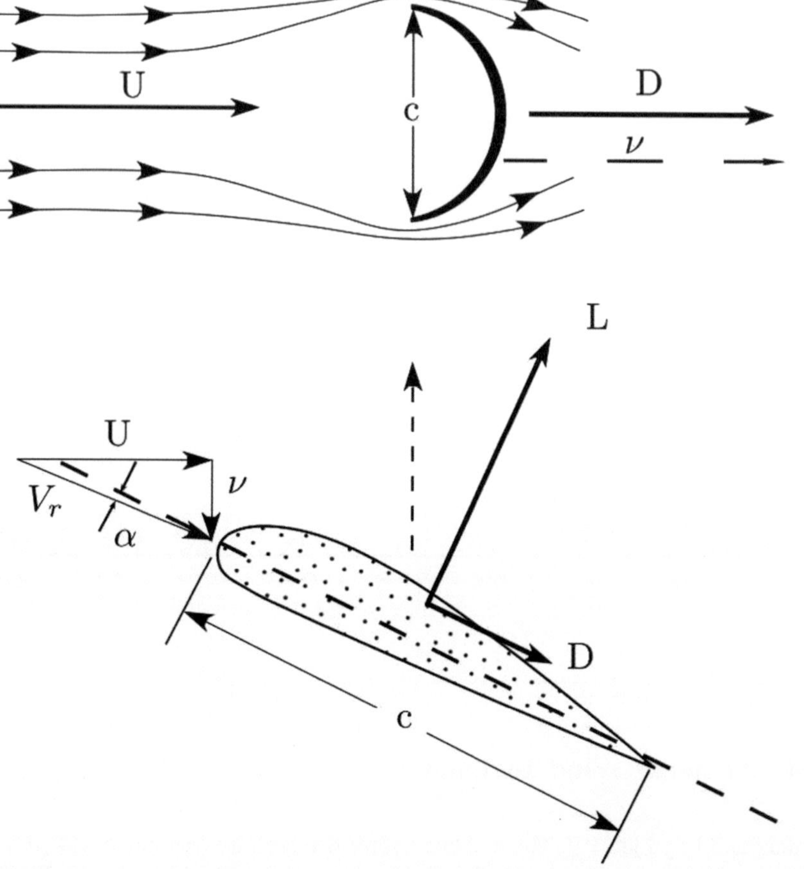

Fig. 2.17 Forces and velocities for purely translating drag device (top) and a translating device using lift (bottom)

Note:

- maximum velocity may be **larger as wind speed**—reached at

$$a = v/U = \frac{2}{3}\frac{C_L}{C_D}\ .$$
(2.14)

- There is a (not negligible) force T parallel to wind which must be compensated.

Figure 2.18 gives an example for $C_D = 2$, $C_L = 1$ and $C_L/C_D = 10$. It has to be noted that [1] confirms these and the investigation of [19] to

$$P = \frac{2}{27}\rho A_r w^3 \cdot \frac{c_L^3}{c_D^2}$$
(2.15)

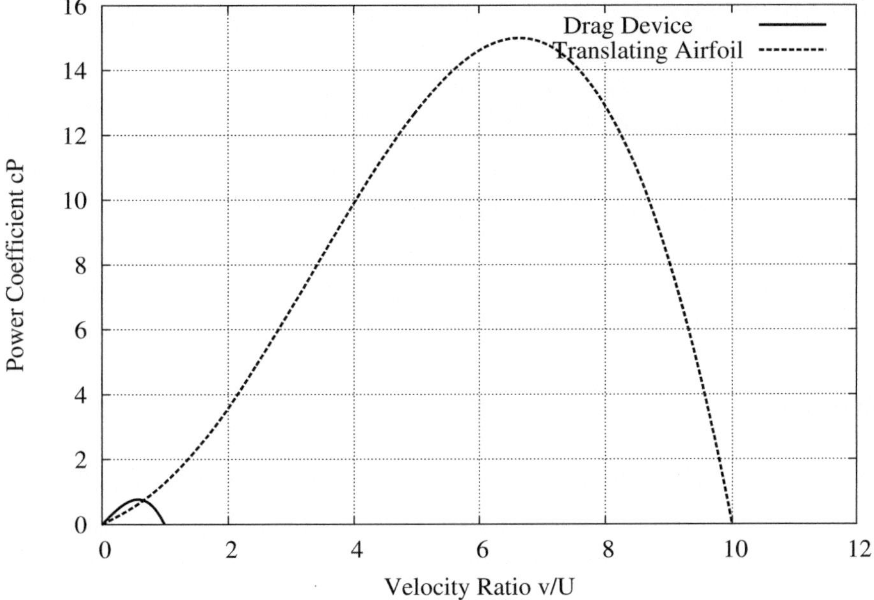

Fig. 2.18 Lifting Device as compared to a pure drag device

which gives

$$c_P = \frac{4}{27} L 2 D^3 c_D \,, \tag{2.16}$$

a value which easily may exceed Betz's limit, because not the swept area but the wind area is used as a reference. To give numbers from a typical example [27] results from a demonstration plant with 4 to 8 m^2 wind area indicate values for c_L/c_D in the order 3, leaving much room for improvement.

2.9 Concluding Remarks

We finish this chapter by noting that still a myriad of other turbine types exist, some of which are based on very specific fluid mechanics principles or ideas. Even a pin-wheel (toy) turbine was investigated experimentally [22]. The reader may consult the older literature [12, 21] for many other highly entertaining examples.

2.10 Problems

Problem 2.1 Derive Eq. 2.10 and find an expression for c as a function of c_{D+} and c_{D-}.

Problem 2.2 Estimate the increase of power for a DAWT with the following properties: $D_{Rotor} = 1m$, $D_{exit} = 1.17$ m and efficiency of diffuser $\eta_{diff} = 0.85$.

Problem 2.3 Determine if a counter-rotating turbine consists of one turbine or two, and give your reasons.

References

1. Ahlers U, Diehl M, Schmehl R (2014) Airborne wind energy. Springer-Verlag, Heidelberg, Berlin
2. A. D. et al (1982) Entwicklung eines 5,5 m durchmesser-windenenergiekonverters mit vertikaler drehachse (phase iii). Technical report, Abschlußbericht zum Forschungsvorhaben T-82-086, Dornier System GmbH, Friedrichshafen, Germany
3. G. B. et al (1978) Entwicklung eines 5,5 m ø-windenenergiekonverters mit vertikaler drehachse (phase ii). Technical report, Abschlußbericht zum Forschungsvorhaben T-79-04, Dornier System GmbH, Friedrichshafen, Germany
4. Bankwitz H et al (1975) Entwicklung einer windkraftanlage mit vertikaler achse (phase i). Technical report, Abschlußbericht zum Forschungsvorhaben ET-4135 A, Dornier System GmbH, Friedrichshafen, Germany
5. Carne TG et al (1982) Finite element analysis and modal testing of a rotating wind turbine. SAND 82–0345
6. Ashwill TD (1992) Measured data for the Sandia 34-meter vertical axis wind turbine. Technical report, SAND91-228. Albuquerque, New Mexico, USA
7. Beurskens J (2023) History of wind energy, Ch. 1 in: understanding wind power technology. In: Schaffarczyk AP (ed), 3rd edn. Springer Nature
8. Brecht B (2008) Life of Galileo. Penguin Classics, London, UK, Reprint
9. Clausen RS, Sønderby IB, Andkjær JA (2006) Eksperimentel og Numerisk Undersøgelse af en Gyro Turbine. The Danish Technical University, Lyngby, Denmark, Institut for Mekanik, Energi og Konstruktion
10. Commission IE (2011) IEC 61400-2 ed. 3, small wind turbines. Technical report, International Electrotechnical Commission
11. Marie DGJ (1931) Turbine having its rotating shaft transverse to the flow field of the current. US Patent, 1(835):018
12. de Vries O (1979) Fluid dynamic aspects of wind energy conversion. Technical report, AGAR-Dograph, No. 243. Neuilly sur Seine, France
13. Ferreira CS (2009) The near wake of the VAWT. PhD thesis, TU Delft, The Netherlands
14. Glauert H (1926) The elements of aerofoil and airscrew theory, Repr, 2nd edn. Cambridge University Press, Cambridge, UK
15. Henseler H (1990) Eole-d mw technologieprgrame darrieus windenergieanlagen anpaßentwicklung, 2. abschlußbericht zum forschungsvorhaben 0328933 p. Technical report, Dornier GmbH, Immenstaad, Germany
16. Herzog R, Schaffarczyk AP, Wacinski A, Zürchner O (2010) Performance and stability of a counter-rotating windmill using a planetary gearing: measurements and simulation. Proc EWEC Warsaw, Poland

17. Homicz GF (1991) Numerical simulation of VAWT stochastic aerodynamic loads produced by atmospheric turbulence: VAWT-SAL code. SAND
18. Lilley GM, Rainbird WJ (1956) A preliminary report on the design and performance of ducted windmills. Technical report, Cranfield College of Aeronautics, Bedford, UK
19. Loyd ML (1980) Crosswind kite power. J Energy 4(3):106–111
20. Mikkelsen R (2003) Actuator disk methods applied to wind turbines. PhD thesis, The Technical University of Denmark, Lyngby
21. Molly J-P (1990) Windenergie - Theorie, Anwendung, Messung, 2nd edn. Verlag C.F. Müller Karsruhe, Germany
22. Nemoto Y, Ushiyama I (2003) Experimental study of a pinwheel-type wind turbine. Wind Eng 27:227–235
23. NN. Technische anlage zum angebot nr. 3026-0-90, lieferung und montage einer 2,25 mw darrieus-windenergieanlage eole-d. Technical report, Dornier GmbH, Friedrichshafen, Germany, 1990
24. Nossen P-O et al (2009) Wind power - the Danish way. The Poul la Cour Foundation, Askov, Denmark
25. Oler JW et al (1983) Dynamic stall regulation of the Darrieus turbine. Technical report, SAND82-7029, Albuquerque, New Mexico, USA
26. Paraschivoiu I (2002) Wind turbine design. With emphasis on Darrieus concept. Polytechnic International Press, Montreal, Canada
27. Ranneberg M, Wolfle D, Bormann A, Breipohl F, Rohde P, Bastigkeit I (2018) Fast power curve and yield estimation of pumping airborne wind energy system. In: Airborne wind energy - advances in technology development and research. Springer, Singapore
28. Savonius SJ (1930) Windrad mit zwei hohlflügeln, deren innenkanten einen zentralen wind-durchlaßspalt freigeben und sich übergreifen. Technical report, Patentschrift Nr. 495 518
29. Schaffarczyk AP (2007) Auslegung einer kleinwindanlage mit mantel, aerodynamischen leistungsdaten, optimierung des diffsors. Technical Report 49, 50 and 51, Kiel University of Applied Sciences, Kiel, Germany
30. Shen WZ, Zakkam VAK, Sørensen JN, Appa K (2007) Analysis of counter-rotation wind turbines. J Phys Conf Ser 75:012003
31. Spera D (ed) (2009) Wind turbine technology, 2nd edn. ASME Press, New York, USA
32. H. M. und Richter B (1988) Messungen an der windkraftanlge dawi 10 und vergleich mit theoretischen untersuchungen, abschlußbericht we-4/88 zum forschungsvorhaben 03e-8384-a. Technical report, Germanischer Lloyd, Hamburg, Germany
33. L. E. und C. Seeßelberg (1990) Analyse und nachweis der 50 kw - windnergieanlage (typ darrieus). Technical report, MEB 55/90, internal report, Dornier GmbH, Immenstaad, Germany
34. van Bussel GJW (2007) The science of making more torque from wind: Diffuser experiments and theory revisited. J Phys Conf Ser 75:012010
35. Wood D (2011) Small wind turbines. Springer-Verlag, London, UK

Chapter 3
Basic Fluid Mechanics

> *Ich behaupte aber, daß in jeder besonderen Naturlehre nur so viel eigentliche Wissenschaft angetroffen werden könne, als darin Mathematik anzutreffen ist (Immanuel Kant, 1786) [30]. (However, I claim that in every special doctrine of nature there can be only as much proper science as there is mathematics therein. (Ref Stanford Encyclopedia of Philosophy).)*

3.1 Basic Properties of Air

Air regarded as an ideal gas may be described by its mass-density $\rho = dm/dV$. To adjust its standard value of $\rho_0 = 1.225$ kg/m^3 to other temperatures (ϑ) and elevations (H) we may use:

$$\rho(p, T) = \frac{p}{R_i \cdot T} \tag{3.1}$$

$$T = 273.15 + \vartheta \tag{3.2}$$

$$R_i = 287 \tag{3.3}$$

$$p(z) = p_0 \cdot e^{-z/z_{ref}} \tag{3.4}$$

$$p_0 = 1015 \text{ hPa} \tag{3.5}$$

$$z_{ref} = 8400 \text{ m.} \tag{3.6}$$

An other important parameter is the velocity of sound:

$$c = \sqrt{\kappa \cdot \frac{p}{\rho}} \; . \tag{3.7}$$

© The Author(s), under exclusive license to Springer Nature Switzerland AG 2024
A. P. Schaffarczyk, *Introduction to Wind Turbine Aerodynamics*, Green Energy
and Technology, https://doi.org/10.1007/978-3-031-56924-1_3

With $\kappa = 1.4$ (meaning that air consists mainly of dual-atom molecules) we have under otherwise standard conditions ($p = p_0$) $c_0 = 340$ m/s. Usually flow is regarded *incompressible* if $v/c \leq 1/3$. This means that the tip speed of a rotating wind turbine blade has to be less than 113 m/s. At the time of this writing (December 2019), the largest rotor has diameter of D $= 220$ m, which means RPM has to be limited to 9.8 to maintain incompressible airflow.

Temperature dependence of dynamic viscosity can be calculated by Sutherland's equation

$$\mu = \frac{C_1 \cdot T^{3/2}}{T + S} \, , \tag{3.8}$$

for air $C_1 = 1.5 \cdot 10^{-6}$ and S $= 110.4$ K. Or if pressure is included as well [27]

$$\mu = (1.82 \cdot (t/298.15))^{0.7}) + 1.6 \cdot 10^{-3} \cdot (P - 1) \cdot 10^{-7} \cdot (79 - (T - (273.15))) \cdot (P - 1)^2 \cdot 10^{-5} \, . \tag{3.9}$$

3.2 The Laws of Fluid Mechanics in Integral Form

3.2.1 Mass Conservation

Continuum mechanics [10] is formulated as a *local* theory for the velocity and (static) pressure field. Unless otherwise indicated, we use the *Eulerian* frame of reference where we assign these fields to points in an absolute (earth-fixed) frame of reference. To arrive at a global solution, we must integrate a set of partial differential equations (see Sect. 3.3) after which we know (in principle) all details of the flow field. In most engineering applications, forces on structures are important and may be derived from the pressure field by further integration.

Fortunately this complicated affair may be circumvented by a direct discussion of the corresponding *conservation laws* which are the basis of the local formulation.

We start with the equation of conservation of mass.

Let V be a finite Volume and ∂V its surface. Then a change of mass may only be possible if there is some flow across the boundary:

$$\dot{m} := \frac{d}{dt} \int_V \rho \cdot dV = - \int_{\partial V} \rho \mathbf{v} d\mathbf{A}. \tag{3.10}$$

Here $\mathbf{v} = (u, v, w)$ is the velocity vector in Cartesian coordinates $\mathbf{r} = (x, y, z)$. The minus sign indicates that we count $d\mathbf{A} = \mathbf{n} dA$ positive *outwards*.

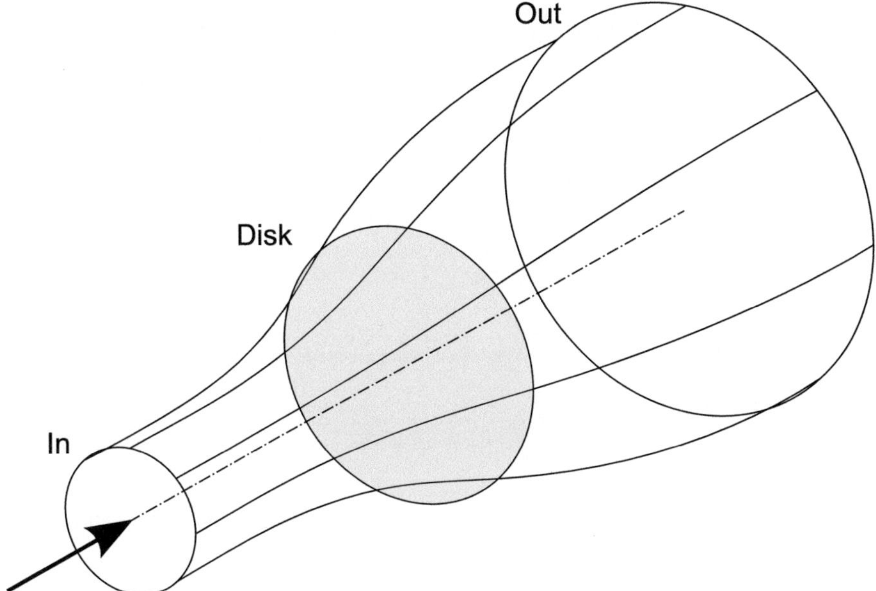

Fig. 3.1 Control volume and conservation of mass

Referring to the example in Fig. 3.1 and assuming no perpendicular flow through the slip stream's surface, we clearly see that along the slipstream (1 = far upstream, 2 = disk and 3 = far downstream):

$$v_1 \cdot A_1 = v_2 \cdot A_r = v_3 \cdot A_3 \tag{3.11}$$

must hold.

3.2.2 Balance of Momentum

Newton's Second Law states that for a point-particle of mass m its momentum $\mathbf{p} := m \cdot \mathbf{v}$ may only be changed by the action of a force:

$$\mathbf{F} = \frac{d\mathbf{p}}{dt} \; . \tag{3.12}$$

Shifting to continuum mechanics, we first have to apply kinematics in an Eulerian frame of Ref. [10]: To get the acceleration

$$\mathbf{a}(t) := \frac{d}{dt}\mathbf{v}(x(t), y(t), z(t), t) \tag{3.13}$$

the *chain rule* of calculus has to be applied:

$$\mathbf{a}(t) = \frac{\partial \mathbf{u}}{\partial x}\dot{x} + \frac{\partial \mathbf{u}}{\partial y}\dot{y} + \frac{\partial \mathbf{u}}{\partial z}\dot{z} + \frac{\partial \mathbf{u}}{\partial t} \ . \tag{3.14}$$

In a more condensed manner, we write

$$\mathbf{a}(t) = \partial_t\mathbf{v} + \mathbf{v} \cdot \nabla\mathbf{v} \ . \tag{3.15}$$

The formal expression

$$\frac{D}{Dt} := \partial_t + \mathbf{v} \cdot \nabla \tag{3.16}$$

is called *convective* or *material* derivative. Within a volume V, the momentum is

$$\mathbf{p} := \int_V \rho\mathbf{v} \cdot dV. \tag{3.17}$$

A slight complication arises because we have now forces acting on the volume (as volume forces $\mathbf{f} = d\mathbf{F}/dV$) as well as on the surface ∂V. If we (for the moment) restrict the analysis to normal stresses (pressure) only, we have

$$\dot{\mathbf{p}} = \frac{d}{dt}\int_V \rho\mathbf{v} \cdot dV = -\int_{\partial V} p\mathbf{n} + \rho\mathbf{v}(\mathbf{v} \cdot \mathbf{n})\mathbf{v}dA + \int_V \rho\mathbf{f}dV \tag{3.18}$$

if—as already remarked—$d\mathbf{A} = \mathbf{n} \cdot dA$.

The second summation term on the right-hand side of Eq. (3.18) may be called *momentum flux per unit area*. Its importance can be seen by the practical application in Fig. 3.1. With no change in total momentum $\dot{\mathbf{p}} = 0$ and no body forces $\mathbf{f} \equiv 0$, we see that if there is no pressure and no velocities outside the jacket:

$$\mathbf{T} + \dot{m}(-v_1 + v_3) \ . \tag{3.19}$$

Here

$$\mathbf{T} = -\int_{A_r} \Delta p \ d\mathbf{A} \tag{3.20}$$

may be used to define the thrust **on** the disk (as a reaction force). It remains to be seen in Chap. 5 how v_1, v_2 and v_3 are related. However, from Eq. 3.19, it is clear that there **must** be forces applied to the disk if $v_3 < v_1$ (Fig. 3.2).

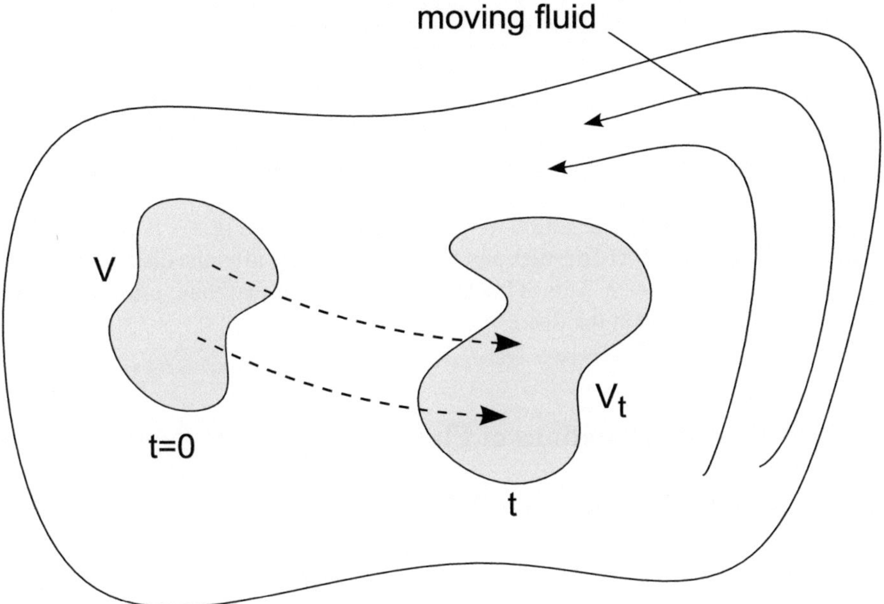

Fig. 3.2 Convected fluid and conservation of energy

3.2.3   Conservation of Energy

Within a volume V, the kinetic energy is

$$E_{kin} := \frac{1}{2} \int_V \rho \mathbf{v}^2 \cdot dV. \tag{3.21}$$

The energy content may only be changed by work done by pressure or body forces:

$$\dot{E}_{kin} = \frac{d}{dt} E_{kin} = -\int_{\partial V_t} p\mathbf{v}d\mathbf{A} + \int_{V_t} \rho \mathbf{v} \cdot \mathbf{f}dV . \tag{3.22}$$

A flow is called *isentropic* if there is a function h, such that

$$\nabla \cdot h = \frac{1}{\rho}\nabla p . \tag{3.23}$$

With this quantity, the integral formulation of energy conservation may formulated as the famous **Bernoulli's Theorem**:

Theorem 3.1 *In stationary isentropic flow, the quantity H*

$$H = \frac{1}{2}\mathbf{v}^2 + h \tag{3.24}$$

remains constant along stream tubes.

Application of this energy equation to our *Actuator Disk* in Fig. 3.1 is then possible from $1 \rightarrow 2^-$ just in front of the disk and from immediately after the disk to a position from downstream $2^+ \rightarrow 3$. This is because we have seen that there must be a force and a pressure drop Δp at the disk.

3.3   Differential Equations of Fluid Flow

3.3.1   Continuity Equation in Differential Form

By applying the *Divergence Theorem* to a somewhat arbitrary vector field $\mathbf{A}$

$$\int_V \nabla \cdot \mathbf{A} \, dV = \int_{\partial V} \mathbf{A} \cdot d\mathbf{A} \tag{3.25}$$

to Eq. (3.10) we have the *continuity equation*

$$\frac{\partial \rho}{\partial t} + \nabla \cdot (\rho \mathbf{v}) = 0 \,. \tag{3.26}$$

If the density ρ is constant in space and time ($\rho(\mathbf{r}, t) = $ const.), this simplifies to a kinematic constraint only:

$$\nabla \cdot \mathbf{v} = \frac{\partial u}{\partial x} + \frac{\partial v}{\partial y} + \frac{\partial w}{\partial z} = 0 \,. \tag{3.27}$$

3.3.2   Momentum Balance

The integral momentum Eq. (3.18) may be put into a differential equation by again applying the Divergence Theorem to the pressure on ∂V. From

$$\mathbf{S}_{\partial V} = -\int_{\partial V} p d\mathbf{A} \,, \tag{3.28}$$

we obtain by using a unit vector $\mathbf{e}$ in any direction:

$$\mathbf{e} \cdot \mathbf{S}_{\partial V} = - \int_{\partial V} p\mathbf{e} \cdot d\mathbf{A} \tag{3.29}$$

$$= - \int_V \nabla \cdot (p\mathbf{e})dV = \int_V (\nabla p)\mathbf{e})dV . \tag{3.30}$$

Together:

$$\frac{\partial \mathbf{v}}{\partial t} + (\mathbf{v} \cdot \nabla)\mathbf{v} = \frac{1}{\rho}(-\nabla p + \mathbf{f}) . \tag{3.31}$$

This is *Euler's equation* of flow.

Several remarks may be made as follows:

- the equation is **non**linear from the convective acceleration;
- it is first order in the velocity field, so only one boundary condition at the solid wall must be applied;
- in case of no volume forces it may be used to calculate the pressure field;
- as we have 1 (from mass conservation) + 3 (momentum balance) = 4 equations and 1 (pressure) + 3 (velocity) unknown fields, the problem is at least not indeterminate.

3.3.3 Differential Energy Equation

For incompressible fluids, or to be more exact, in flow conditions where the compressibility may be neglected ($v/c \leq 1/3$), mass and momentum balance provide a sufficient number of equations. Therefore the differential energy equation is trivial:

$$\frac{D\rho}{Dt} = 0 . \tag{3.32}$$

This situation changes only if $\rho = \rho(\mathbf{r}, t)$ becomes variable. In addition thermodynamics enter the picture, as air then must be regarded at least as an ideal gas and $\rho = \rho(T, p)$ as mentioned earlier.

3.4 Viscosity and Navier–Stokes Equations

The early description of fluid motion starting with D. Bernoulli's *Hydrodynamics* [13] was hopelessly wrong in describing practical applications, as shown by the aforementioned Paradox of D'Alembert Theorem 2.1. One important step toward a more *realistic* description, in the form of accurate numerical prediction, becomes possible with the inclusion of internal friction or viscosity. Under one-dimensional flow conditions, it simply states there must be shear stress to maintain cross-gradients:

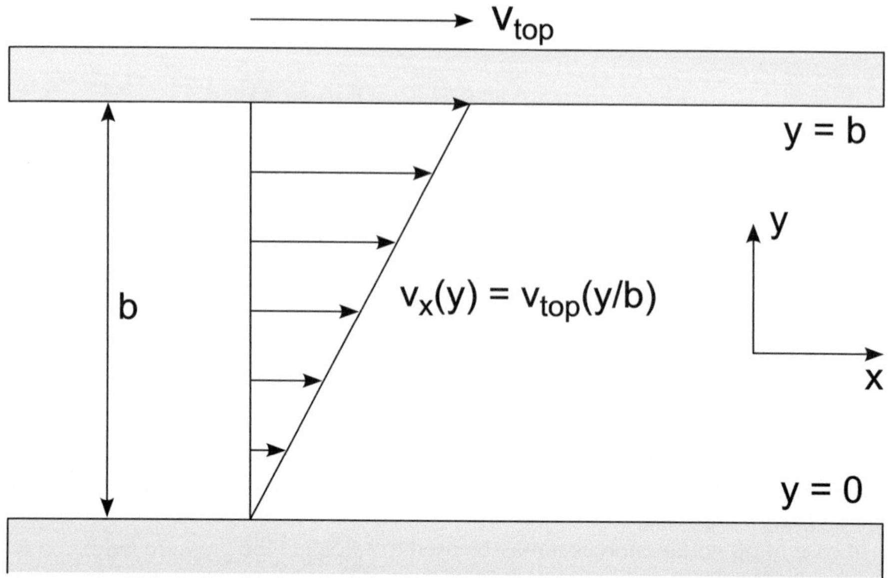

Fig. 3.3 On the definition of viscosity

$$\tau := \eta \cdot \frac{du}{dy} \, . \tag{3.33}$$

See Fig. 3.3. A fluid fulfilling this property is called a *Newtonian* fluid. A generalization to three dimensions is straightforward. Instead of (x, y, z) for the Cartesian components, we will include indices for the coordinates (x_1, x_2, x_3). Latin indices (i, j, k, …) then obey $1 \leq i, j, k \leq 3$ and $(i, j) \in \mathbb{N}$. The Cauchy stress tensor

$$\sigma = (\sigma_{ij}) = \begin{pmatrix} \sigma_{11} & \sigma_{12} & \sigma_{13} \\ \sigma_{21} & \sigma_{22} & \sigma_{23} \\ \sigma_{31} & \sigma_{32} & \sigma_{33} \end{pmatrix} \tag{3.34}$$

simply gives forces (index 1) in direction i with reference to directions (index 2) j. With $\mathbf{v} = (u, v, w) \rightarrow (v_1, v_2, v_3)$ we define a deformation tensor[1]

$$\mathbf{D} = (D_{ij}) = \frac{1}{2}\left(\frac{\partial v_i}{\partial x_j} + \frac{\partial v_j}{\partial x_i}\right). \tag{3.35}$$

---

[1] A tensor may be represented by a n × n matrix. However, like scalars and vectors it is defined by its transformation rules under change of coordinates.

Now the 3D generalization of Eq. (3.33) is

$$\sigma = 2\eta \cdot \mathbf{D} \,. \tag{3.36}$$

Now after some algebra we finally arrive at the *Navier–Stokes Equations* (NSE).

$$\frac{\partial \mathbf{v}}{\partial t} + (\mathbf{v} \cdot \nabla)\mathbf{v} = \frac{1}{\rho}\left(-\nabla p + \mathbf{f}\right) + \nu \Delta \mathbf{v} \,. \tag{3.37}$$

Here $\Delta = \nabla^2$ is the Laplacian Operator which acts on each component and

$$\nu = \frac{\eta}{\rho}, \tag{3.38}$$

the *kinematic viscosity*. By now it is generally believed that all types of flow—as long as the fluid is Newtonian—may be described by Eq. 3.37. We will come back to this discussion when talking about *turbulence* in Sect. 3.8. Again we close with some remarks:

- NSE are **non**linear;
- they are of **second** order, so two boundary conditions at the solid wall must be provided. This defines the so-called *slip condition*, which means that normal (as in the case of Eulerian flow) as well as tangential components of velocity have to vanish at solid walls;
- the limit $\nu \to 0$ is highly nontrivial as it changes the order of the differential equation. As this limit is at the heart of *Boundary Layer Theory*, we will discuss it in more detail in Sect. 3.7.

3.5 Potential Flow

3.5.1 General 3D Potential Flow

Another very important quantity is *vorticity*. It is defined by

$$\omega := \nabla \times \mathbf{v} \tag{3.39}$$

and is closely related to *circulation*:

$$\Gamma_{C_t} := \oint_{C_t} \mathbf{v} \cdot d\mathbf{r} \tag{3.40}$$

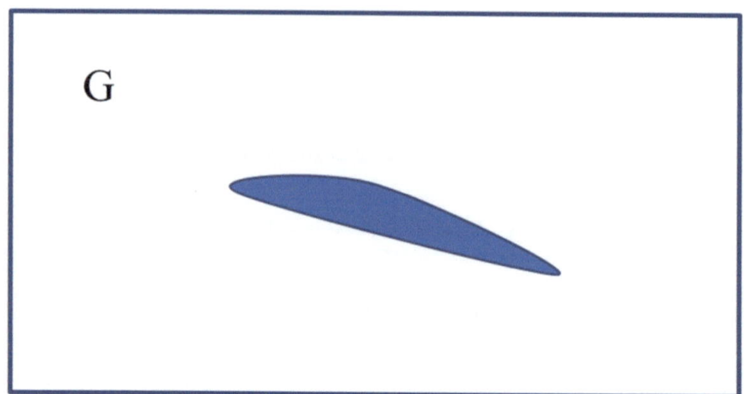

Fig. 3.4 Example for a multi-connected region

which may be seen using *Stokes Theorem* for a vector field **B** on a surface A with boundary $\partial A = C_t$:

$$\int_{\partial A} (\nabla \times \mathbf{B}) \cdot d\mathbf{A} = \oint_A \mathbf{B} \cdot d\mathbf{r} \,. \tag{3.41}$$

This establishes the close relationship between both quantities. A flow is called *irrotational* if $\nabla \times \mathbf{v} = 0$ everywhere. If we assume there exists a scalar function φ with

$$\mathbf{v} := \nabla \Phi \,, \tag{3.42}$$

we call Φ a *velocity potential*. With some weak assumption on the differentiability on Φ we get from $\nabla \times \nabla \varphi = 0$ directly $\nabla \times \mathbf{v} = 0$, which means that such a flow is *irrotational*. This holds only if the region in which the flow is irrotational is simply connected, which means in simple terms that the region does not have holes. Note that even one point may be regarded as a hole (Fig. 3.4).

In such cases

Theorem 3.2 *In isentropic flow, the circulation, Γ_{C_t}, Eq. (3.40) is a conserved quantity—constant in time holds [3, 10, 40]. Potential theory is especially useful and beautiful in two dimensions. Therefore a subsection on this topic is appropriate. Because of*

$$\nabla \cdot (\nabla \times \mathbf{A}) \equiv 0 \,, \tag{3.43}$$

the Continuity Equation is fulfilled identically if we set

$$\mathbf{v} \equiv \nabla \times \mathbf{A} \,. \tag{3.44}$$

A is then called *vector potential*. We will come back to this useful quantity in Sect. 6.7. It has to be noted that 3D potential theoretical calculations were very popular [31] in the early days of *computational fluid mechanics*. As they predict forces resulting only from vortex-induced flow (see Sect. 3.6) they may serve only as to give an impression of the general flow field. Nevertheless it will be seen in the context of *boundary layer flow* (see Sect. 3.7) that—in case of non-*separated* flow—except for a thin layer, most of the flow may be regarded as potential flow.

3.5.2 2D Potential Attached Flow

In this subject we will explain how useful and easy potential flow methods are in two dimensions. To obey the 2D continuity equation (from Eq. 3.27 by omitting z and w), we have

$$\frac{\partial u}{\partial x} + \frac{\partial v}{\partial y} = 0 \, , \tag{3.45}$$

a stream function

$$\psi = \psi(x, y) \tag{3.46}$$

is introduced. It is easily seen that after canceling terms the only remaining component of the 3D vector potential, Eq. 3.44:

$$u = \frac{\partial \psi}{\partial y} \tag{3.47}$$

$$v = -\frac{\partial \psi}{\partial x} \, . \tag{3.48}$$

Now ψ as well as ϕ (the vector potential) obey

$$\Delta \Psi = \Delta \Phi = 0 \, . \tag{3.49}$$

Now introducing *complex variables* $z := x + i \cdot y \in \mathbb{C}$: together with complex velocity $\mathbb{C} \ni w = u - i \cdot v$ and *complex potential* $F(z) := \Phi + i \cdot \Psi$. An important result from complex calculus is that (by the action of *Cauchy–Riemann's* Theorem [31].

Theorem 3.3 *Every complex function* $A(z) = u(x, z) + i \cdot v(x, y)$ *which is* homomorphic *(complex differentiable) broken down to real and imaginary parts satisfies the potential equation*

$$\Delta u = \Delta v = 0 \, . \tag{3.50}$$

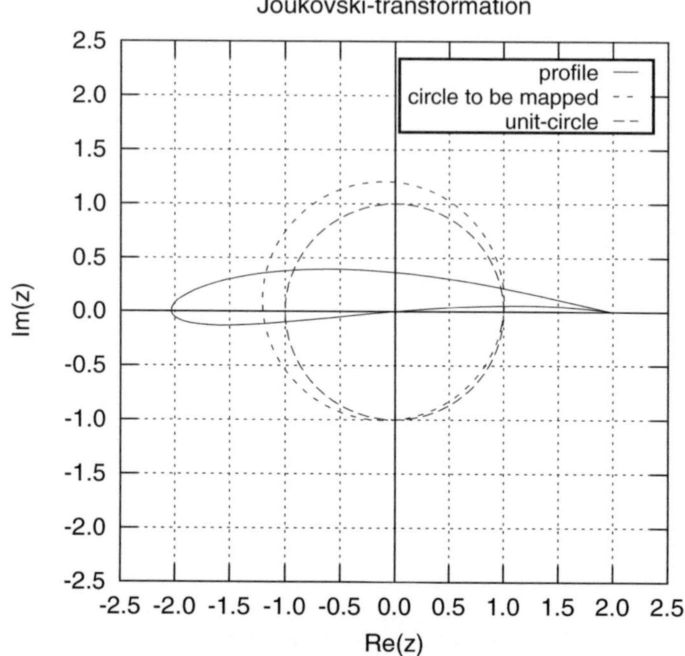

Fig. 3.5 Generation of a Joukovsky profile by conformal mapping

Now let $\mathbb{C} \to \mathbb{C}$ with $\zeta \to zf(\zeta)$ via

$$\zeta \to := z + \frac{c^2}{z} \tag{3.51}$$

the *Joukovsky Transformation* (see Problem 3.4 for details) which maps a circle $z(\phi) = z_m + R \cdot e^{i \cdot \phi}, 0 \le \phi \le 2\pi$ to the profile in Fig. 3.5.

The linearity of the potential equation allows us to superimpose various solutions. For a cylinder of radius in uniform flow with $w = u + i \cdot 0$, we have

$$F(z) = u \cdot z + \frac{R^2 \cdot u}{z} \tag{3.52}$$

which can be transformed to a corresponding pressure distribution on the profile; see Fig. 3.6. As Eq. 3.51 is holomorphic except for $\mathbb{C} \ni z = 0$ its inversion is as well. Therefore a one-to-one relationship between pressure distribution and profile shape exists, which may be extended to a design tool for aerodynamic profiles [14, 16] (Fig. 3.7).

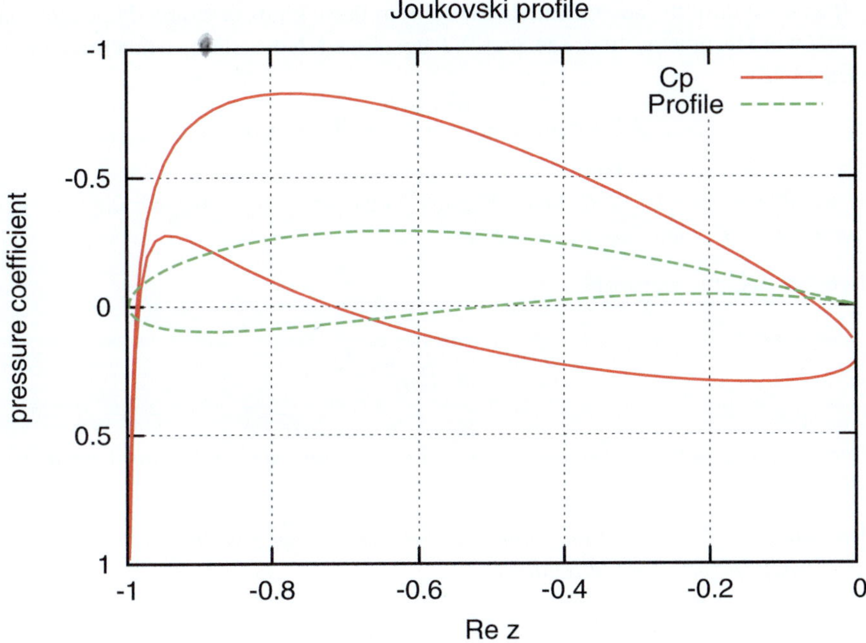

Fig. 3.6 Pressure distribution of a simple Joukovsky profile

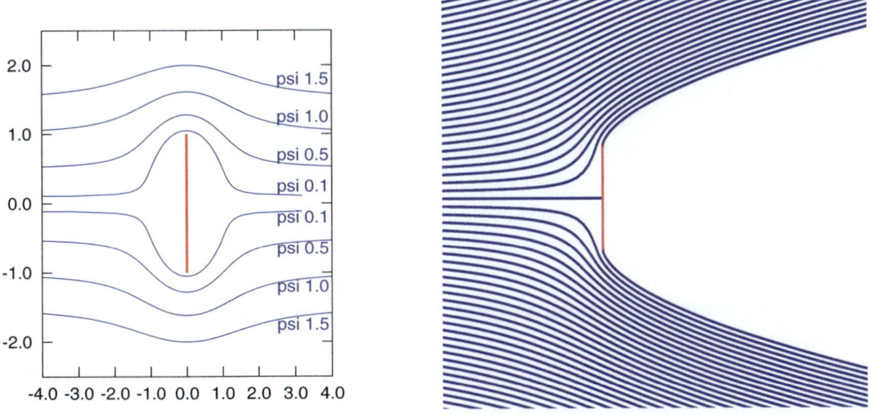

Fig. 3.7 Flow normal to a plate (red). Attached (left $c_D = 0$) and separated (right, $c_D = 0.88$) potential-theoretic flow

It was not until the advent and distribution of these kinds of simple PC-codes that special wind turbine profiles, see Sect. 9.2, developed. Now adding a potential vortex to Eq. (3.52):

$$F(z) = u \cdot z + \frac{R^2 \cdot u}{z} - i \cdot \Gamma \, log(z) \, , \qquad (3.53)$$

we are able to generate *Lift*. From Blasius'[2] theorem [10] using a complex force $\mathcal{F} = F_x + i \cdot F_y$ and $\overline{\mathcal{F}} = F_x - i \cdot F_y$:

Theorem 3.4 *In exterior flow:*

$$\overline{\mathcal{F}} = -\frac{i\rho}{2} \int_{\partial \mathcal{B}} w^2 \, dz \, . \qquad (3.54)$$

From this very general theorem, we obtain if we include circulation (positive in z-direction):

Theorem 3.5 *If there is circulation* $\Gamma = \oint \mathbf{v} \cdot d\mathbf{r}$ *around* $\mathcal{B}$ *with inflow velocity* $\mathbf{v}_\infty$, *then a force (positive in y-direction)*

$$L = \rho \cdot \mathbf{v}_\infty \cdot \Gamma \qquad (3.55)$$

emerges.

The amount of circulation may be determined by the following corollary:

Corollary 3.1 *The flow around an airfoil has to be of finite (but not continuous) velocities [49].*

This more or less mathematical statement[3] gave a first explanation of lift which for a long time was a mystery. Even today many incorrect explanations exist. For an easy-to-read version, see [2] and [39] of Chap. 1. An extension of these ideas gave rise to *Thin Airfoil Theory*, originated by *Max Munk* [42], *Hermann Glauert* [24] and others (see Sect. 3.6.2).

One of the fundamental results is with c_L being the *lift coefficient* and α the *angle of attack*; see Fig. 3.8

$$c_L = 2\pi \cdot \alpha \, . \qquad (3.56)$$

---

[2] Heinrich Blasius, * 09.08.1883, † 24.04.1970, was one of *L. Prandtl's* first Ph.D. students. His work on forces and *Boundary Layer* flat plate theory is discussed in almost every textbook of fluid mechanics even today. Less known is that he taught over 50 years (1912–1970) at *Ingenieuerschule* (Polytechnic—now University of Applied Sciences) Hamburg.

[3] It may interest to note that here the essential argument is to avoid singularities in terms of infinite velocities, whereas in other applications this does not apply, for example the infinite pressure at an inclined flat plate.

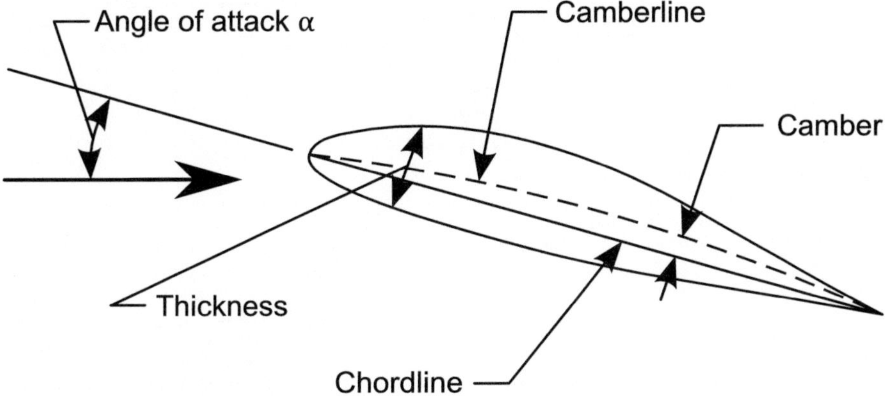

Fig. 3.8   Angle of attack and other geometrical properties regarding an airfoil section

One important ingredient is the *Kutta Condition* which fixes the otherwise indeterminate amount of circulation by a simple mathematical condition that all velocities have to be bounded.

3.5.3   2D Potential Flow Behind a Semi-infinite Set of Lamina

Prandtl and Betz, see [5, 59] and Fig. 5.10, used the complex potential

$$F(z) = -v \, \frac{d}{\pi} \, arccos(exp(\pi z/d))$$
(3.57)

with v the inflow and d the spacing of the plates.

3.5.4   2D Separated Flow

As the result of further progress in complex calculus, Kirchhoff and Helmholtz [40] were also able to investigate separated flow. One success was the approximate formulation of drag on a inclined flat plate:

$$c_D = \frac{2\pi \cdot sin(\alpha)}{4 + \pi \cdot sin(\alpha)} \, ,$$
(3.58)

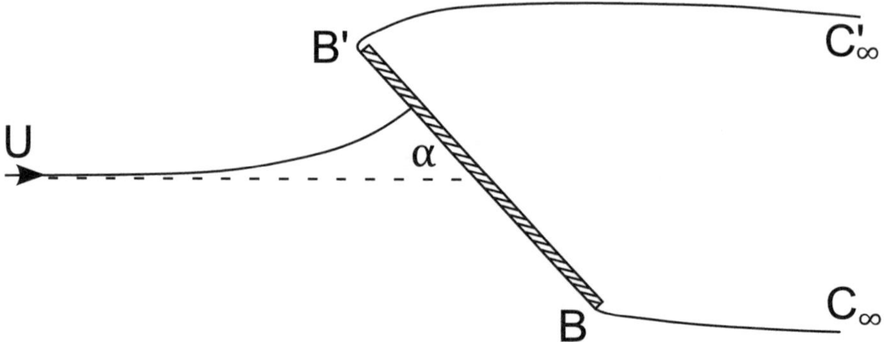

Fig. 3.9 Flow pattern of Helmholtz–Kirchhoff Flow on an inclined flat plate, adapted from this chapter [40]

which at $90°$ gives $c_{D,90} = 0.88$. The experimental value varies according to Reynolds number and an aspect ratio between 1.2 and 2.0. An [36, 44] application of this model pertaining to *dynamic stall* is available (Fig. 3.9).

3.6 Formulation of Fluid Mechanics in Terms of Vorticity and Vortices

The pressure may be eliminated from the body-force-free NSE, Eq. (3.37) by taking the *curl* of **v**

$$\frac{D\omega}{Dt} = \omega \cdot \mathbf{v} + \nu \Delta \omega \,. \tag{3.59}$$

As an outcome of pressure, there may be computed only from *Poisson's equation* which is a non-homogeneous Laplace equation:

$$\Delta p = -\rho \frac{\partial^2 u_i u_j}{\partial x_i \partial x_j} \equiv S(\mathbf{r}). \tag{3.60}$$

Formally[4] this equation then may be solved by the introduction of *Green's function*:

$$p(\mathbf{r}) = p^{harmonic}(\mathbf{r}) + p(\mathbf{r}) + \frac{\rho}{4\pi} \int_{\mathbb{R}^3} \frac{S(\mathbf{r})}{|\mathbf{r} - \mathbf{r}'|} \, d\mathbf{r}' \,, \tag{3.61}$$

where $p^{harmonic}(\mathbf{r})$ is a solution of the homogenous pressure equation:

---

[4] Here we use Einstein's summation convention, to sum over all dual-indexed variables: $u_i \cdot u_i := \sum_{i=1}^{N} u_i \cdot u_i$.

$$\Delta p^{harmonic}(\mathbf{r}) = 0 \; . \tag{3.62}$$

Obviously the situation becomes much easier if vorticity is concentrated. If so, these compact distributions then may be regarded as the *sinews and muscles of fluid motion* [49]. The simplest case is a so-called *point-vortex* (2D) and *line vortex* (3D). From Eq. (3.42) we get

$$\omega = \nabla \times \mathbf{v} = -\nabla^2 \mathbf{A} \tag{3.63}$$

if $\nabla \cdot \mathbf{A} = 0$. Again by using *Green's functions*:

$$\mathbf{A} = \frac{1}{4\pi} \int_{\mathbb{R}^3} \frac{\omega}{|\mathbf{r} - \mathbf{r}'|} \, dV \tag{3.64}$$

which is called *Biot–Savart-Law*. Inserted into the velocity equation we get

$$\mathbf{v} = \frac{1}{4\pi} \int_{\mathbb{R}^3} \nabla \times \frac{\omega}{|\mathbf{r} - \mathbf{r}'|} \, dV \; . \tag{3.65}$$

We apply this now to a specific situation in which ω is only non-zero along a line, for example the z-axis in a 3D Cartesian system. Performing the integration (see Problem 3.2) and referring to Fig. 3.10 we arrive at (Fig. 3.11):

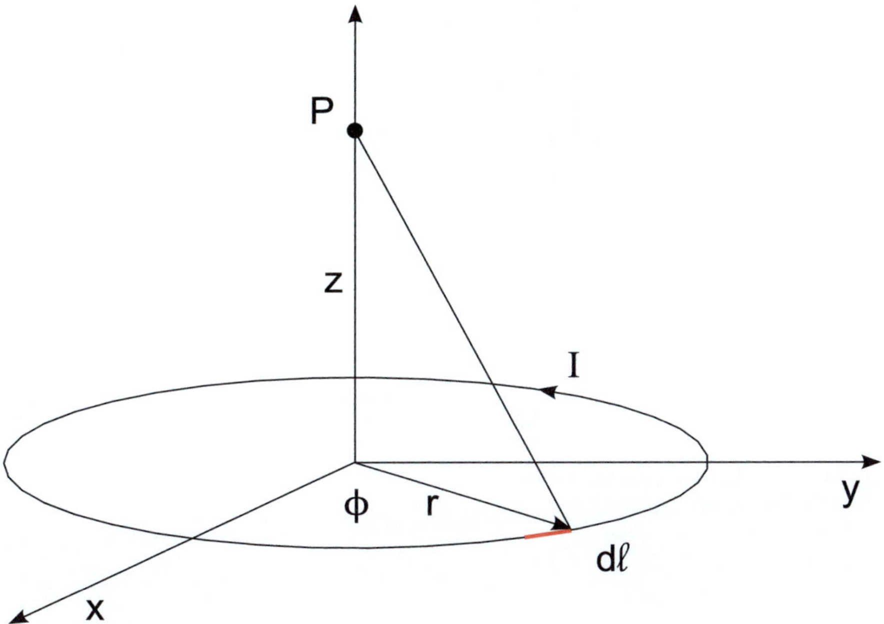

Fig. 3.10 Biot–Savart-law

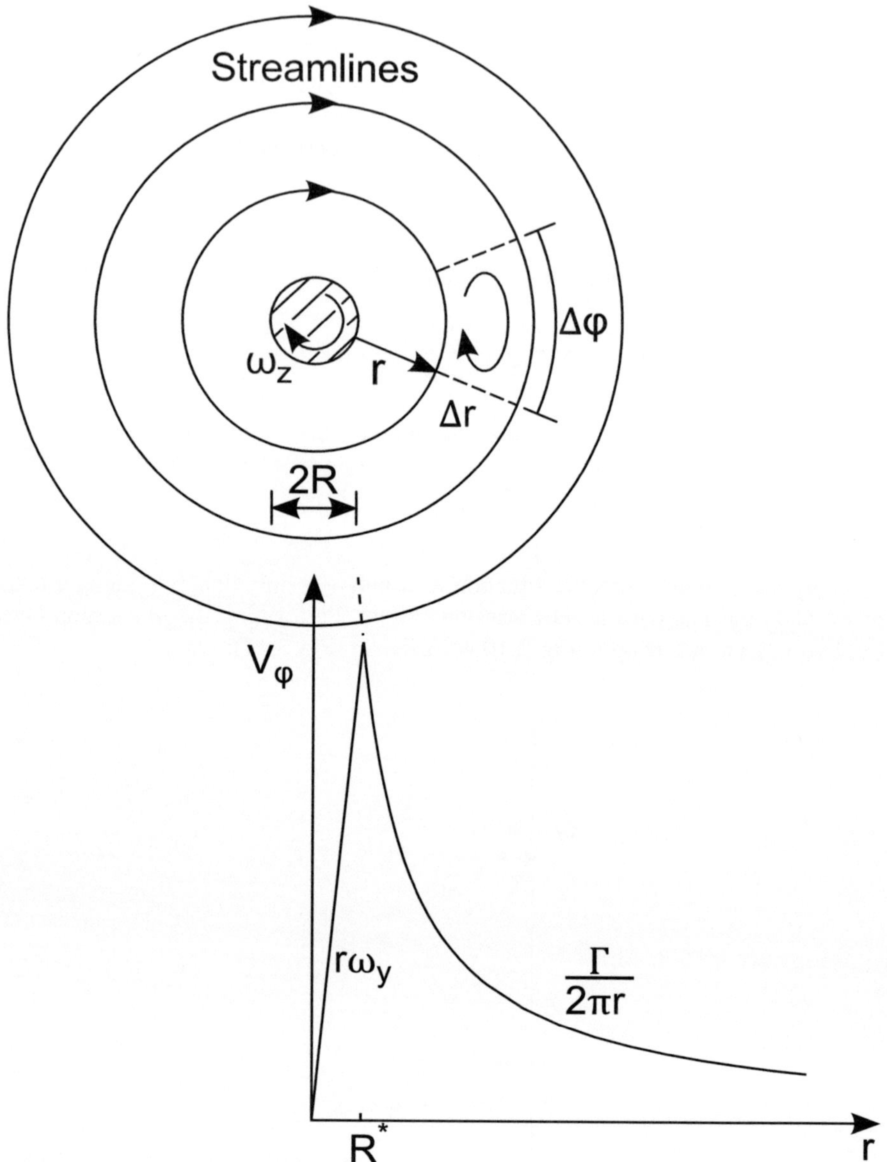

Fig. 3.11 Point(2D)- or Line(3D) vortex

$$v_r = 0 \tag{3.66}$$

$$v_\phi = \frac{\Gamma}{2\pi r} \tag{3.67}$$

with Γ from Eqs. (3.39) to (3.41) (Figs. 3.12 and 3.13).

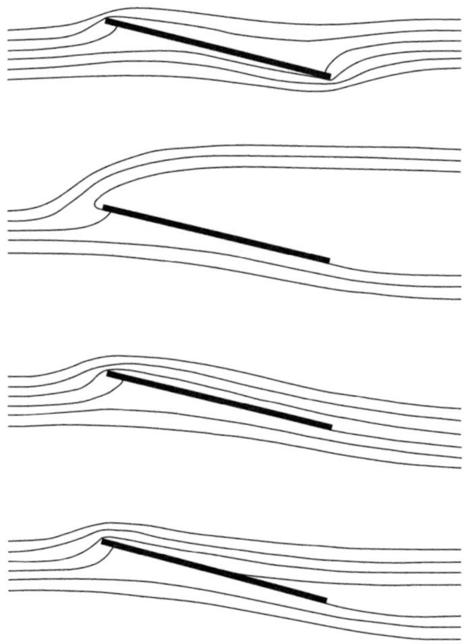

Fig. 3.12 Origin of forces induced by an inclined flat plate depending on the flow pattern. Adapted from Jones [29]

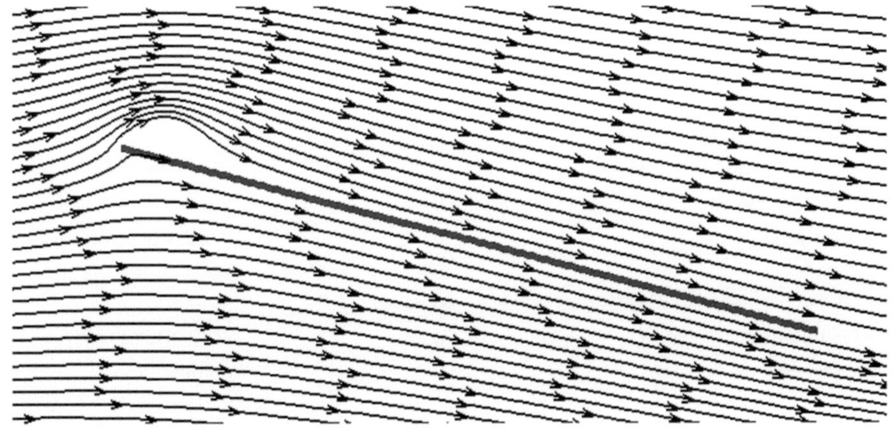

Fig. 3.13 CFD (RANS) simulation of flat plate flow, according to Fig. 3.12

3.6.1   Flow and Forces

3.6.2   Thin Airfoil Theory

With [24] we have a 2D section of chord c where the circulation is distributed along a so-called *skeleton line* $y(x)$, $0 \leq x \leq c$ of (Table 3.1)

$$\Gamma = \int_0^c \gamma(x) \, dx \; . \tag{3.68}$$

According to Biot–Savart's Law Eq. (3.65) we have induced velocities:

$$v(x') = \int_0^c \frac{\gamma(x) \, dx}{2\pi \, (x - x')} \; . \tag{3.69}$$

If the inflow velocity is designated by V then by noting that no normal flow through the airfoil is possible, the important constraint is

$$\alpha + \frac{v}{V} = \frac{dy(x)}{dx} \; . \tag{3.70}$$

Glauert [24] solved this equation by expanding $\gamma(x)$ in terms of a *Fourier Series*:

$$\gamma(x) = 2V \left(A_0 \cdot cot \left(\frac{\theta}{2} + \sum_{n=1}^{\infty} A_n sin(n\theta) \right) \right) \tag{3.71}$$

$$x(\theta) = \frac{1}{2} c(1 - cos(\theta) \; 0 \leq \theta \leq \pi \; . \tag{3.72}$$

Table 3.1 Numerical values

Type	c_L	c_D	Remark
Potential theory	0.00	0.00	
Helmholtz–Kirchhoff	0.33	0.09	
Kutta–Joukovsky	1.63	0.00	
BL-theory (1)	1.53	0.02	NACA0015 XFoil
BL-theory (2)	0.70	0.25	NACA0005 XFoil
RANS laminar	0.92	0.25	FLUENT
RANS turbulent	1.03	0.30	FLUENT $k - \epsilon$
Measurement	0.81	0.22	Föppl [20]

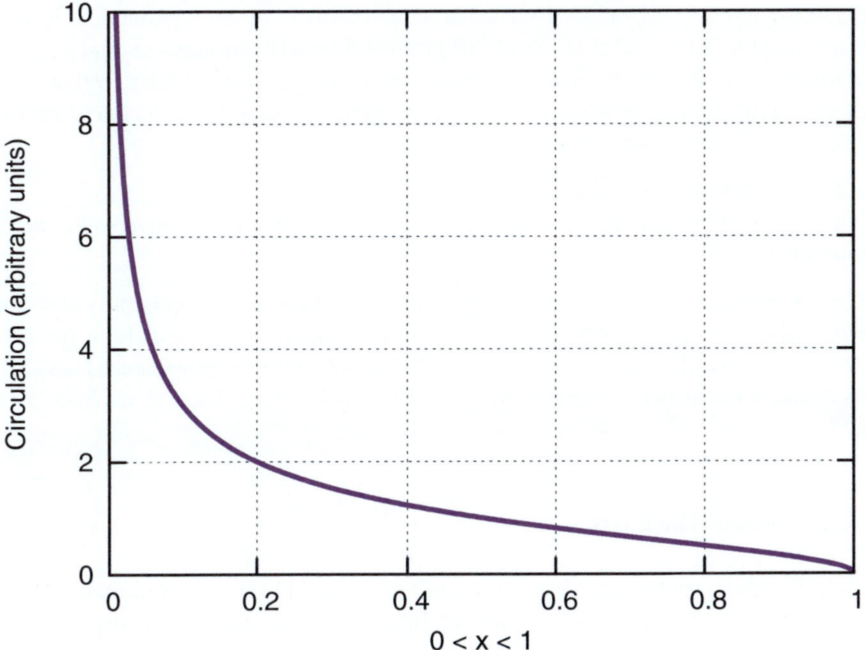

Fig. 3.14 Circulation distribution for an inclined flat plate

He found:

$$c_L = 2\pi \left(A_0 + \frac{1}{2} A_1 \right) \tag{3.73}$$

$$A_0 = \alpha - \frac{1}{\pi} \int_0^\pi \frac{dy(x)}{dx} d\theta \text{ and} \tag{3.74}$$

$$A_n = \frac{2}{\pi} \int_0^\pi \frac{dy(x)}{dx} \cos(n\theta) d\theta \ . \tag{3.75}$$

For a flat plate $A_0 = 1$ and $A_1 \approx 0$. The corresponding circulation distribution $\gamma(x)$ is shown in Fig. 3.14. It has a square-root singularity for $x \to 0$. The Kutta(–Joukovsky) condition is implemented via $\gamma(x = 1) = 0$, as can be seen in Fig. 3.14.

3.6.3 Viscous Thin Airfoil Theory

TAT is based on inviscid models of fluid flows (only density as a material enters) and as a consequence, circulation—among others—is a conserved quantity, i.e. it can neither be created nor destroyed. Therefore, more sophisticated models (and equations) must be included if the emergence of lift is to be explained. It is well known that the Navier–Stokes Equations provide this basis, adding a second material parameter, viscosity. In a series of journal and technical papers Yates [60], among others, was able with the help of a Oseen-type approximation (in fact a linearization) using these Navier–Stokes equations to

1. derive the Kutta condition and
2. to give asymptotic corrections to the lift-curve slope in terms of inverse Reynolds number.

This is somewhat surprising as Oseen-type-Flow [3], Sect. 4.10, is generally assumed to be valid in low-Re (RN $<$ 1, creeping) flow only, whereas in high-Re flow (RN $>$ 10^5) boundary layer theory Sect. 3.7 should be more appropriate. Meanwhile this discussion has been extended by so-called *Triple–Deck-Theory*; see Sect. 3.7.4 and [50].

3.6.3.1   Finite Thickness

A lot of authors including Abbot and von Doenhoeff [1] tried to improve TAT by investigating the influence of thickness on lift-curve-slope which typically results in equations like [1],

$$c_L = 2\pi \left(1 + \tau\right) , \tag{3.76}$$

$$\tau = \frac{\epsilon}{a} = \frac{4\sqrt{3}}{9} \cdot \frac{t}{c} . \tag{3.77}$$

Yates [60] further gave Reynolds number corrections

$$c_L = 2\pi \left(1 + \tau\right) \cdot \left(1 - \frac{4}{\log\left(64Re\right) + \gamma_E}\right) \tag{3.78}$$

$\gamma_E = 0.57722$ being Euler's constant which shows a decrease of about 10% at t/c $=$ 0.3. from RN effects which—at least—is partly compensated by the first (thickness) term. No negative lift-curve slope values are likely. McLean [39], Chap. 7.4, pp 313/314 gives further details.

Not included in all these discussions is the influence of the flow-state of the boundary layer, whether it is laminar or turbulent. In our discussions, we assume that lift (in the linear part) is not influenced as strong as drag. It is well known that drag can be much higher when most parts of the boundary layer are turbulent.

Another important phenomenon, flow separation, the starting point defined by

$$c_f = \mu \cdot \frac{dv_t}{dn} \leq 0 , \tag{3.79}$$

is closely related to limiting c_L to values around 1.5 (of course with remarkable exceptions).

3.6.4 Vortex Sheets

A special kind of 2D extended vortical objects are called *vortex sheets*. As can be shown [46, 49] this is more or less equivalent to a discontinuity in tangential velocity only. These sheets may be created, for example, by sudden movement of a *Kaffelöffel* (coffee spoon) [32]. A major application is Prandtl's *lifting line theory* . We will come back to this point later in Sect. 6.6.

3.6.5 Vorticity in Inviscid Flow

As will be presented in more detail in Chap. 6, Theorem 1 (*Helmholtz' Theorem*), circulation is a conserved quantity in inviscid flow. In addition, along streamlines energy is also conserved when expressed in terms of enthalpy H, Eq. (3.23). H varies when changing streamlines according to *Crocco's equation*

$$\mathbf{u} \times \omega = \nabla \cdot H \ . \tag{3.80}$$

3.7 Boundary Layer Theory

3.7.1 The Concept of a Boundary Layer

As was remarked earlier, NSE and the inviscid Euler Equation are not connected by an analytical limit $v \to 0$, because the order of both differential equations is not the same. However, a very useful approximation is possible as will be seen in the following. We start with one of the simplest cases, the flat plate, which is described by uniform flow $u(y) = u_\infty = const, x < 0$ and $u(x, y = 0) = v(x, y = 0) = 0$, $x > 0$. We follow closely, but not completely, the path presented in [52]. We start with the stationary 2D Cartesian NSE $\mathbf{v} = (u,v)$ and set $\rho = 1$. Then we have the following set of equations:

$$\frac{\partial u}{\partial x} + \frac{\partial u}{\partial y} = 0 \tag{3.81}$$

$$u\frac{\partial u}{\partial x} + u\frac{\partial u}{\partial y} = -\frac{\partial}{\partial x}p + v\Delta u \tag{3.82}$$

$$u\frac{\partial v}{\partial x} + v\frac{\partial v}{\partial y} = -\frac{\partial}{\partial y}p + v\Delta v. \tag{3.83}$$

As usual in 2D we obey continuity by introducing a stream function ψ:

$$u = \frac{\partial \psi}{\partial y} \tag{3.84}$$

$$v = -\frac{\partial \psi}{\partial x} \, . \tag{3.85}$$

We try a *separation approach* via

$$\psi := \sqrt{2\nu x u_\infty} \cdot f(y) \, . \tag{3.86}$$

In addition, we use a *similarity approach* for the vertical coordinate (η now not to be confused with the earlier defined dynamic viscosity)

$$\eta := y \sqrt{\frac{u_\infty}{2\nu x}} \, . \tag{3.87}$$

Therefore,

$$u = x \cdot f', \ v = -f \, . \tag{3.88}$$

Inserting into Eqs. (3.82) and (3.83) we may drop the pressure to arrive at *Blasius' Equation*:

$$f''' + f f'' = 0 \, . \tag{3.89}$$

At the wall $y = 0$ which means $f(0) = f'(0) = 0$ to obey the *no-slip* condition $u = v = 0$. We now have a third-order differential equation, so we have to add one boundary condition. We demand asymptotically for $y \to \infty$ or equivalently $\eta \to \infty$

$$f'(\eta) = 1 \ as \ \eta \to \infty \, . \tag{3.90}$$

Blasius originally [6] solved Eq. (3.89) by a power-series approach. By now the mathematical understanding is much more complete [8], and very easy numerical schemes exist. The boundary condition (3.90) may be shifted to $f''(0) = const$ and looking for a constant *const* which leads to $lim_{\eta \to \infty} f(\eta) = 1$. As may be seen from Fig. 3.15 at $\eta = 3.8\,(5)$ we have $f'(\eta) = 0.996\,(0.99994)$ so that we may define a *boundary layer thickness* by

$$\delta_{99\%} := 3.5 \cdot \sqrt{2\nu x u_\infty} \, . \tag{3.91}$$

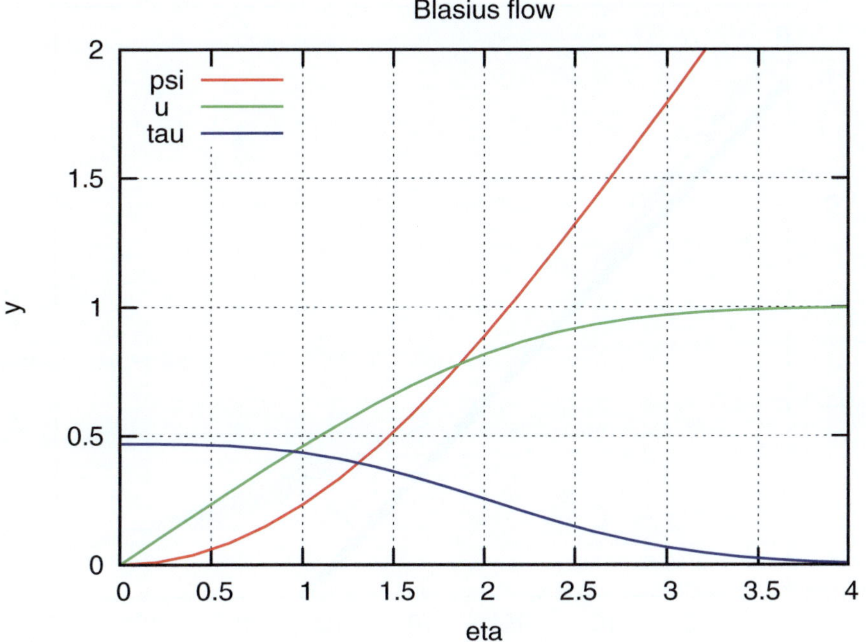

Fig. 3.15 Boundary layer at a flat plate—Blasius' solution

Two other variables, displacement and momentum thickness, are also used:

$$\delta^\star := \int_0^\infty \left(1 - \frac{u}{u_\infty}\right)\, dy \tag{3.92}$$

$$\theta := \int_0^\infty \frac{u}{u_\infty} \cdot \left(1 - \frac{u}{u_\infty}\right)\, dy \,. \tag{3.93}$$

Now we are able to calculate shear stresses on the wall:

$$c_f := \frac{2\tau_w}{\rho u_\infty^2} = \frac{0.664}{\sqrt{Re_x}} = \frac{\theta}{x} \,. \tag{3.94}$$

Rather accidentally we introduced a very important number the *Reynolds number*

$$Re_x := \frac{u_\infty \cdot x}{\nu} \,. \tag{3.95}$$

With typical numbers from wind turbines (x = 1 …10 m, u = 10 …50 m/s, $\nu = 1.5 \cdot 10^{-5}\,\mathrm{m^2/s}$) we have $Re_x \sim 0.6 \ldots 35 \cdot 10^6$, therefore BL-thicknesses on the

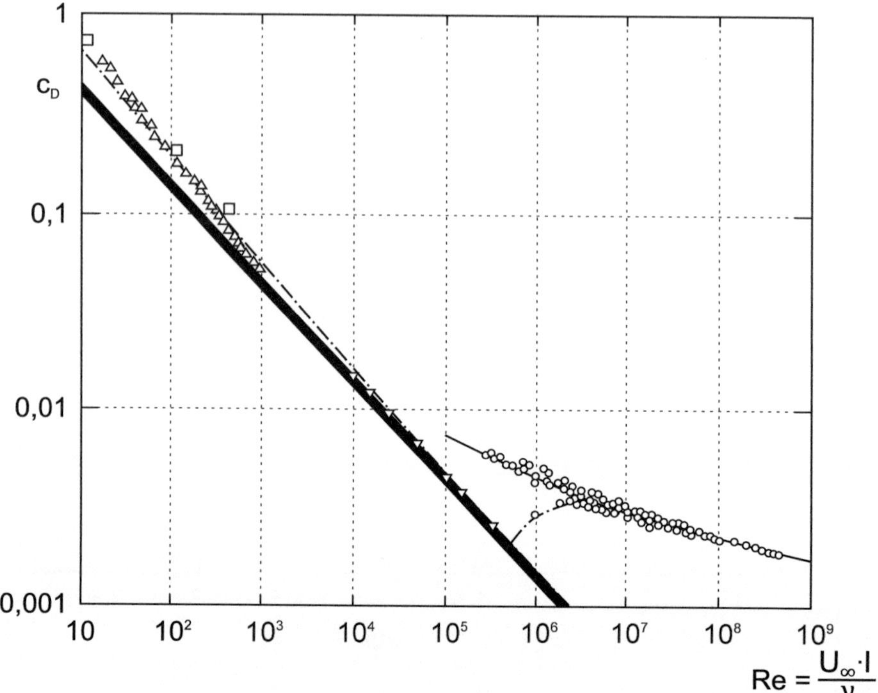

Fig. 3.16 Comparison of measurement and boundary layer theory

order of cm. Figure 3.16 shows a comparison of measurement of friction with the theory developed so far. Two things are remarkable:

- the calculated c_F value fit very well in the range $Re < 10^5$ but starts to deviate at values less than 10^2;
- for $Re > 10^5$ c_F is much larger, but a theory is also available.

3.7.2   *Boundary Layers with Pressure Gradient*

As has been seen for the case of a simple Joukovsky airfoil, pressure is not constant on these important 2D sections. Therefore is it very useful to look for a simple (potential) flow which also may be calculated by BL-methods. This kind of flow is the so-called wedge flow or *Falkner–Skan-Flow*; see Fig. 3.17. The pressure may be characterized by a dimensionless number Λ and the profiles within the boundary layer are shown in Fig. 3.18. As may be expected, beginning separated flow ($\Lambda \approx 12$) exhibits a profile with a vertical tangent.

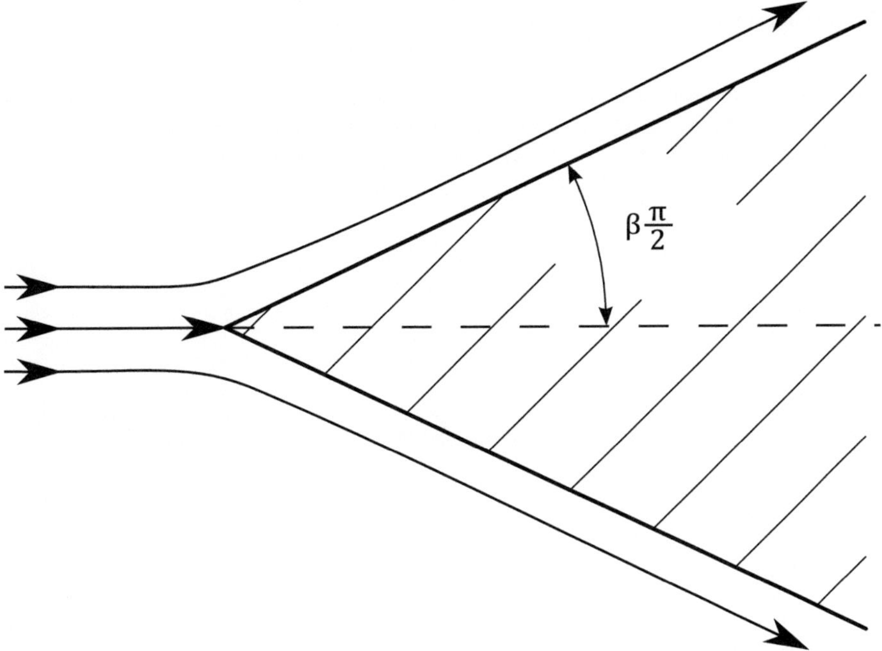

Fig. 3.17 Faulkner–Skan flow

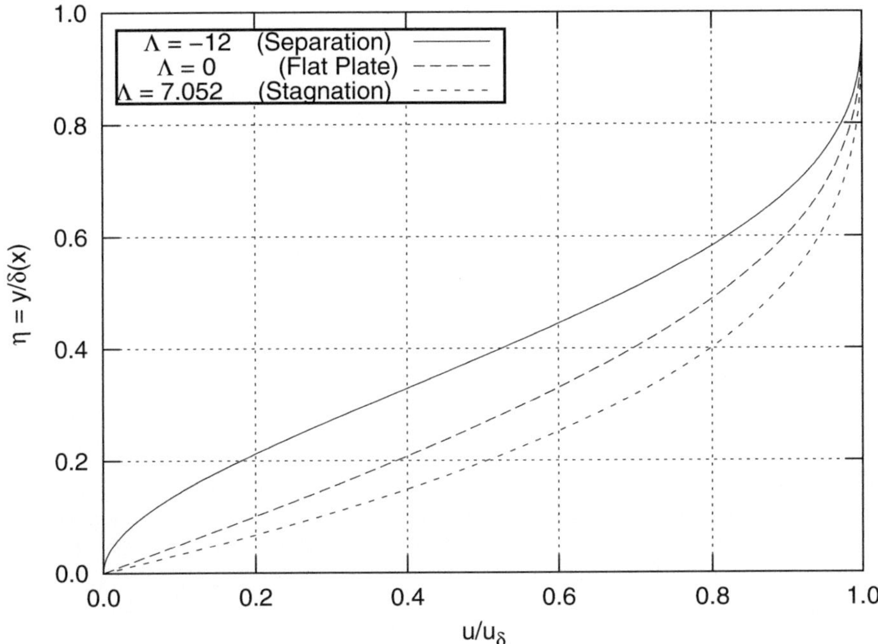

Fig. 3.18 Faulkner–Skan flow: velocity profiles

3.7.3 Integral Boundary Layer Equations

Differential equations are useful if all details of the flow are sought. In most cases, however, only the shear stress at the wall is needed. Considerable simplification can be achieved when integrating the BL-DEQ, Eq. (3.82) [31]. After some algebra one arrives at an ordinary DEQ for the momentum thickness Eq. (3.93)

$$\frac{d\theta}{dx} + (H+2)\frac{\theta}{u_e}\frac{du_e}{dx} = \frac{C_f}{2} \ . \tag{3.96}$$

And using Eqs. 3.92 and 3.93,

$$C_f := \frac{\tau_w}{\frac{1}{2}\rho u_e^2} \ , \tag{3.97}$$

$$H := \frac{\delta^\star}{\theta} \ , \tag{3.98}$$

u_e is the edge velocity which also predicts the pressure at the BL-edge. As may be readily seen, this DEQ contains three unknown (functions): $\delta^\star$, θ and the wall shear stress τ_w via the friction coefficient. To solve for the velocity profile, families of polynomials are used:

$$\frac{u}{u_e} = a_1\eta + a_2\eta^2 + a_3\eta^3 + a_4\eta^4 + \cdots 0 \leq \eta \leq 1 \ . \tag{3.99}$$

Applying the boundary conditions for the flat plate (Blasius) case, this reduces to

$$\frac{u}{u_e} = 2\eta - 2\eta^3 + \eta^4 \ . \tag{3.100}$$

Integration of Eq. (3.96) can be performed easily with the result for the friction coefficient, e.g.

$$C_f \cdot \sqrt{Re_x} = 0.6854 \ , \tag{3.101}$$

instead of

$$C_f \cdot \sqrt{Re_x} = 0.664 \ , \tag{3.102}$$

for the exact Blasius solution. This method has been successfully implemented into XFoil (open-source, [14]) and Rfoil [58], the latter being enhanced for rotational effects which may become important at the root sections of WT-blades. Figure 3.19 shows as an example the results of lift and drag for the Joukovsky airfoil from Fig. 3.5.

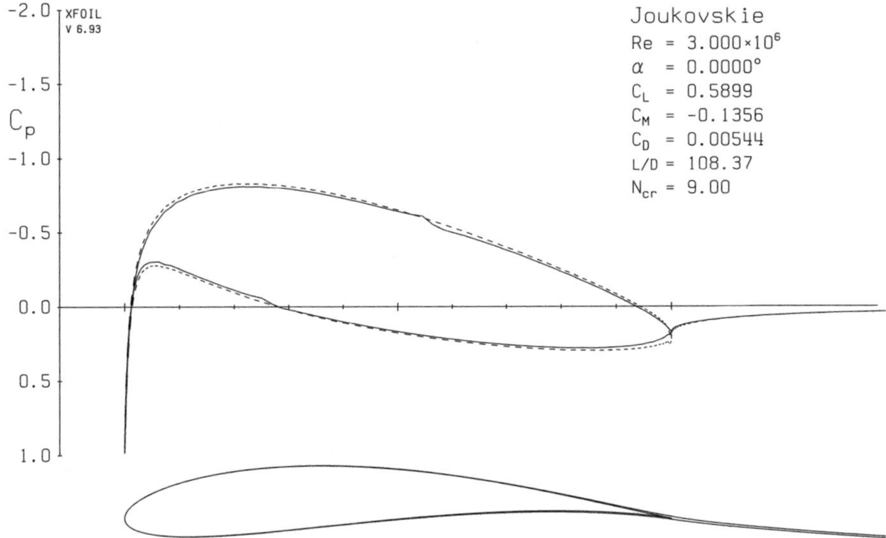

Fig. 3.19 Lift and drag data from integral BL theory (XFoil) for a Joukovsky airfoil

Where do boundary layers occur in wind turbine aerodynamics? Two places are to be noted first and foremost: the lower atmosphere may be regarded as a boundary layer (with a thickness of several hundred meters, however) and when investigating flow over a wind turbine blade in more detail. Unfortunately, the latter BLs are of a very different kind—they are *turbulent*, which will be explained in Sect. 3.9.

3.7.4   *Higher Order Boundary Layer Theory*

Very gradually the original Boundary Layer Theory, see first paper of [46], was understood as an asymptotic expansion in $RN^{-(1/n)}$, $n = 8$. Improvements in terms of higher-orders have been looked for; however, this proved to be more difficult than expected; see [11, 47, 54–56]. Nevertheless, as already stated [50] this model seems to be far enough developed to **explain** the physical origin of the Kutta Condition and derive viscous correction to the famous 2π slope of the lift coefficient.

3.8   Stability of Laminar Flow

Instability of laminar flow is one pre-condition for the emergence of a very different other type of flow, turbulent flow (Figs. 3.20, 3.21 and 3.22).

It is somewhat surprising that a stability analysis was not performed for wind turbine blade flow until recently. Hernandez [25] modified the stability equations to

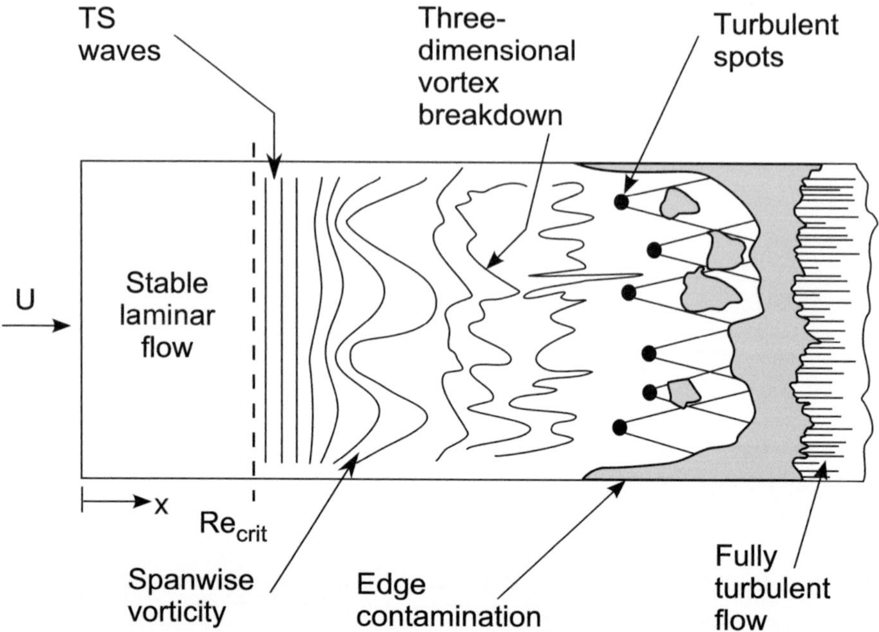

Fig. 3.20 Transition from unstable laminar to fully turbulent flow

account for rotational effects. More about his results applied to transition prediction will be found in Sect. 8.7. Further, very recently [17] stability investigations were used to gain theoretical insights into the laminar–turbulent transition mechanism on a wind turbine blade under atmospheric inflow conditions; see Sect. 8.7.4 for further details.

3.9 Turbulence

3.9.1 Introduction

The NSE may be written in a non-dimensional version if reference values of time (T), length (l) and velocity (U) are introduced: $x \rightarrow x/l, t \rightarrow t/T, (u, v, w) \rightarrow u/U, v/u, w/U$:

$$\frac{\partial \mathbf{v}}{\partial t} + (\mathbf{v} \cdot \nabla)\mathbf{v} = (-\nabla p + \mathbf{f}) + \frac{1}{Re}\Delta \mathbf{v}\,, \tag{3.103}$$

with $Re := \frac{U \cdot L}{\nu}$ and pressure in non-dimensionalized form.

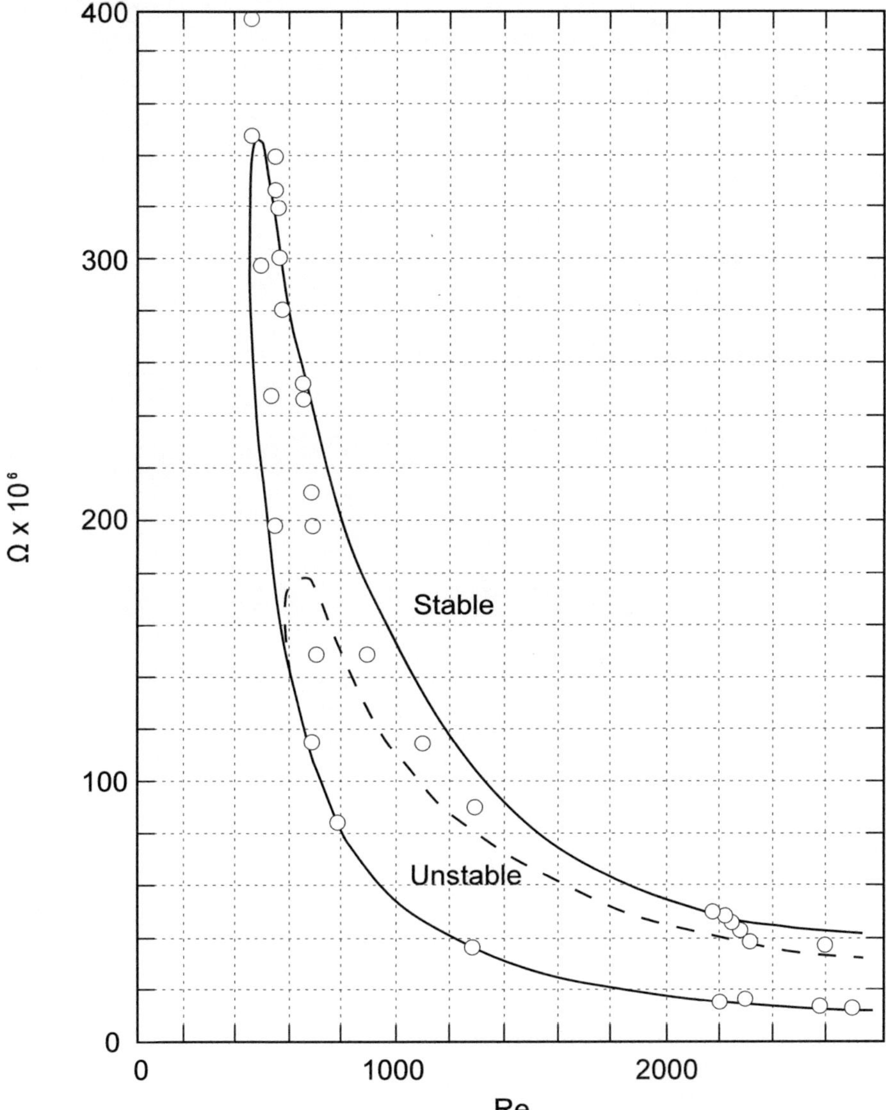

Fig. 3.21 Results from linear stability theory (lines) compared to measurements (dots) for flat plate flow, $Re_{\Theta}^{cr} \sim 520$

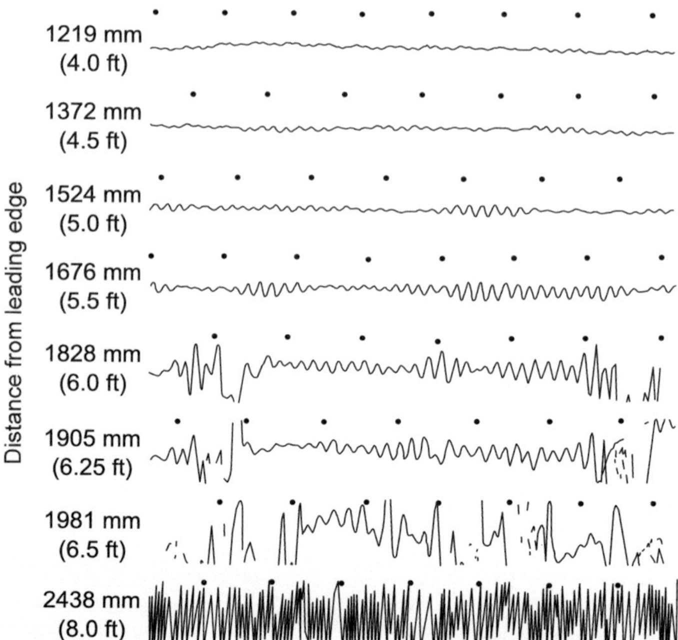

Oscillograms of u-fluctuations showing laminar boundary-layer oscillations in boundary layer of flat plate. Distance from surface, 0.5842 mm; U_∞= 24384 mm per second; time interval between dots, 1/30 second.

Fig. 3.22 First measurements of Tollmien–Schlichting-waves in a flat plate boundary layer, data from [53]

It was recognized very early by *Leonardo da Vinci* that a special (highly disordered—or as he called it *hairy*) state of fluid flow exists and he actually introduced the term (see Fig. 3.23) more than 500 years ago.

But it was not before the end of the nineteenth century that a scientific fluid-dynamical investigation was undertaken by *Osborne Reynolds*. He injected dye into a pipe of circular cross section and found that in case of high flow velocity, the shape and structure changed into an irregular and rapidly changing pattern (see Fig. 3.24). In addition to that he introduced *Reynolds averaging* so that a special kind of NSE is now abbreviated as *RANS* (Reynolds Averaged Navier–Stokes Equation).

Fig. 3.23 One of the first sketches of turbulence *turbolenza* from Leonardo da Vinci [12] which was drawn around 1510. It includes remarkably many details. Reproduced with Permission of Royal Collection Trust/©Her Majesty Elizabeth II 2013

Fig. 3.24 O. Reynolds classical experiments

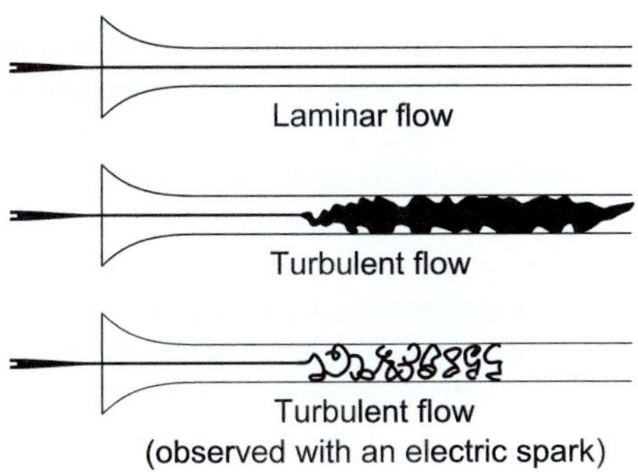

3.9.2 *Mathematical Theory of Turbulence*

3.9.2.1 Existence and Properties of Solutions

Many people believe that such investigations are of minimal importance. This is not at all true. A sound mathematical model serves in all known cases as a rich source for improving computation efforts. As explained in [18, 19, 23, 37] the mathematical theory of the properties of NSE is of utmost importance, because there is a still unproven conjecture that turbulence might be defined as the *breakdown of smooth solutions* of NSE after a finite time. By smooth we mean:

$$p, \mathbf{v} \in C^{\infty}(\mathbb{R}^3 \times [0, \infty)) \,. \tag{3.104}$$

An important tool—and intermediate step—in investigating existence and regularity [18] of NSE is the concept and distinction of *weak* and *strong* solutions. By that the smoothness of the derivatives (particularly important for the existence and boundedness of vorticity ω) is tamed to some extent. Strong solutions then are genuine solutions.

3.9.2.2   Dynamical System Approach

Another important tool from mathematical investigations arises from the question: how it is at all possible that a *deterministic system*[5] is capable of becoming *random*.[6] An important concept is that of an *attractor* or *attracting set* in phase space [23]. It seems that within this framework Kolmogorov's picture (see Sect. 3.9.4) of turbulence may be connected to NSE rigorously and, in addition, a specific route or *scenario* was introduced [48].

3.9.3   The Physics of Turbulence

Through the development of the atomic model during the nineteenth century in physics, an improvement in the mechanics of heat (thermodynamics) was gained. Knowing that many ($\sim 10^{23}$) atoms must be present in ordinary amounts of matter, only statistical (averaged) quantities seemed to be of interest. Here statistics emerged as a tool for the reduction of information. Even more fundamentally, quantum mechanics showed in the first half of the twentieth century that probabilities not only emerge by omitting known mechanical information, but also as the only way of describing a quantum mechanical system, a hydrogen atom for example. During the Second World War, these methods were generalized to systems of infinite numbers of degrees of freedom (fields), including classical (non-quantum) systems.

It appeared more than natural to apply these methods to the turbulence problem.[7] What has been achieved by purely physical methods? A summary of results up to the end of the 1980s may be found in [38]. At that time application of so-called *Renormalization Group (RNG)* methods for derivation of improved engineering turbulence models was heavily debated. A broad variety of theoretical field methods like those presented in the work of R. Kraichnan, e.g.[8] has been tried but with limited success. A notable exception is the exactly analytically solvable model of an advective *passive scalar* [33], sometimes called the *Ising-model* of turbulence [9].

---

[5] Sometimes called chaos theory.

[6] By random we mean that there exist only probabilities for the field quantities, at least in the sense of classical statistical physics.

[7] Many well-known physicists worked on turbulence: W. Heisenberg, L. Onsager, Carl-Friedrich von Weizsäcker, to name a few of them. An often repeated quote is that of Richard Feynman, that *turbulence is the most important unresolved problem in classical physics.*

[8] Especially on 2D-turbulence for which he was awarded the Dirac-medal in 2003.

3.9.4 Kolmogorov's Theory

3.9.4.1 Length Scales

Turbulence may be defined in terms of time averages. With

$$< u >:= \int_{t_0}^{t_e} u(t) \, dt \overset{!}{=} \int_o^\infty P(v) \cdot v \, dv \tag{3.105}$$

and $t_0 = 0$ and $t_e = 600\,\text{s}$ in wind energy we may separate the fluctuation u'

$$u(t) = < u > + u' . \tag{3.106}$$

Clearly $\sigma^2 := < u'^2 > \neq 0$. The last equality in Eq. (3.105) is connected to the so-called *ergodicity* hypothesis, which means that time averages and ensemble-averages are equal. In the case of wind this is simply a question of how long to measure to obtain a reliable probability-density function of wind speed. In most practical cases this is reduced to an estimation of two parameters (A,k) of Eq. (1.2). Let P and P' be two points with location $\mathbf{x}$ and $\mathbf{x}'$ and $\mathbf{r} = \mathbf{x} - \mathbf{x}'$. The correlation tensor then takes a special form:

$$< u'_\alpha(\mathbf{x})u'_\beta(\mathbf{x}') > = Q_{\alpha\beta}(\mathbf{r}) = < u'^2 > (f(r) - g(r))\frac{r_\alpha r_\beta}{r^2} + < u'^2 > g(r) \cdot \delta_{\alpha\beta} \tag{3.107}$$

only if the flow is homogenous and isotropic [4]:

- Homogeneity: correlations only depend on $\mathbf{r}$ and not on $\mathbf{x}, \mathbf{x}$';
- Isotropy: no direction is preferred.

With that at least three length scales may be defined:

1. an integral correlation length L,
2. a micro-scale of *Taylor* λ_T,
3. a dissipation length of *Kolmogorov* η_L

with (Fig. 3.25)

$$L := \int_0^\infty f(r)dr \tag{3.108}$$

$$\frac{1}{\lambda_T^2} := -f''(0). \tag{3.109}$$

Equation (3.109) implies that λ_T may be easily derived from a Fourier Transform of a wind time series (see Fig. 3.26)

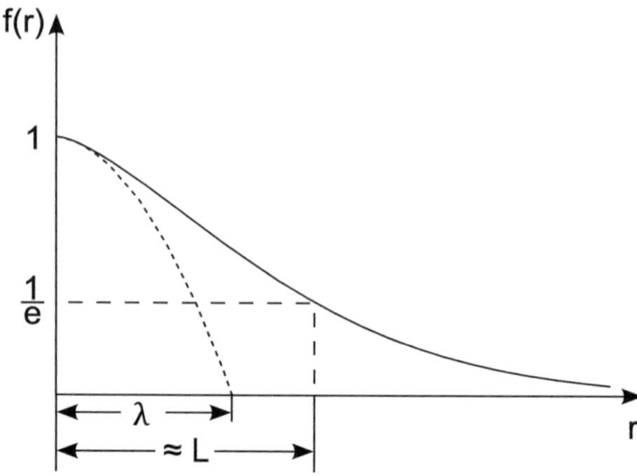

Fig. 3.25 Global and Taylor length scales

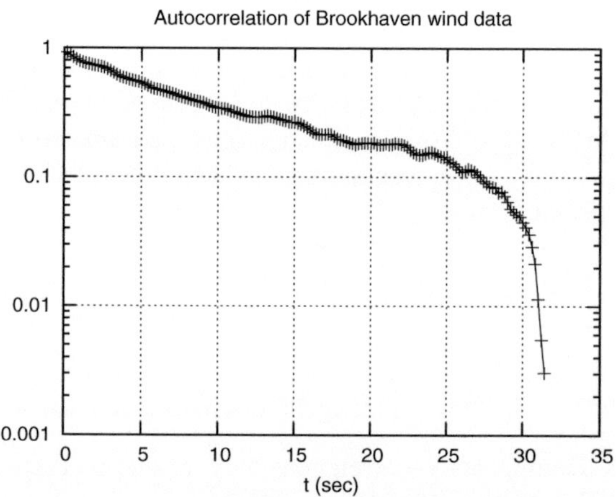

Fig. 3.26 Example of an autocorrelation, Eq. (3.110). Data from [35]. Note that the slope at t $= 0$ appears to be non-zero because of the violation of non-stationarity

$$\rho(s) :=< u(0) \cdot u(s) > / < u'(0)^2 > . \tag{3.110}$$

Originally Taylor thought that λ_T gives the size of the smallest eddies within a turbulent flow. This is not the case, however, as Kolmogorov showed.

With (3.21) for the kinetic energy within a volume V, it can easily be shown from NSE at $\rho \equiv 0$ that

$$\frac{dE_k}{dt} = \int_V u_\alpha \cdot f_\alpha - \int_V \epsilon \, dV \tag{3.111}$$

introducing

$$2 \cdot \epsilon = \nu \left(u_{\alpha,\beta} + u_{\beta,\alpha} \right)^2 = \nu \cdot (\nabla \times \mathbf{v})^2 \tag{3.112}$$

and using again a notation of Einstein:

$$F_{\alpha,\beta} := \frac{\partial F_\alpha}{\partial x_\beta} \tag{3.113}$$

λ_T may be used to connect [45] u'^2 and ϵ:

$$\epsilon = 15\nu \frac{<u'^2>}{\lambda_t^2} . \tag{3.114}$$

In flows with no obvious length scale, a turbulent Taylor–Reynolds number is often used:

$$Re_{\lambda_T} := \sqrt{<u'^2>} \cdot \lambda_T/\nu. \tag{3.115}$$

3.9.5 *Dissipation Scales*

+ Kolmogorov introduced in 1941 [22, 45] a theory of turbulence with a very small set of assumptions and—remarkably—without reference to the NSE. He assumed that at *sufficiently high Reynolds numbers* only two parameters ϵ (dissipation) and ν (kinematical viscosity) determine all the properties of the flow. As we already know, $\nu \sim 10^{-5}$ for air and ϵ may be estimated via $\sim u'^3/L$. From that scales for length, time and velocity may be defined as follows:

$$\eta = (\nu^3/\epsilon)^{1/4} \tag{3.116}$$
$$\tau_\eta = (\nu/\epsilon)^{1/2} \tag{3.117}$$
$$u_\eta = (\nu \cdot \epsilon)^{1/4} . \tag{3.118}$$

At length scales less than η, dissipation into heat starts. Therefore the ratio

$$L/\eta \sim Re^{3/4} \tag{3.119}$$

gives an estimate of the number of points to be resolved in a numerical solution of the NSE. In 3D and assuming a $Re_L \sim 10^6$ we then arrive at $N_{points} \sim 3 \cdot 10^{13}$. Resolving turbulent flow down to this scale is called *Direct Numerical Simulation (DNS)*. In [26] such a HIT (= **H**omogeneous **I**sotropic **T**urbulence) Simulation is reported for λ_T-values up to 1200.

From Table 3.3 a rough estimate for η from wind measurement is shown. It gives values on the order of 1 mm.

Another important parameter is the kinetic energy density as well as the standard deviation. Both have the dimension $(m/s)^2$.

Defining the *turbulent kinetic energy* by

$$k := \frac{1}{2} < u_i' \cdot u_i' > \tag{3.120}$$

we may look for the energy contained in a specific wavenumber range,[9] Slightly generalizing Eq. (3.110)

$$R_{ij}(\mathbf{r}, \mathbf{x}, t) := < u_i'(\mathbf{x}, t) \cdot u_i'(\mathbf{x} + \mathbf{r}), t > \tag{3.124}$$

we now transform to

$$\Phi_{ij}(\kappa, t) := \frac{a}{(2\pi)^3} \int e^{-i\kappa \cdot \mathbf{r}} R_{ij}(\mathbf{r}, \mathbf{x}, t) \cdot d\mathbf{r} . \tag{3.125}$$

The *energy spectrum* function now reads

$$E(\kappa, t) := \frac{1}{2} \int \Phi_{ii}(\kappa, t)\delta(|\kappa| - \kappa)d\kappa . \tag{3.126}$$

It may be easily shown that

$$\int_o^\infty E(\kappa, t)d\kappa = \frac{1}{2}R_{ii}(0, t) = \frac{1}{2} < u_i \cdot u_i > . \tag{3.127}$$

Now with the same class of arguments, Kolmogorov showed:

$$E(\kappa) = \epsilon^{2/3} \cdot \kappa^{-5/3} \cdot \Psi(\kappa \cdot \eta) . \tag{3.128}$$

Figure 3.27 shows measurement [28] at a site close to the North Sea indicating validity of Kolmogorov's *five-thirds-law* for at least five orders of magnitude.[10]

------

[9] It is well known from the theory of linear partial differential equations that a number of problems are simplified if formulated by Fourier-transform into wavenumber–frequency co-space. The NSE equation then reads as

$$k_\beta \cdot u_\beta(\mathbf{k}, t) \tag{3.121}$$

$$\left(\frac{\partial}{\partial t} + v \cdot k^2\right) u_\alpha(\mathbf{k}, t) = M_{\alpha\beta\gamma}(\mathbf{k}, t) \sum_{\mathbf{j}} u_\beta(\mathbf{j}, t) \cdot u_\gamma(\mathbf{k} - \mathbf{j}, t) \ and \tag{3.122}$$

$$M_{\alpha\beta\gamma}(\mathbf{k}, t) = (2i)^{-1} \left(k_\beta D_{\alpha\gamma}(\mathbf{k}) + k_\gamma D_{\alpha\beta}(\mathbf{k})\right) . \tag{3.123}$$

[10] Taylor's frozen turbulence hypothesis has been used here. It states that time series may be used instead of spatially varying values.

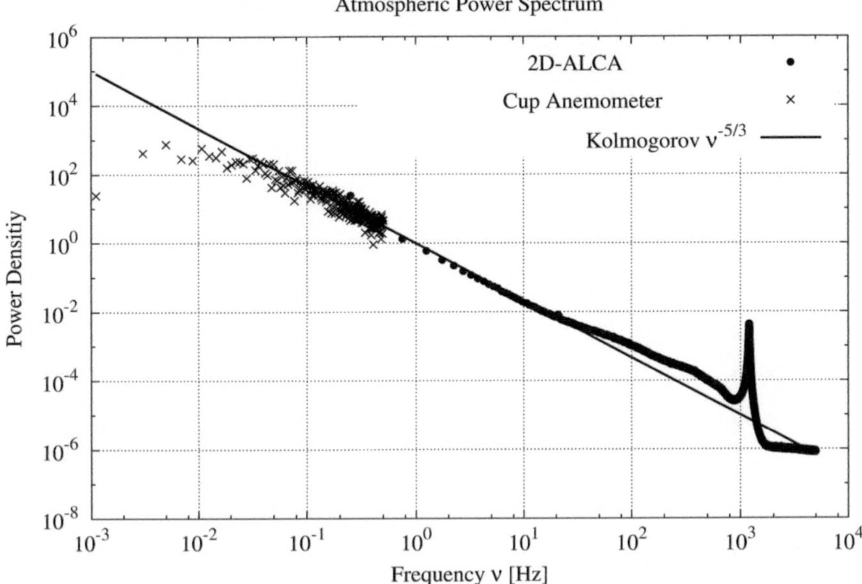

Fig. 3.27 Energy density spectra close to the North Sea. The peak at 1 kHZ is an artifact of the measurement system. 10^{-2} Hz seems to be an approximate maximum

In wind energy, semi-empirical spectra have to be used for generating *synthetic wind*. Equation 3.130 and Fig. 3.28 show one commonly used, originating from von Karman.

$$E(\kappa) = C_K \cdot \epsilon^{2/3} \cdot \kappa^{-5/3} \cdot f_L(\kappa \cdot L) f_\eta(\kappa \cdot \eta) \tag{3.129}$$

$$f_L(\kappa \cdot L) = \left(\frac{\kappa L}{[(\kappa L)^2 + c_L]^{1/2}} \right)^{5/3 + n_0} . \tag{3.130}$$

3.9.6 *Turbulence as a Stochastic Process*

Stochastic Processes [57] were first used in description of *Brownian motion* where the motion of particles is described by classical mechanics and is subjected to additional *random forces*. Of course specifics must be applied to define the meanings and differences of *random, statistical* and *stochastic*. As *Statistical Fluid Mechanics* [41] often refers to the *Mechanics of Turbulence*, it seems natural to apply these methods to turbulent flows. Starting with the work of Friedrich and Peinke [21] attempts were made to derive Fokker–Planck-Equations for the description of turbulence. We will

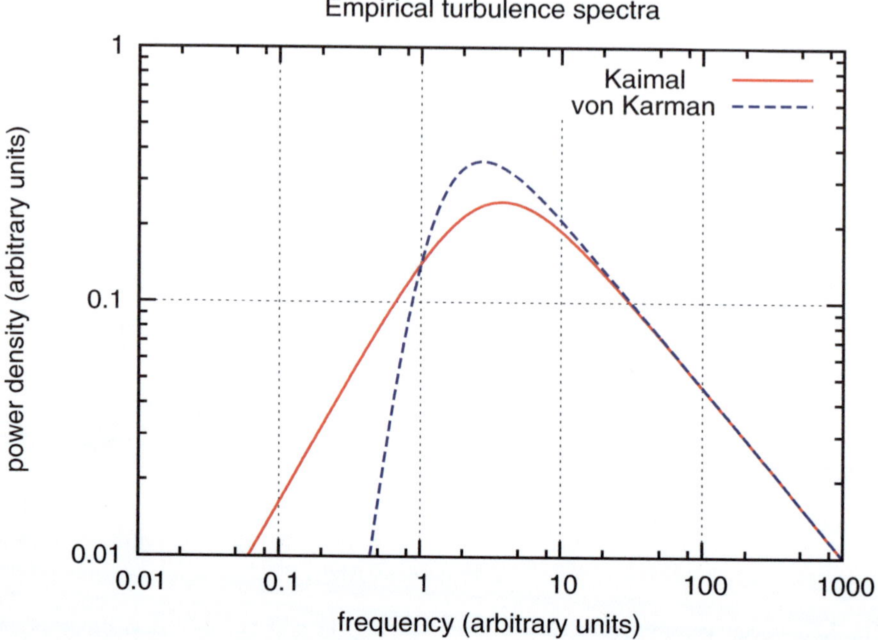

Fig. 3.28 Semi-empirical turbulence spectra

discuss this important issue of *synthetic wind* in more detail in Sects. 4.5 and 4.6;
see [32, 51]. As an outcome of this approach (see Fig. 3.29) we see that at least the
important starting points are *velocity increments* and *structure functions*:

$$\delta \mathbf{u}(\mathbf{x}, \mathbf{r}) := \mathbf{u}(\mathbf{x} + \mathbf{r}) - \mathbf{u}(\mathbf{r}) \tag{3.131}$$

which according to Kolmogorov's theory [22] scales like

$$< (\delta \mathbf{u}r)^2 > = C \cdot \epsilon^{2/3} \cdot r^{2/3} \tag{3.132}$$

$$< (\delta \mathbf{u}r)^3 > = -\frac{4}{5} \cdot \epsilon \cdot r \tag{3.133}$$

$$S_n(r) = < (\delta \mathbf{u}r)^n > = C_n \cdot \epsilon^{n/3} \cdot r^{n/3} . \tag{3.134}$$

An interesting parameter called *intermittency* originally means the occurrence of
intermittent bursts of high activity. In addition it may cause a deviation of the Kolmogorov scaling $log(S_n) \sim n$ [22] and may serve as a model of *gusts* [7] (Fig. 3.30).

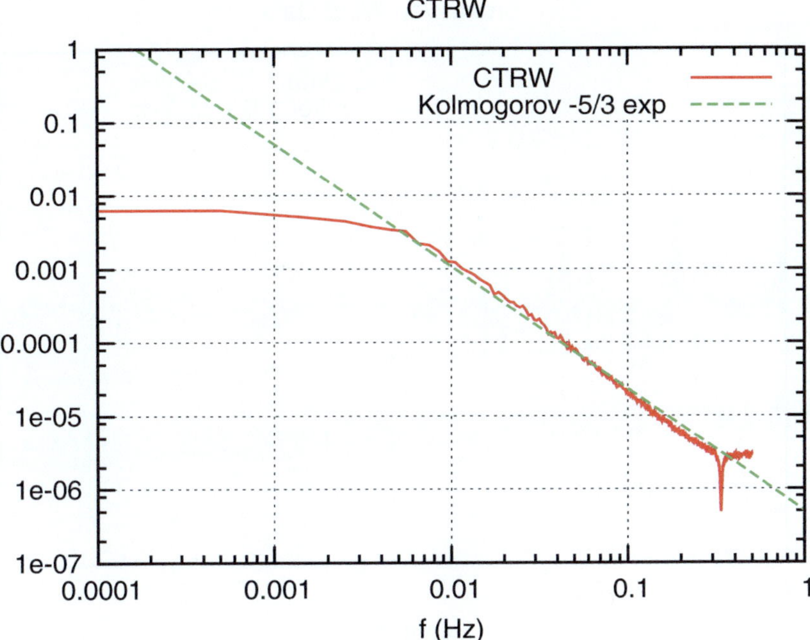

Fig. 3.29 Energy density spectra from a CTRW-model

3.9.7 The Turbulent Boundary Layer

Boundary layer theory successfully describes the velocity profile coming from the outer inviscid flow to the wall. In its simplest form it is the Blasius profile for a flat plate. If turbulence comes into play, unfortunately things become much more complicated even for a simple flat plate [45].

Nevertheless this type of flow is important both in application to wind (as will be seen in the next section) and to CFD (= numerically solved Navier–Stokes equations).

With the term *fully developed* we loosely define a type of a flow in which at all locations large fluctuations (over time) are present. We define *Reynolds averaging (RA)* as separating averages from fluctuations

$$u(t) = U + u'(t) \tag{3.135}$$

with

$$U := <u(t)> := \int_{-\infty}^{\infty} u(t)\, dt \; . \tag{3.136}$$

Therefore $<u'> = 0$. Now we consider a simple 2D planar flow. Applying RA and BL-approximations to the appropriate NAST, we get

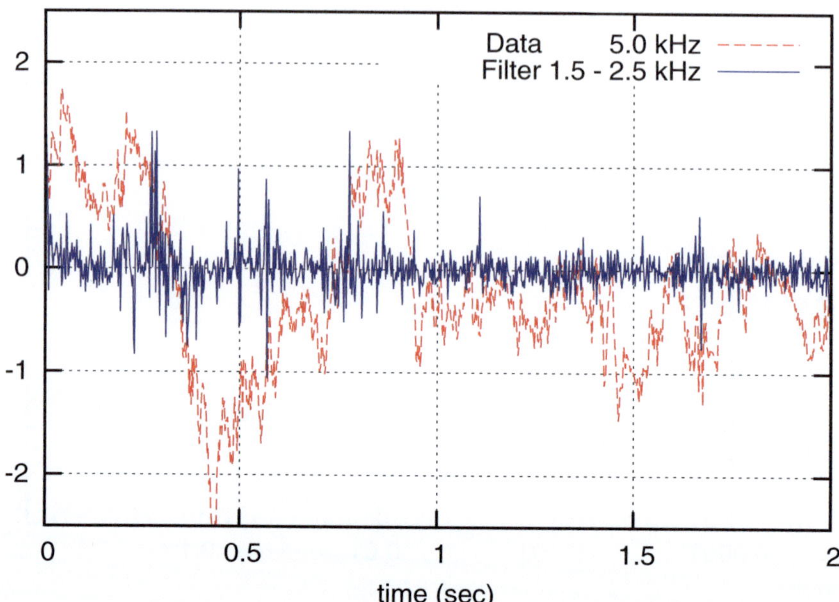

Fig. 3.30 Intermittent time signal from high pass filtering of wind data

$$\frac{\partial <U>}{\partial x} + \frac{\partial <V>}{\partial x} = 0, \tag{3.137}$$

$$<U> \frac{\partial <U>}{\partial x} + <V> \frac{\partial <U>}{\partial y} = v\frac{\partial^2 <U>}{\partial y^2} - \frac{\partial}{\partial y} <u' \cdot v'> \tag{3.138}$$

$$\frac{1}{\rho}\frac{\partial <p>}{\partial x} = \frac{1}{\rho}\frac{dp_0}{dx} - \frac{\partial <v'^2>}{\partial x}. \tag{3.139}$$

The shear stress then becomes

$$\tau = \rho v \frac{\partial <U>}{\partial y} - \rho <u' \cdot v'> . \tag{3.140}$$

Expressions like $<u_i' \cdot u_j'>$ are called *Reynolds stresses*. The *wall shear stress* is particularly important

$$\tau_w := \tau(0) . \tag{3.141}$$

Using this quantity as the primary one we may define a *friction velocity*

$$u_\tau := \sqrt{\frac{\tau_w}{\rho}} \tag{3.142}$$

and a *viscous length scale* [45]

$$\delta_v := v\sqrt{\frac{\rho}{\tau_w}} \ . \tag{3.143}$$

Wall distance and velocity now may be measured with reference to these values using dimensionless numbers only:

$$y^+ := \frac{u_\tau y}{v} \tag{3.144}$$

$$u^+ := \frac{U}{u_\tau}. \tag{3.145}$$

3.9.8 The Log Law of the Wall

Boundary layer theory successfully describes the velocity profile for laminar flow coming from outer inviscid flow to the wall. In its simplest form, this is the Blasius profile for a flat plate, as a no-slip boundary condition. If turbulence comes into play, unfortunately things become much more complicated even for a simple flat plate [45]. A law of the wall now is an equation of the type

$$u^+ = f_w(y^+) \ . \tag{3.146}$$

In total we may distinguish about 5 [45] sub-layers within the turbulent boundary layer, but two are of particular importance:

(a) Very close to the wall ($y^+ < 5$) we must have

$$f_w(y^+) = y^+ \tag{3.147}$$

because $u^+(0) = 0$ (no-slip condition), and $f'_w(0) = 1$ because of the special normalization we used. Sometimes this sub-layer is called *viscous sub-layer*.

(b) Now if $y^+ > 100$ Prandtl and von Karman [45] assumed that viscosity is of no importance and

$$\frac{du^+}{dy} = const. \cdot \frac{u_\tau}{y} \tag{3.148}$$

simply because of dimensional analysis. The constant $const. := \kappa \approx 0.4$ is called *von Karman constant*. If we integrate Eq. (3.148) we get the famous *logarithmic law of the wall* :

$$u^+ = u_\tau \left(\frac{1}{\kappa} \, ln(y) + C \right) \ . \tag{3.149}$$

Figure 3.31 shows an example from the German off-shore measurement station FINO1, about 40 km north of the island of Borkum. All data up to 90 m seem to

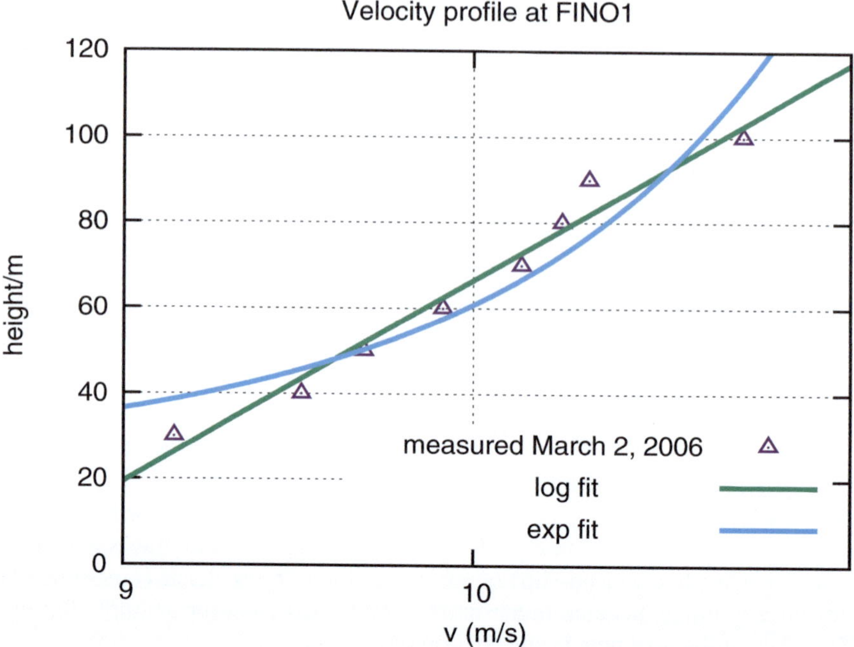

Fig. 3.31 Example of the validity of the log law of the wall. Data from FINO1 [15]

fit rather well to the logarithmic law, with the exception of the highest anemometer at $H = 100$ m above sea level.

3.10 Wind as Turbulent Flow in the Lower Atmosphere

The methods applied above may also be applied to the *Atmospheric Boundary Layer* (ABL) which is sometimes also called *Planetary Boundary Layer*. Two important distinctions have to be noted:

- geometry is that of the surface of a sphere;
- there is no inertial system of reference, the earth is rotating with ω_{earth} $= 7 \cdot 10^{-5}$/s.

The outer-flow is called *geostrophic wind* and is induced by the combined action of pressure-gradients and Coriolis forces (Fig. 3.32).

A naive estimation assuming laminar flow produces far too small values for the thickness of the ABL of only few meters [61]. More realistic estimations must include effects from turbulence and thermal radiation. Putting everything together, one can say that the thickness of the ABL varies from approximately very few hundreds of meters (at night) to about 1500 m in the daytime.

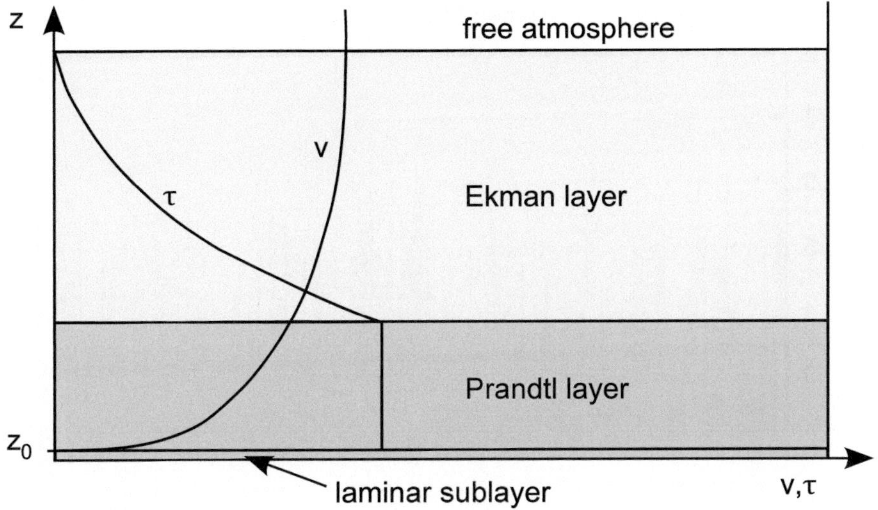

Fig. 3.32 Structure of the atmospheric boundary layer (ABL), adapted [61]

One special mention has to be made of *roughness length* Eq. (3.149) which is used for estimating wind velocities at different heights, when one measurement z_r, v_r is known:

$$v(z) = v_r \cdot \frac{ln(z) - ln(z_0)}{ln(z_r) - ln(z_0)} \, .$$ (3.150)

Obviously $v(z_0) \equiv 0$ (Table 3.2).

We will learn more in Sect. 4.2 about vertical profiles.

As already shown, wind may be used as an example to show the validity of the Kolmogorov theory across more than five decades (Fig. 3.27). Figure 3.33 shows an example of the underlying time series.

Many more turbulence characteristics may be derived from these types of measurements. Table 3.3 shows comparison of the turbulent dissipation rate in wind (from [28, 34, 35]).

Table 3.2 Some examples of roughness heights

Surface	$z_0(m)$
Water	$0.1–1 \ 10^{-3}$
Shrub	0.1–0.2
Forest	0.5
Cities	1–2
Mega cities	5
Mountains	1–5

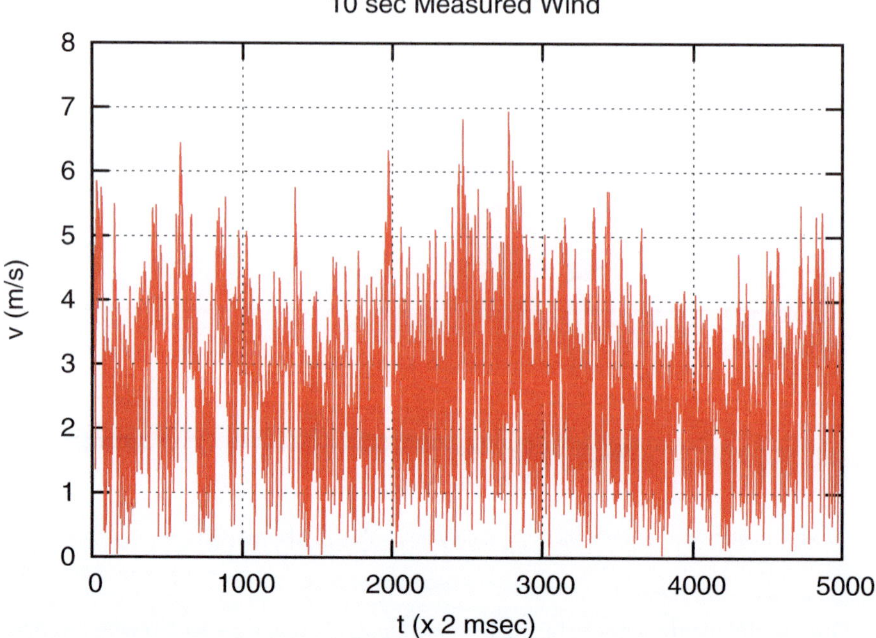

Fig. 3.33 Typical time series of a high-frequency wind speed measurement

Table 3.3 Measurements of turbulent dissipation rates in atmospheric winds

Author	Source	ϵ	$\sqrt{<u'^2>}$ (m/s)
Kunkel et al.	[34]	0.1 to $1 \cdot 10^{-2}$	~1.0
Srinivasan et al.	[35]	$1.0 \cdot 10^{-2}$	~1.5
Jeromin, Schaffarczyk	[28]	$\sim 3 \cdot 10^{-2}$	~2.0

3.11 Problems

Problem 3.1 Derive Bernoulli's equation from the Euler Equation, Eq. (3.31).

Problem 3.2 Derive the velocity distribution of a line vortex from integration and calculate the kinetic energy contained.

Problem 3.3 Use [4] to find the complex velocity potential flow of Fig. 3.7, write and calculate the dividing (free) streamline.

Problem 3.4 Use [43] or [40] to investigate the Joukovsky transformation (3.51) in more detail. Find an expression for the thickness and the camber as function of z_m.

Problem 3.5 (a) Derive from Table 3.3 corresponding values for the Taylor- and Kolmogorov length scales, Eqs. (3.114) and (3.116).

(b) Discuss your results with reference to the size of the active measurement equipment.

References

1. Abbott I, von Doenhoff A (1959) Theory of wing sections: including a summary of airfoil data. Dover
2. Babinsky H (2003) How do wings work? Phys Edu 8(6)
3. Bachelor GK (1967) An introduction to fluid dynamics. Cambridge University Press, Cambridge, UK
4. Bachelor GK (1993) The theory of homogeneous turbulence. Cambridge University Press, Cambridge, UK, reprinted
5. Betz A (1919) Schraubenpropeller mit geringstem energieverlust mit einem zusatz von l.prandtl. In: Nachr. d. Königl. Gesell. Wiss. zu Göttingen, Math.-phys. Klasse, pp 198–217. Königl. Gesell. Wiss. zu Göttingen
6. Blasius H (1908) Grenzschichten in füßigkeiten mit kleiner reibung. Z Math Phys 66:1–37
7. Böettcher F, Barth S, Peinke J (2006) Small and large scale fluctuations in atmospheric wind speeds. Stoch Environ Res Risk Ass
8. Boyd JP (1999) The blasius function in the complex plane. Exp Math 8:1–14
9. Chen S et al (2008) Obituary of R. Kraichnan. Phys Today 71:70–71
10. Chorin A (1993) A mathematical introduction to fluid mechanics. Springer, New York, Berlin, USA
11. Cousteix J, Mauss J (2007) Asymptotic analysis and boundary layers. Springer, Berlin Heidelberg
12. da Vinci L, Verso: studies of flowing water, with notes. Royal Collection Trust, Windsor, UK, 1266, 1510
13. Darrigol D (2005) Worlds of flow. Oxford University Press, Oxford, UK
14. Drela M (1990) XFOIL: an analysis and design system for low Reynolds number airfoils, vol 54. Springer lecture notes in engineering. Springer, Berlin, Heidelberg, Germany
15. Emeis S (2012) Private communication
16. Eppler R (1990) Airfoil design and data. Springer, Berlin, Heidelberg, Germany
17. Fava T, Lobo B, Schaffarczyk A, Breuer M, Henningson D, Hanifi A (2023) Numerical investigation of transition on a wind turbine blade under free-stream turbulence at $re_c = 10^6$. Flow, turbulence and combustion, page to be published
18. Fefferman C (2000) Existence & smoothness of the Navier–Stokes equations. Clay Mathematics Institute, The Millennium Prize, Problems, Navier–Stokes Equations, Providence, RI, USA
19. Foias C, Manley O, Rosa R, Temam R (2001) Navier–Stokes equations and turbulence. Cambridge University Press, Cambridge, UK
20. Föppl O (1911) Windkräfte an ebenen und gewölbten platten. Jahrbuch der Motorluftschiff-Studiengesellschaft, Berlin, Germany (in German)
21. Friedrich R, Peinke J (1997) Description of a turbulent cascade by a Fokker–Planck equation. Phys Rev Lett 78
22. Frisch U (1995) Turbulence. Cambridge University Press, Cambridge, UK
23. Gallavotti G (2002) Foundations of fluid mechanics. Springer, Berlin, Heidelberg, Germany
24. Glauert H (1926) The elements of aerofoil and airscrew theory, Repr, 2nd edn. Cambridge University Press, Cambridge, UK
25. Hermandez GGM (2011) Laminar-Turbulent transition on wind turbines. PhD thesis, Technical University of Copenhagen, Denmark

26. Ishihara T, Gotoh T, Kaneda Y (2009) Study of high-reynolds number isotropic turbulence by direct numerical simulation. Annu Rev Fluid Mech 41:165–180
27. Jacobs M (2015) High reynolds number airfoil test in dnw-hdg. DNW-GUK-, 2014
28. Jeromin A, Schaffarczyk AP (2012) Advanced statistical analysis of high-frequency turbulent pressure fluctuations for on- and off-shore wind. In: Höelling M et al (eds) Proceedings of Euromech coll 528
29. Jones RT (1990) Wing theory. Princeton University Press, Princeton, NJ, USA
30. Kant I (1786) Metaphysische anfangsgründe der naturwissenschaft. Königsberg, Königreich Preussen
31. Katz J, Plotkin A (2001) Low-Speed Aerodynamics, 2nd edn. Cambridge University Press, Cambridge, UK
32. Klein F (1910) über die bildung von wirbeln in reibungslosen flüssigkeiten. Z. f. Mathematik u. Physik 58:259–262
33. Kraichnan R (1994) Anomalous scaling of a randomly advected passive scalar. Phys Rev Lett 72:1016–1019
34. Kunkel KE, Eloranta EW, Weinman JA (1980) Remote determination of winds, turbulence spectra and energy dissipation rates in the boundary layer from lidar measurements. J Atmos Sci 37:6
35. Kurien S, Sreenivasan KR (2001) Measures of anisotropy and the universal properties of turbulence. In: Lesieur M et al (eds) Les Houches summer school in theoretical physics. Springer, Berlin, Heidelberg, New York
36. Leishman JG (2002) Challenges in modeling the unsteady aerodynamics of wind turbines. AIA, pp 2002–0037
37. Lions P-L (1996) Mathematical topics in fluid mechanics, vol 1. Incompressible models. Clarendon Press, Oxford, UK
38. McComb WD (1992) The physics of fluid turbulence. Clarendon Press, Oxford, UK
39. McLean D (2013) Understanding aerodynamics. Boeing, Wiley, Chichester, UK
40. Milne-Thomson LM (1996) Theoretical hydrodynamics, 5th edn. Dover Publications, New York, USA
41. Monin AS, Yaglom AM (2007) Statistical fluid mechanics, vol 2. Dover Publications, New York
42. Munk MM (1922) General theory of thin wing sections. Technical report, NACA, Report No. 142
43. Panton RL (1996) Incompressible flow, 2nd edn. Wiley, New York
44. Pereira R, Schepers G, Pavel MD (2013) Validation of the Beddoes-Leishmann dynamic stall model for horizontal axis wind turbines using Mexico data. Wind Energy 16(2):207–219
45. Pope SB (2000) Turbulent flows. Cambridge University Press, Cambridge, UK
46. Prandtl L, Betz A (2010) Vier Abhandlungen zur Hydrodynamik und Aerodynamik. Universitätsverlag Göttingen, Germany (in german)
47. Ruban T, A.I
48. Ruelle D, Takens F (1971) On the nature of turbulence. Comm Math Phys 20:167–192
49. Saffman PG (1992) Vortex dynamics. Cambridge University Press, Cambridge, UK
50. Schaffarczyk A (2021) An explanation and understanding of aerodynamic lift by triple deck theory. Preprints
51. Schaffarczyk AP et al (2010) A new non-gaussian turbulent wind field generator to estimate design-loads of wind-turbines. In: Peinke J, Oberlack M, Talamelli A (eds) Progress in turbulence III. Springer proceedings in physics, vol 131. Springer, Dordrecht
52. Schlichting H, Gersten K (2000) Boundary layer theory. Springer, Berlin, Heidelberg, Germany
53. Schubauer GB, Skramstad HK (1943) Laminar-boundary-layer oscillations and transition on a flat plate. Technical report, NACA-TR-909, 1943/47
54. Sobey I (2000) Introduction to interactive boundary layer theory. Oxford University Press, Oxford, UK
55. Sychev V, Ruban A, Sychev V, Korolev G (2008) Asymptotic theory of separated flows. Cambridge University Press, Cambridge, UK

56. Van Dyke M (1975) Perturbation methods in fluid mechanics. The Parabolic Press, Stanford, CA, USA
57. van Kampen N (2007) Stochastic processes in physics and chemistry, 3rd edn. Elsevier, Amsterdam, The Netherlands
58. van Rooij RPJOM (1996) Modification of the boundary layer calculation in rfoil for improved airfoil stall prediction. Technical report, Internal report IW-96087R, TU Delft, The Netherlands
59. Wald QR (2006) The aerodynamics of propellers. Prog Aero Sci 42:85–128
60. Yates YE (1991) A unified viscous theory of lift and drag of 2-d thin airfoils and 3-d thin wings. Technical report, NASA report, CR-4414
61. Zdunkowski W, Bott A (2003) Dynamics of the atmosphere. Cambridge University Press

Chapter 4
Inflow Conditions to Wind Turbines

4.1 Importance of Inflow Conditions to Rotor Performance

In Chap. 3, some of the necessary fluid mechanics background was introduced. Now looking closer at the wind turbine, we may ask which details of the turbulent wind field are generally more important and in which order the properties of this field should be discussed. Certainly one may start with the easiest flow field: a spatially and temporally constant one. This model is used in most analytical studies.

Nevertheless, wind is as it is, so we will approach reality in steps of increasing sophistication. Starting with the simplest spatial inhomogeneity, we consider at first a gradual vertical increase of wind speed called wind shear.

4.2 Wind-Shear

As we have already seen in Sect. 3.9.8, and as exemplified in Fig. 3.31, a turbulent boundary layer shows a logarithmic increase of wind speed with height. The slope of this curve, described in Eq. 3.150, uses one parameter only, the roughness length. At least two measurements and two different heights are therefore necessary for extrapolation to other heights. Figure 4.3 shows an example of a location in Hamburg, Germany. Any decision on hub heights and the resulting tower investment relies heavily on these data (see Table 3.2 and [21]). Figures 4.1 and 4.2 give two more examples of locations in cities where z_0 is particularly large, giving steep velocity gradients.

Due to the change (in most cases decrease) in temperature with height, there is an energy flux in upward direction which influences the profile of the mean horizontal wind speed and the turbulence characteristics. A (Monin-Obukhov) length scale

© The Author(s), under exclusive license to Springer Nature Switzerland AG 2024

A. P. Schaffarczyk, *Introduction to Wind Turbine Aerodynamics*, Green Energy
and Technology, https://doi.org/10.1007/978-3-031-56924-1_4

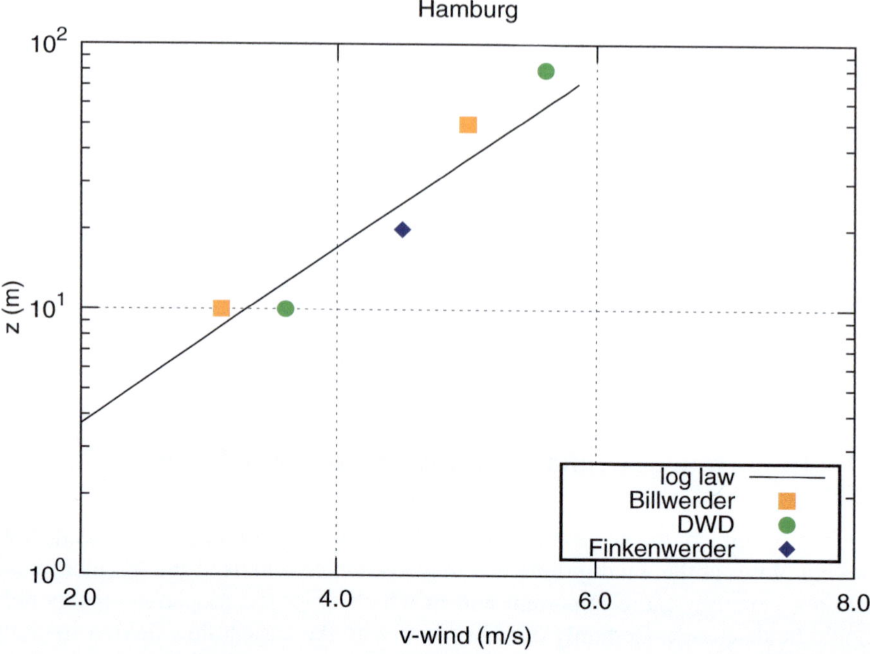

Fig. 4.1 Wind profile at locations (Billwerder, Finkenwerder) close to Hamburg, Germany. DWD means German Weather Service

$$L := \frac{u^{\star 3} c_p \rho T}{k g H} \tag{4.1}$$

has to be introduced with $u^\star$, the friction velocity, c_p, specific heat of the air, ρ, density, T, temperature, k von Karman's constant ≈ 0.4, g, earth acceleration and H, the flux. The profile then is expressed [21] as

$$u(z) = \frac{u^\star}{k} \left[ln \frac{z}{z_0} - \psi_m \left(\frac{z}{L} \right) \right] \tag{4.2}$$

with a universal function $\psi_m(z/L)$. As a result, the profiles change significantly diurnally, and the profile is fundamentally different over water surfaces, which is important for offshore wind energy applications (see Fig. 4.4).

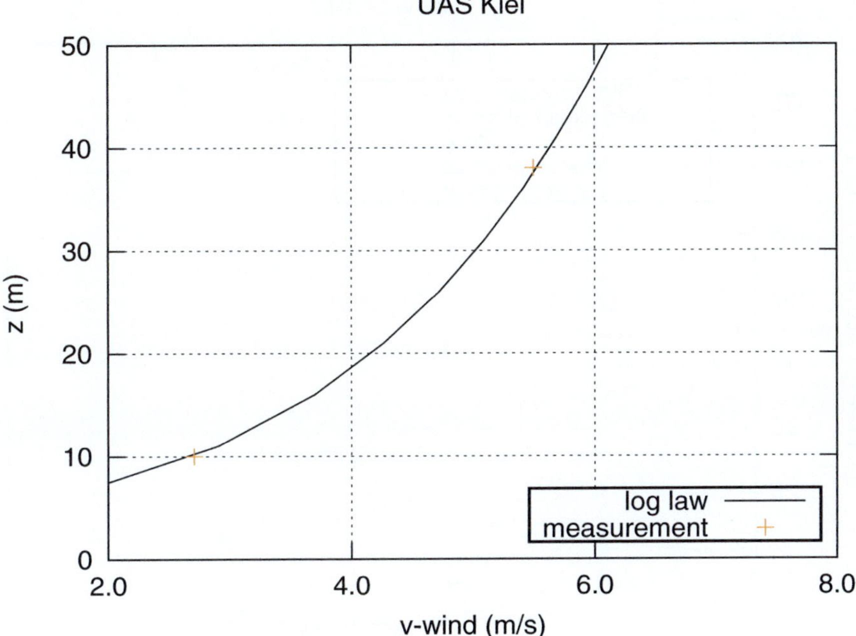

Fig. 4.2 Wind profile at Kiel, Germany

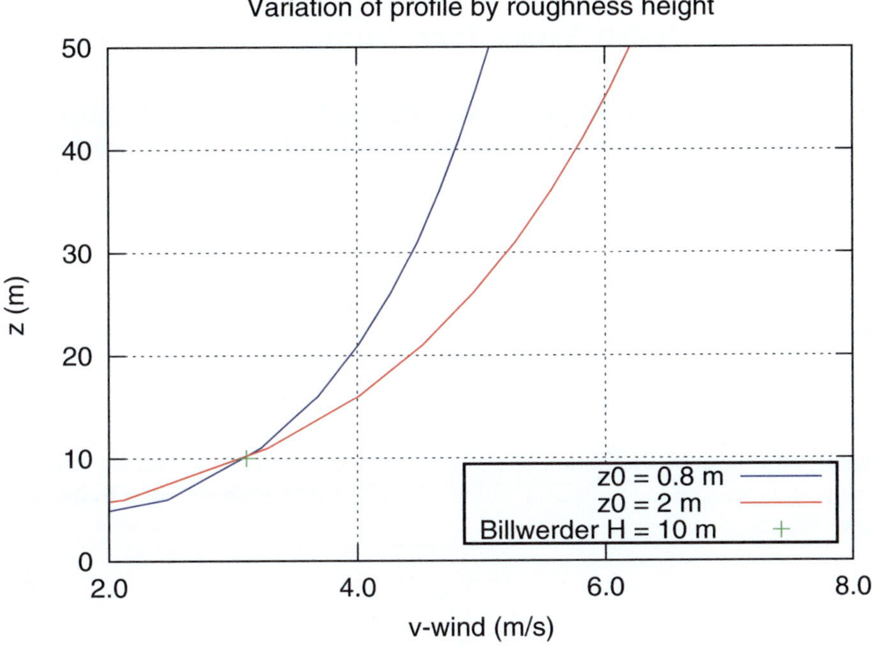

Fig. 4.3 Dependence of wind profile shape (City) on z_0

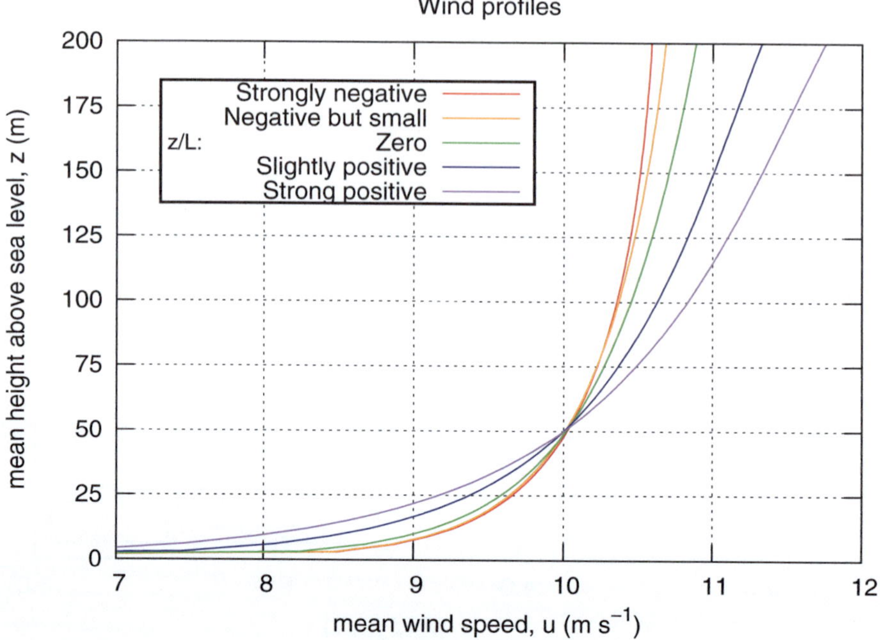

Fig. 4.4 Wind profiles as function of temperature differences

4.3 Unsteady Inflow Versus Turbulence

As wind turbines are flexible structures, they may be able to react to instationary inflow conditions up to a certain frequency. These limit frequencies are on the order of a few Hertz, as may be seen from a sample *Campbell* Diagram. Therefore, from a structural point of view, the transition between instationary flow and turbulence may be drawn (without any rigorous reasoning) at 10 Hz (Fig. 4.5).

From a more aerodynamic standpoint the situation is not as clear. Starting with a laminar Boundary Layer, Stokes and others [4] have shown, that starting from a simple oscillation flow

$$u(x, t) = u_0(x) \left(1 + B \cdot cos(\omega t)\right) \tag{4.3}$$

the response is spatially limited due to viscous damping. For the laminar case, the boundary layer is able to respond within a length $\sim 1/k$ perpendicular to the main flow only for

$$k = \sqrt{\frac{\omega}{2\nu}} . \tag{4.4}$$

For air and $\omega = 60/s$, we have approx. 4 mm for this length scale.

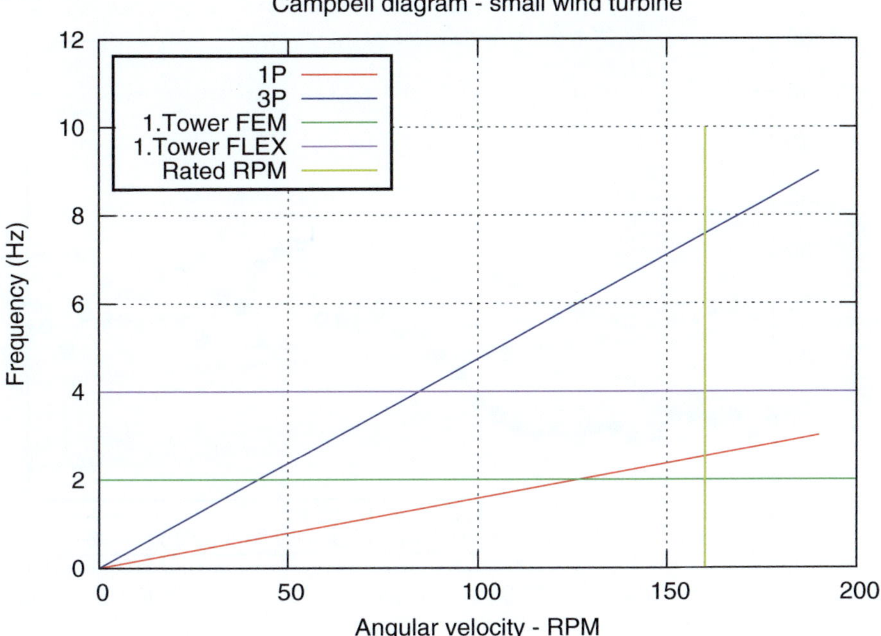

Fig. 4.5 Sample resonance (Campbell) diagram of a small wind turbine. FEM means results from a Finite Element Method code and FLEX is an often used Aero-Elastic tool

In turbulent flow, there is even a characteristic frequency called the *bursting frequency* [5] which may be defined as

$$f_B = U/5\delta_T \tag{4.5}$$

where δ_T is the turbulent boundary layer thickness. On a wind turbine blade with U = 40 m/s we may estimate $\delta_z \approx 7$ mm, and therefore, $f_B \approx 200$ Hz. To conclude, we may designate a very blurred border between structural and aerodynamic turbulence somewhere between 10 and 200 Hz.

4.4 Measuring Wind

Measuring the wind is an important part of site assessment. As special examples, we show in Figs. 4.6, 4.7, 4.8 and 4.9, the measurements for storm *Christian* occurring in Northern Germany in October 2013. KWK is located close to the North Sea, whereas FINO3 is located about 80 km west of the German island of Sylt.

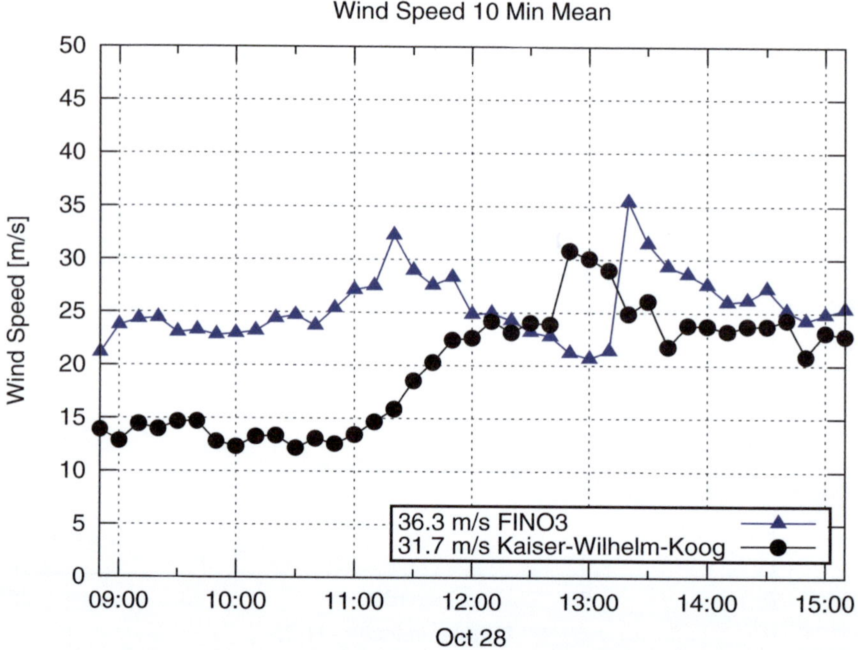

Fig. 4.6 Storm Christian on Oct 28, 2013, at FINO3 and KWK, 10 min average

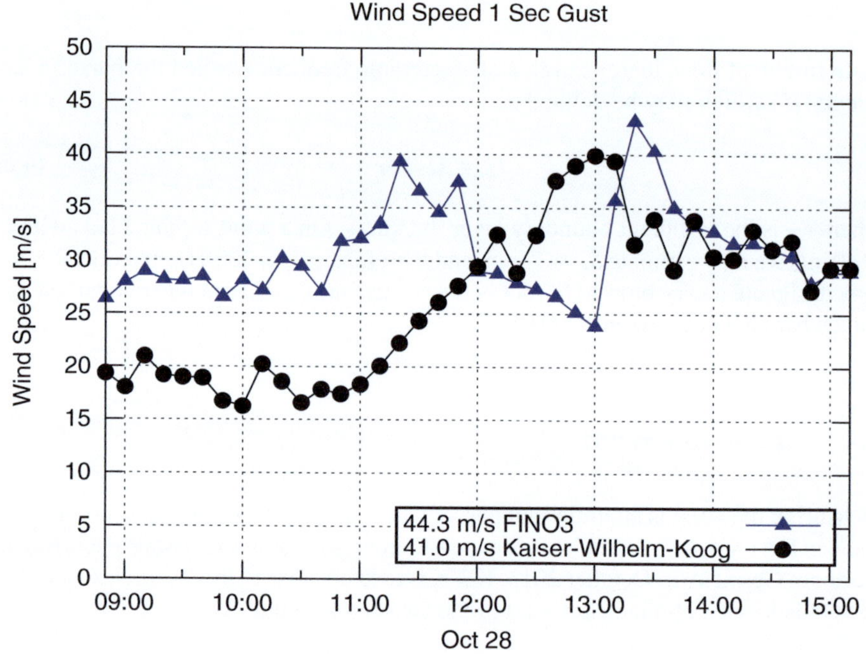

Fig. 4.7 Storm Christian on Oct 28, 2013, at FINO3 and KWK, 1 s gust

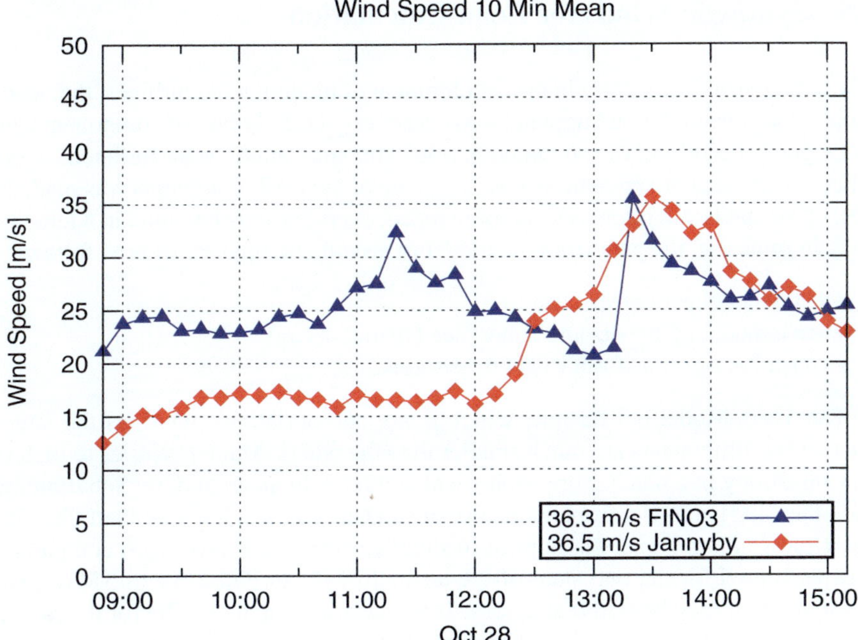

Fig. 4.8 Storm Christian on Oct 28, 2013, at FINO3 and Janneby, 10 min average

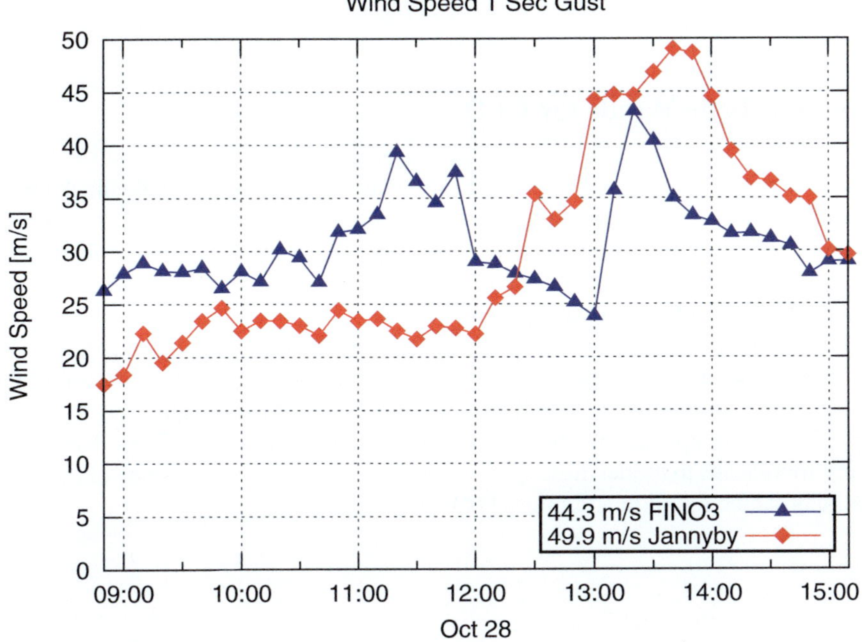

Fig. 4.9 Storm Christian on Oct 28, 2013, at FINO3 and Janneby, 1 s gust

4.5 Synthetic Winds for Load Calculation

It was recognized very early in the development of modern wind turbines, that instationary loads from fluctuating wind may cause very early failure of components due to *fatigue*.[1] Therefore, the so-called aero-elastic simulations were initiated, which relied on artificial or synthetic winds [27]. One of the earliest attempts was made by Veers [26] and is still often used. A more recent overview may be found in Kleinhans [14]. In principle, the path from the model to a synthetic time series is as follows:

- select a model energy spectrum,
- select a model of anisotropic coherence (spatial decay),
- perform Fourier transform to get time series.

The Veers Model is isotropic, which is not appropriate for wind, as the lateral fluctuation components are much smaller than the others. Mann [19] tries to include this anisotropy in a much more sophisticated model. In an attempt by Schaffarczyk and others [23], a model from the theory of stochastic processes (so-called *Continuous Time Random Walks* was applied to simulate turbulent wind [13]. A comparison was performed for the two state-of-the-art multi-MW turbines (3 and 5 MW), and in summary, it may be said that at least 10% deviations in the results for the fatigue loads may be seen. It was even observed that some completely wrong result, see [25]; therefore, it is clear that many more efforts must be made to get reliable and more sophisticated synthetic wind models which are appropriate for the certification process [20].

4.6 Synthetic Winds for CFD

A similar problem appears if one tries to simulate wind farms on a scale much larger than the typical atmospheric length scale of several hundred meters. Then specific inflow conditions are definitely needed if a special kind of transient calculations (Large Eddy Simulation—LES, see Chap. 7) are performed. An impressive description of the state of the art may be found in [22].

In case of turbulent inflow conditions, it is necessary that spatially and temporally resolved turbulent flow fields are set-up such that they statistically resemble the real world or experimental conditions as closely as possible. This results in the need for setting up the so-called turbulent boundary conditions. There have been many approaches to model these spatiotemporal inflow conditions, some of the early methods can be found in Kleinhans [14].

---

[1] A well-known example is the German early Multi-MW turbine GROWIAN (1976–1987) which had to be shut down after only about 400 h of operation [24].

4.6.1 General Approach: From Experimental Data to Synthetic Time Series

A common approach for the generation of artificial inflow turbulence data that satisfies certain statistical properties known from experimental data such as the energy spectra, mean velocity, fluctuations, cross correlations, length and time scales, etc. is as follows:

1. For each velocity component, a 3D velocity signal is generated such that the two-point statistics are fulfilled.
2. The cross correlations between the different velocity components are considered for the generation of the velocity signal in case of anisotropic inflow conditions.
3. An inverse Fourier transform is then used on the obtained velocity signal generating velocity data that satisfies the specified energy spectrum.

To successfully generate time series data using this approach, one would need 3D energy spectra which is experimentally difficult to obtain. To overcome this problem, Lee et al. [15] uses a model spectrum which represents isotropic turbulence:

$$E(k) \sim k^4 exp(-2(k/k_0)^2) \text{ with wave number vector k}$$
$$k = (k_1^2 + k_2^2 + k_3^2)^{1/2} k_0 = \text{peak wave number} \tag{4.6}$$

Equation 4.6 is one of the many possible spectra that can be employed. As seen in Klein et al. [12] for example, the question on how to choose k_0 cannot be uniquely answered. To overcome some of the challenges faced with the general approach outlined above, a method based on digital filtering of random data was developed.

4.6.2 Digital Filters

The basis for the development of the digital filter method by Klein et al. was its practicability. This means that only those statistical quantities that can be obtained with reasonable expenses or from heuristical estimates would be the necessary input. It was concluded that this method requires the definition of certain statistical quantities such as the mean velocity, Reynolds stress terms and the definition of one integral time scale and two integral length scales.

The flow signal can be described as the sum of the average velocity and a fluctuating part. Using the Cholesky decomposition a_{ij} of the Reynolds stress tensor and synthetically generated velocity fluctuations u_m, Klein et al. [12] used the method by Lund et al. [17] to ensure the correct cross correlations. The instantaneous velocity signal is as follows:

$$u = \overline{u} + a_{ij}(u_m) \tag{4.7}$$

The definition of the tensor a_{ij} can be found in Lund et al. [17] and Klein et al. [12]. To satisfy the two-point correlations, a series of random numbers r_m is defined such that $\overline{r_m} = 0$ and $\overline{r_m r_m} = 1$, then the convolution or a digital non-recursive filter with filter coefficient b_n and filter length N is defined as described in Klein et al. [12]

$$u_m = \sum_{n=-N}^{N} b_n r_{m+n} \tag{4.8}$$

Since $\overline{r_m r_n} = 0$ for $m \neq n$, it follows the autocorrelation:

$$\frac{\overline{u_m u_{m+k}}}{u_m u_m} = \sum_{j=-N+k}^{N} b_j b_{j-k} \bigg/ \sum_{j=-N}^{N} b_j^2 \tag{4.9}$$

This gives a relation between the filter coefficients and the autocorrelation function. A 3D filter can be constructed through the product of three 1D filters:

$$b_{ijk} = b_i . b_j . b_k \tag{4.10}$$

Klein et al. proposed in contrast to the full autocorrelation function $R_{uu}(x, r)$ where r is a distance vector and $r = |r|$, a length scale should be prescribed. This leads to a special shape of R_{uu}. With the assumption of homogeneous turbulence and a fixed time as seen in Batchelor [1],

$$R_{uu}(r, 0, 0) = exp\left(-\frac{\pi r^2}{4 L^2}\right) \tag{4.11}$$

With Δx as the grid spacing and $L = n\Delta x$ as the length scale, the autocorrelation function in discretized form is:

$$\frac{\overline{u_m u_{m+k}}}{u_m u_m} = R_{uu}(k\Delta x) = exp\left(-\frac{\pi (k\Delta x)^2}{4(n\Delta x)^2}\right) = exp\left(-\frac{\pi k^2}{4n^2}\right) \tag{4.12}$$

The 3D filter coefficient is:

$$b_k \approx \tilde{b}_k \bigg/ \left(\sum_{j=-N}^{N} \tilde{b}_j^2\right)^{1/2} \text{ and } \tilde{b}_k := exp\left(-\frac{\pi k^2}{2n^2}\right) \tag{4.13}$$

This by convolution gives the 3D filter as seen in Eq. 4.10. The support of the filter must be large enough to capture twice the length scale, i.e. $N \geq 2n$ and $n \geq 2$. Furthermore, using the method discussed above, different length scales L_x, L_y and L_z are possible in the different spatial directions with n being replaced by n_x, n_y and n_z, respectively.

A more efficient procedure to reduce computational effort and memory requirement as well as increase parallel scaling performance is proposed by Kempf et al. [11].

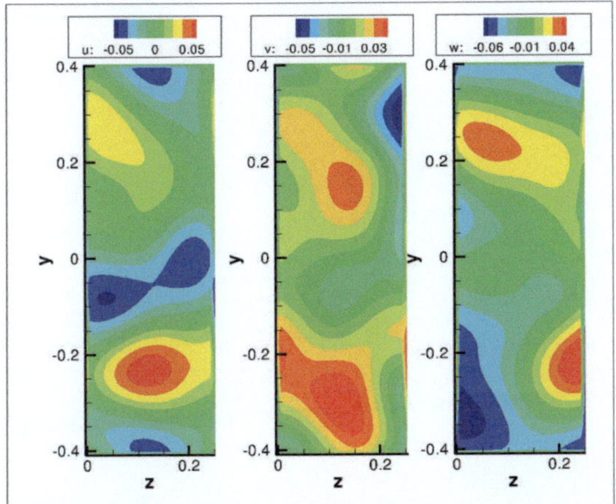

Fig. 4.10 Cartesian velocity components of the instantaneous velocity fluctuations normalized by u_∞ and generated by the digital filter method

Table 4.1 Kaimal length scales and standard deviation ratios from IEC61400-1 [20]

	Velocity component		
	Longitudinal	Lateral	Vertical
Standard deviation σ	σ_1	$0.8\,\sigma_1$	$0.5\,\sigma_1$
Integral scale, L_k	$8.1\,\Lambda_1$	$2.7\,\Lambda_1$	$0.66\,\Lambda_1$

Here an exponential filter kernel is applied to the field of random noise which is generated such that any parallel process may generate the same random number for any given location within the domain and the filter is chosen such that the required integral length scale is recoverable.

Figure 4.10 shows the Cartesian velocity components of the instantaneous velocity fluctuations in a plane normal to the inflow and normalized by the inflow velocity for a turbulence intensity of 2.8%. An isotropic case with a length scale per unit chord of 0.118 was considered based on an experiment by Hain et al. [8].

In order to generate anisotropic inflow turbulence for studies on wind turbines, the input parameters, such as the relationship between the standard deviations in the three principal directions and the length scales used for the generation of the inflow turbulence could be based on those suggested by the IEC-61400-1 standard as shown in Table 4.1.

According to the IEC standard, for hub heights greater than 60 m, the relative length scale is $\Lambda_1 = 42$ m which, according to Table 4.1, leads to length scales of 340.2 m, 113.4 m and 27.72 m in the longitudinal, lateral and vertical directions, respectively. It may be necessary to scale down these length scales for use within the employed computational domain. Furthermore, to allow for a streamwise and spanwise $(u - w)$ correlation, a non-zero Reynolds shear stress is included. As discussed

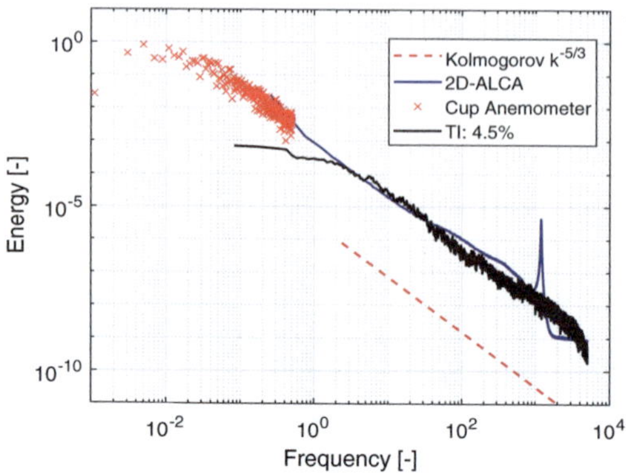

Fig. 4.11 Comparison of the generated inflow turbulence for $TI = 4.5\%$ with measurements from the lower atmosphere [10]

by Jonkman [9], this can be set as $\overline{u'_x u'_z} = -U^2_{star}$, where U_{star} is the friction or shear velocity and typically ranges from 0.05 to 0.1 depending on the ground roughness scale.

The drawback of the standard digital filter method described above is that it only allows the definition of one length scale per direction. However, this disadvantage can be compensated by superimposing the solutions of different length scales given by the maximal length scale divided by the factor 2^{n-1} where n refers to the index of the signals to be superimposed. Here n can be chosen based on the smallest length scale that can be resolved by the computational domain. For the inflow turbulence generated by superposition of several length scales to decay according to Kolmogorov's 5/3 law, a scaling according to $2^{(-5/6)\times(n-1)}$ is necessary as discussed in Lobo [16]. Figure 4.11 shows a comparison of the generated inflow turbulence using the method of superposition with measurements from the atmosphere as employed in large-eddy simulations by Lobo [16]. The spectra are in agreement, especially for $f > 10$ dimensionless units. For an agreement at even lower frequencies up to the order of 10^{-2}, an increase in computational time by an order of magnitude would be necessary, which was not computationally feasible at the time these simulations were carried out.

4.6.3   Adding Turbulence Close to the Region of Interest

A common problem associated with the introduction of turbulent boundary conditions at the inflow plane of the computational domain is that the larger grid resolution leads to the damping out of essential higher frequencies. This problem can be overcome by introducing the generated turbulence close to the region of interest which typically has a finer grid resolution. However, care must be taken to ensure that the divergence-free criterion is maintained.

One way to accomplish this is through a source term formulation where an additional source term is added to the momentum equation. A detailed description using the integral form of a general conservation equation as used in the finite volume

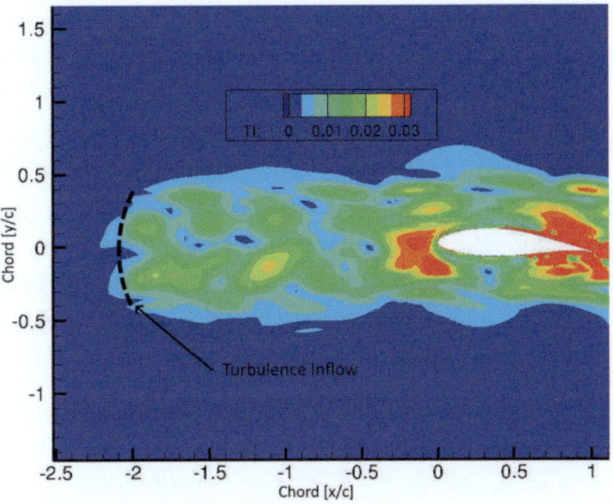

Fig. 4.12 Instantaneous spatial distribution of turbulent fluctuations with an inflow turbulence intensity of 1.4%

scheme is found in Schmidt and Breuer [3]. It is shown that despite the superposition of the synthetically generated velocity fluctuations as a source term, the subsequent iterations within their predictor-corrector scheme ensure a divergence-free formulation at the end of the time step.

Figure 4.12 shows an instantaneous view of the computational domain with the added inflow turbulence with the injection area located about two chord lengths upstream of the leading edge, whereas the inflow plane is located further upstream at about eight chord lengths upstream of the leading edge.

4.7 Problems

Problem 4.1 Imagine a rotor of diameter $\varnothing = 175\,$m and hub height $z_1 = 100\,$m Assume a wind speed $u_1 = 12\,$m/s and a logarithmic profile with $z_0 = 0.001$ (0.01) m.

Now calculate:

(a) (area)averaged wind speed

and

($\star$ b) averaged power

in comparison to the constant values.

Problem 4.2 Turbulence intensity

$$I := \frac{\sigma_v}{\bar{v}} \tag{4.14}$$

may be used for a simple engineering correction [6] of average power when influenced by I.

Show:

$$\overline{P(v)} = P(\bar{v}) + \frac{1}{2} P''(\bar{v})\sigma_v^2 \, . \tag{4.15}$$

Problem 4.3 Getting familiar with measured wind.

Download 1 h wind data measured with 25 Hz resolution and use some data analysis software (for example TISEAN from http://www.mpipks-dresden.mpg.de/~tisean/TISEAN_3.0.1/index.html) to get basic information:

(a) Plot a histogram and compare it with a Gaussian distribution.

(b) Calculate the energy spectra which is defined by Eq. (3.126).

(c) finally calculate the (auto)correlation function which is defined by

$$R(\tau) = \frac{1}{\sigma_u^2} \int_0^\infty < u(0) \cdot u(\tau) > \cdot dt. \tag{4.16}$$

Estimate a global correlation time and Taylor's microscale, Eq. (3.109).

Problem 4.4 Gust prediction—return time

(a) From Eq. (5) of [2], a normalized maximum gust wind speed $U_s = \frac{U_{max} - \bar{U}}{\sigma}$ during a large *return time T* may be estimated by

$$< U_s > = (2 \cdot ln(\nu T_R))^{1/2} \, . \tag{4.17}$$

Parameter ν (not to be mixed up with kinematic viscosity) is related to Taylor's microscale 3.109 by

$$\lambda_T = 1/(\nu \pi \sqrt{(2)}) \tag{4.18}$$

By using an estimated value for $\lambda_T \approx 20 \ msec$, $\bar{U} = 6$ m/s and $\sigma = 1$ m/s, the corresponding gust speed for return times of 1, 10, 50 and 100 years was calculated.

(b) Compare to a Rayleigh distribution

$$r(v, \bar{v}) = \frac{\pi}{2} \frac{v}{\bar{v}^2} \cdot \exp -\frac{\pi}{4} \left(\frac{v}{\bar{v}}\right)^2 \tag{4.19}$$

Problem 4.5 Time series averaging

For typhoon Haiyan from November 2013, a wind speed of 65.3 (87.5) m/s during 600 (60) s averaging time were measured. Try to give assumptions for the estimation of a 3 s value.

Problem 4.6 (a) Download *TurbSim* from https://wind.nrel.gov/designcodes/preprocesors/turbsim and create sample time series of artificial wind. Produce plot of the time series, a histogram and the energy spectra.

(b) The same with a sample time series from the so-called anisotropic Mann-Model [18, 19].

(* c) The same with the so-called *continuous time random walk* model [7].

References

1. Bachelor GK (1993) The theory of homogeneous turbulence. Cambridge University Press, Cambridge, UK, reprinted
2. Beljaars ACM (1987) The influence of sampling and filtering on measured wind gusts. J Atmos Oceanic Technol 4:613–626
3. Breuer M, Schmidt S (2017) Source term based synthetic turbulence inflow generator for eddy-resolving predictions of an airfoil flow including a laminar separation bubble. Comput Fluids 146:1–22
4. Cebeci T, Cousteix J (1999) Modeling and computation of boundary-layer flows. Springer, Berlin
5. el Hak MG (1989) Feasibility of generating an artificial burst in a turbulent boundary. Technical report, NASA-18292, Washington
6. K, K et al (2007) Turbulence correction for power curves. In: Peinke PS, Barth S (Eds) Proceedings of the euromech colloquium. Wind energy. Springer, Berlin, Heidelberg
7. Gontier H et al (2007) A comparison of fatigue loads of wind turbine resulting from a non-gaussian turbulence model vs. standard ones. In: DTU (Ed) The science of making torque from wind, Lyngby, Danmark, IoP, p. 012070
8. Hain R et al (2009) Dynamics of laminar separation bubbles at low-reynolds-number aerofoils. J Fl Mech 630:129–153
9. J JB (2009) TurbSim user's guide: Version 1.50, NREL/TP-500-46198. Technical report, National Renewable Energy Laboratory
10. Jeromin A, Schaffarczyk AP, Puczylowski J, Peinke J, Hölling M (2014) Highly resolved measurements of atmospheric turbulence with the new 2D-atmospheric laser cantilever anemometer. J Phys: Conf Ser 555:012054
11. Kempf AM et al (2012) An efficient, parallel low-storage implementation of klein's turbulence generator for les and dns. Comput Fluids 60:58–60
12. Klein M et al (2003) A digital filter based generation of inflow data for spatially developing direct numerical or large eddy simulations. J Comput Phys 186(2):652–665
13. Kleinhans D (2008) Stochasitische Modellierung komplexer Systeme. PhD thesis, Dissertation Universität Münster in German
14. Kleinhans D et al (2010) Synthetic turbulence models for wind turbine applications. In: Talamelli J, Peinke A, Oberlack M (Eds) Progress in turbulence III. Springer, Dordrecht, London, New York, pp 111–114
15. Lee S et al (1992) Simulation of spatially evolving compressible turbulence and the application of taylors hypothesis. Phys Fluids A 4(7):1521–1530
16. Lobo B (2023) Investigation into boundary layer transition on a wind turbine airfoil using wall-resolved large-eddy simulations and modeled inflow turbulence. PhD thesis, Helmut-Schmidt Universitä/UniBwH
17. Lund TS et al (1998) Generation of turbulent inflow data for spatially-developing boundary layer simulations. J Comput Phys 140(2):233–258
18. Mann J (1994) The spatial structure of neutral atmospheric surface-layer turbulence. J Fluid Mech 273:141–168
19. Mann J (1998) Wind field simulation. Prob Eng Mech 13(4):269–282
20. NN (2019) Iec 61400, wind turbines, design requirement. Technical report, International Electrotechnic Commision
21. Panofsky HA, Dutton JA (1984) Atmospheric turbulence. Wiley-Interscience, New York
22. Schaffarczyk A (2013) Pressure. In: Shen W (Ed) International conference on aerodynamics of offshore wind energy systems
23. Schaffarczyk AP et al (2010) A new non-gaussian turbulent wind field generator to estimate design-loads of wind-turbines. In: Peinke MO, Talamelli A (Eds) J. Progress in turbulence III, Springer proceedings in physics, vol 131, Springer, Dordrecht, London, New York
24. Seeger T, Köttgen V, Oliver R (1990) Schadensuntersuchung growian: Schlussbericht. Technical report, FB-7-1990,0328949A, Darmstadt

25. Steudel D (2007) Private communication. priv comm
26. Veers PS (1988) Three-dimensional wind simulation. Technical report, SAND88-0152, Albuquerque, NM, USA
27. Veltkamp D (2006) Chances in wind energy, a probabilistic approach to wind turbine fatigue design. PhD thesis, TU Delft, The Netherlands

Chapter 5
Momentum Theories

In the meantime, several relevant textbooks have been published:

- Reference [58] summarizes advances in momentum theory with special emphasis to connection to vortex theory and compares Joukovsky's and Betz' Vortex model,
- Reference [7] starts from *vorticity-based methods* and gives a very detailed compilation of applications to wind turbine aerodynamics,
- Reference [68] discusses the path from basic (Euler's) differential equations to the Actuator Disk model,
- Reference [50] this most recent book gives a *self-contained basis* for design. Special emphasis is given to various approaches to aerodynamic design.

It may be noted that Wood and Hammam [71] recently re-examined conditions for optimum (max. c_P) Actuator Disks and used Variational calculus. As a result, circumferential (i.e., a') remains free of singularities but $c_T(\lambda)$ becomes non-monotonic and shows a maximum at $\lambda \approx 2$.

5.1 1D Momentum Theory

5.1.1 Forces

We now apply our knowledge from Chap. 3 to the situation shown in Fig. 3.1. To make it quantifiable by hand calculations, we start by using integral momentum theory. We choose three special locations and assume pure 1D flow (in X-direction: u) with no spatial or temporal variation.

- 1: far upstream: In, u_1,
- 2: the location of the wind turbine, Disk, u_2,
- 3: far downstream, Out, u_3.

© The Author(s), under exclusive license to Springer Nature Switzerland AG 2024
A. P. Schaffarczyk, *Introduction to Wind Turbine Aerodynamics*, Green Energy
and Technology, https://doi.org/10.1007/978-3-031-56924-1_5

We assume that the action of the turbine (now also called *actuator disk*) reduces the inflow wind $v_{wind} = u_1$ to u_2 at the disk and further to u_3 far downstream. That there is a further decrease after the disk seems to be somewhat surprising, but this will become clear after proper application of all three theorems of conservation.

Starting with the momentum theorem, we see that at *In*, we have momentum flow $\dot{p}_1 = \dot{m} \cdot u_1$, whereas at *out*, it is only $\dot{p}_3 = \dot{m} \cdot u_3$. From that, we have to conclude that there must be a force T exerted by the turbine on the passing air. From Newton's Third Law, we know there is an opposing force of the same magnitude $-T$ applied by the air to the disk. Together

$$\dot{m}(u_1 - u_3) - T = 0 \ . \tag{5.1}$$

Now we use the Energy Equation (Bernoulli's law). Because we have seen that there is a force at location 2, this may be converted to a pressure drop $\Delta p_{2^- 2^+} = T/A_r$. As already remarked, the reference area A_r is the *swept area* of the rotor $\frac{\pi}{4} D^2$. This implies that we may use Bernoulli only from $1 \to 2^-$ and then further from $2^+ \to 3$. Here $-$ refers to a location immediately upstream of the disk and $+$ to a location immediately downstream of the disk. From $1 \to 2^-$:

$$p_0 + \frac{\rho}{2} u_1^2 = p_2^- + \frac{\rho}{2} u_2^2 \ . \tag{5.2}$$

and the same from $2^+ \to 3$:

$$p_2^+ + \frac{\rho}{2} u_1^2 = p_0 + \frac{\rho}{2} u_3^2 \ . \tag{5.3}$$

Figure 5.1 shows the development of the pressure along the slipstream. As there is a pressure drop across the disk to negative values (if $p_0 = 0$), there must be a further decrease of velocity for recovering the pressure to p_0.[1]

p_0 denotes the ambient pressure and may be set to 0 without loss of generality. We assume that it acts on the whole surface of our *control volume*.

Solving Eqs. (5.2) and (5.3) for $\Delta p = p_2^+ - p_2^-$, we get

$$\Delta p = \frac{\rho}{2} \left(u_1^2 - u_3^2 \right) \ . \tag{5.4}$$

Now we have to find an expression for the mass flow which must be constant from $1 \to 2 \to 3$. Taking the definition of density into account and assuming—as already mentioned—constant velocity across A_r, we have

$$\dot{m} = \rho \dot{V} = \rho A_r \cdot u_2 \ . \tag{5.5}$$

---

[1] In [69], Fig. 12 it is discussed that the pressure jump is not symmetrical, at least for $c_T = 8/9$.

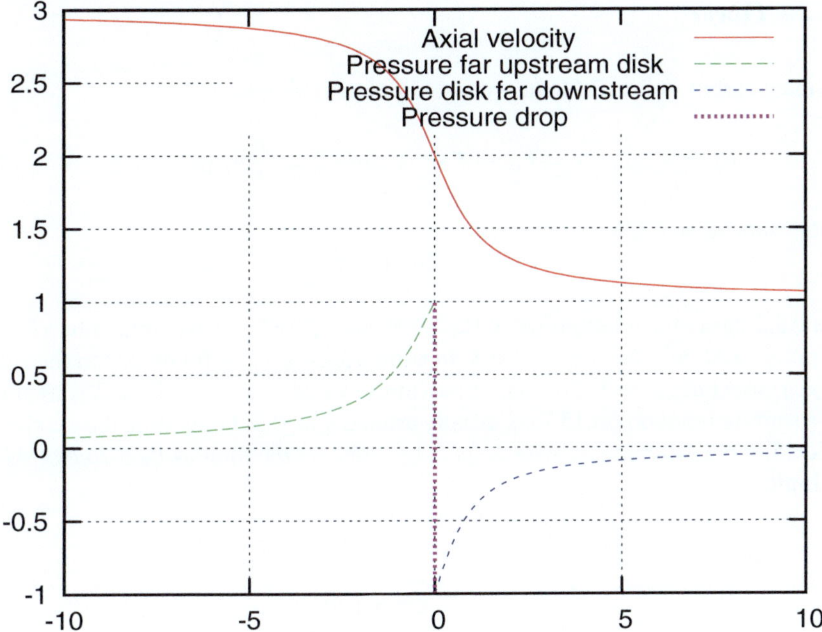

Fig. 5.1 Development of velocity and pressure along the slipstream (arbitrary units)

Equivalently, this equation may be regarded as the *definition* of averaged velocity at the disk. Comparing now Δp from Eq. (5.1) with that from Eq. (5.4), we arrive at the famous *Froude's Law*:

$$u_2 = \frac{1}{2}\left(u_1 + u_3\right) . \tag{5.6}$$

If we define a reference force as the product of the stagnation pressure of the wind $q_0 := \frac{\rho}{2}u_1^2$ times our reference area A_r, $F_{ref} := q_0 \cdot A_r$ and the so-called *(axial) induction factor* by $u_2 := (1 - a)u_1$, then it is immediately seen that if $c_T := T/F_{ref}$ we may express non-dimensional thrust by only one variable a.

$$c_T(a) = 4\left(1 - a\right)a . \tag{5.7}$$

A slipstream can only be existing if $u_3 \geq 0$. Therefore, the inequality $0 \leq a \leq 1$ must hold.

5.1.2 Power

As stated earlier in Eq. (2.1), the extracted power is simply

$$P_T = P_{extracted} = P_{in} - P_{Out} + \Delta p \cdot \dot{V} = \frac{\rho}{2} A_r \left(u_1^3 - u_3^3 \right) . \qquad (5.8)$$

In non-dimensional form:

$$c_P(a) = 4a \left(1 - a \right)^2 . \qquad (5.9)$$

Figure 5.2 shows the relationship of Eqs. (5.9) and (5.7). Note that momentum theory is strictly valid only for $0 \leq a \leq 0.5$. In some cases, it is useful to extend the range of $c_T(a)$ to values $a \approx 1$. For that, a maximum value of $c_T(a = 1) = 2$ is assumed and a starting point on Eq. (5.7) is usually around $a = 0.3$. For getting the maximum of Eq. (5.9) we differentiate with respect to a and get the famous *Betz-Joukovski* [6, 34] limit.

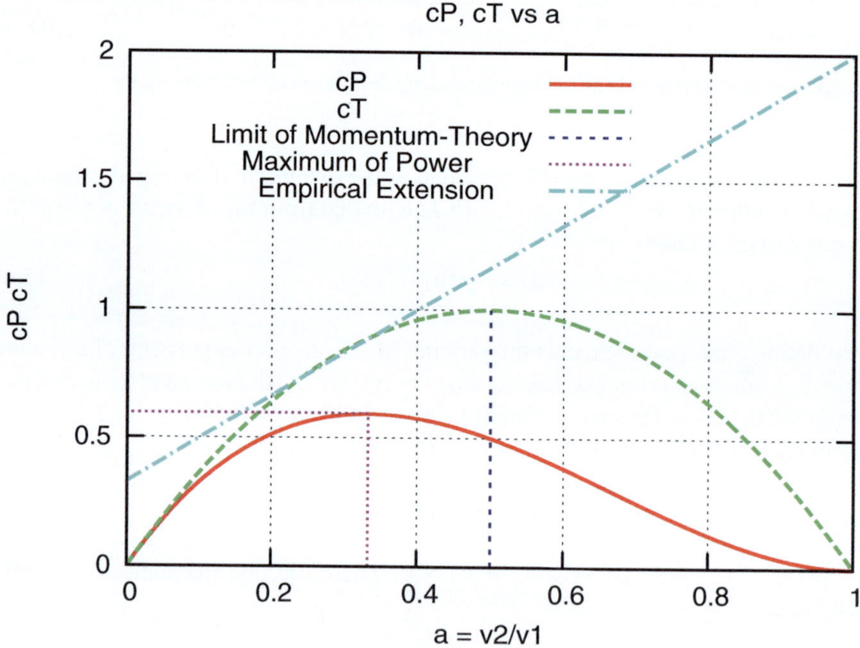

Fig. 5.2 c_P and c_T as function of induction factor a

$$c_P^{max} = \frac{16}{27} \tag{5.10}$$

$$at \; a = \frac{1}{3} \tag{5.11}$$

$$and \; c_T \left(a = \frac{1}{3} \right) = \frac{8}{9}. \tag{5.12}$$

5.1.3 Remarks on Beating Betz

There is an ongoing discussion about whether it might be possible to get closer to a c_P^{max} of 1. In a lesser-known paper—probably because it is published in German only—Betz [5] discussed already in 1926 the possibility of exceeding the $\frac{16}{27}$ value if one relaxes the underlying assumptions, the most restrictive of which is the 2D actuator disk.[2] See Problem 5.2 for a short introduction to the matter.

This is particularly obvious in the case of more 3D-like wind turbines known as DAWTs, see Sect. 2.5. Here most of the confusion comes from the fact that an appropriate definition of the reference area is less obvious. Jamieson [24, 25] proposed the furthest upstream sectional area A_1 to be used as a reference. He further showed that

$$c_P \leq \frac{16}{27}(1 - a_0) \tag{5.13}$$

if a_0 is defined by $v_2 = v_1(1 - a_0)$ when there is no energy extraction from the diffuser—the empty case. It is clear that such a device should have a_0 negative with low additional costs compared to the original turbine.

It is clear then now the discussion can be limited to the diffuser only. (Compare to Ref. [65] from Chap. 2).

5.2 General Momentum Theory

The 1D model of the preceding chapter is a very simple one and may be regarded as purely academic, as it was shown that such a 1D model does not exist at all [59]. One reason for this is that, the DEQ of fluid mechanics couples the velocity components so rigidly that, from $v \equiv w \equiv 0$, it must follow $u = const$. In addition, no hints on how to design blades can be deduced from this 1D theory.

---

[2] Page 14: Because the ratio $\frac{L_{max}}{L_0}$ plays an exceptionally important role, it is appropriate to consider the extent to which the derived value of 16/27 depends on specific assumptions and if it is possible to increase it intentionally. The energy behind the first wind turbine can easily be exploited by the use of a second wind turbine. Our original derivation applies only to a single disk-shaped wind turbine. If one allocates the space behind the first wind turbine for energy use, it is possible to achieve $L_{max/L_=} \approx 1$, assuming the prescribed diameter is not exceeded (translated by John Thayer).

Therefore, Glauert [19] formulated a 2D theory in which circumferential flow is allowed. So instead of a Cartesian coordinate system (x, y, z), we now use a cylindrical one (x, r, φ); one with the same notations for the velocity components.

To extract power, we now use *torque* Q

$$P = F \cdot v = Q \cdot \Omega , \tag{5.14}$$

with Ω being the rotational velocity of the turbine

$$\Omega = \frac{N \cdot \pi}{30} = RPM \cdot 0.1047 . \tag{5.15}$$

From Newton's third law for axial momentum, there must be an (angular) momentum balance as well. Therefore, introducing torque and tangential forces, we quantify tangential induced velocities

$$a'(r) \cdot \Omega r, \ x = 0 \tag{5.16}$$
$$2a'(r) \cdot \Omega r, \ x > 0 . \tag{5.17}$$

The total velocity at the disk now consists of two components perpendicular to each other:

$$u_2 = (1 - a)u_1 \tag{5.18}$$
$$u_\varphi = (1 + a') \cdot \Omega r \tag{5.19}$$
$$in \ total : \ w^2 = u_2^2 + v_t^2 \tag{5.20}$$

We divide the disk into annuli of thickness dr and area $dA = 2\pi r \cdot dr$. (See Fig. 5.3).

The change of angular momentum is

$$dQ := d\dot{m} \cdot v_\varphi \cdot r = \tag{5.21}$$
$$\rho dA u_1(1 - a) \cdot 2a'\Omega r \cdot r . \tag{5.22}$$

As $P = Q \cdot \Omega$ also $dP = dQ \cdot \Omega \stackrel{!}{=} dT \cdot u_2$.

The last equation seems a little bit surprising as one may think that axial and tangential forces should be independent. This is obviously not the case if we use the energy (Bernoulli's) equation across the disk. Specific enthalpy (Bernoulli constant) was already introduced in Sect. 3.2.3, Eq. 3.24. It may be shown[3] [19, 60] that

$$\Delta p = \rho u_\varphi \cdot \left(\Omega r + \frac{1}{2} u_\varphi \right) \tag{5.23}$$

---

[3] Sharpe [11, 52] disagreed with Glauert on how the energy equation for the rotational part is transferred. We come back to this below in Sect. 5.3.

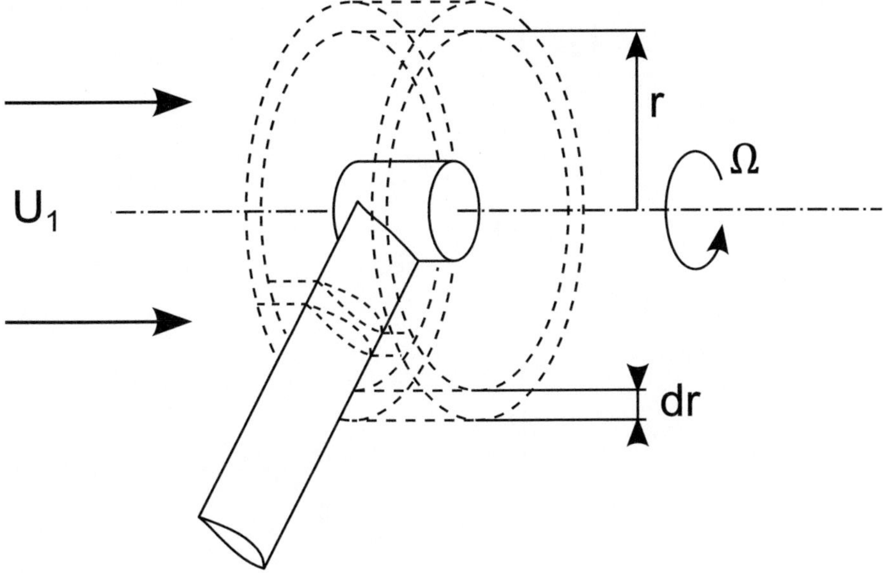

Fig. 5.3 Annuli of extension dr from $0 \leq r \leq R_{Tip}$

with u_φ from Eq. (5.19). If we now equate

$$dT = 4\pi \rho u_1^2 (1 - a) a r \, dr \qquad (5.24)$$

$$and \qquad (5.25)$$

$$dT = \Delta p \cdot 2\pi r dr = 2\pi r u_\varphi \left(\Omega r + \frac{1}{2} u_\varphi \right) r dr, \qquad (5.26)$$

we get

$$(1 - a)a = \lambda^1 x^2 (1 + a')a' \qquad (5.27)$$

with $x := r/R_{Tip}$, $0 \leq x \leq 1$ and $\lambda = \Omega R/u_1$ as already defined in Eq. (2.3) (Fig. 5.4). The increment per annulus for the power coefficient is

$$dc_p = 8(1 - a)a'\lambda^2 x^3 \, dx \,, \qquad (5.28)$$

or if we integrate over all annuli

$$c_p = 8\lambda^2 \int_0^1 (1 - a)a'x^3 \, dx \,, \qquad (5.29)$$

To maximize the total power Eq. (5.29), we have to optimize locally Eq. (5.28). For that, we define a local efficiency

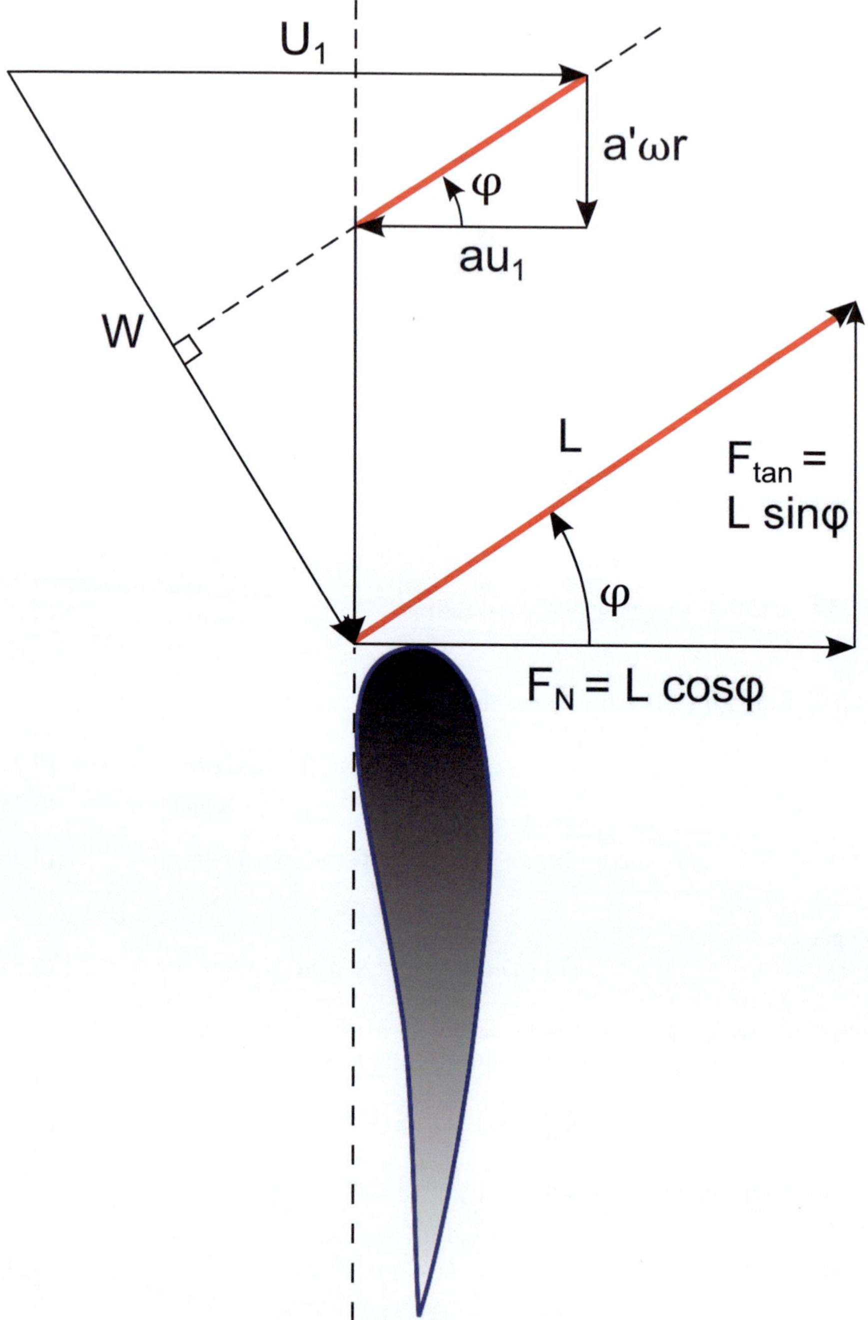

Fig. 5.4 Velocity and force triangles

$$\eta(a, a', x) := a'\,(1 - a)\,x^3 \tag{5.30}$$

and differentiate $\partial\eta/\partial a'$ to obtain

$$\frac{da(a')}{da'} = \frac{1 - a}{a'} \;. \tag{5.31}$$

Doing the same with Eq. (5.27), we have

$$\frac{da(a')}{da'} = \frac{(\lambda \cdot x)^2}{1 - 2a} \;. \tag{5.32}$$

or

$$\frac{1 - a}{a'} = \frac{\lambda^2 x^2}{(1 - 2a)} \;. \tag{5.33}$$

Combining Eqs. (5.27) and (5.33), we may drop $\lambda \cdot x$ to obtain a relation $a'(a)$ only:

$$a' = \frac{1 - 3a}{4a - 1} \;. \tag{5.34}$$

As always, $a' > 0$, this implies $1/3 > a > 1/4$.

To solve for numbers, one may start with the cubic equation

$$16a^3 - 24a^2 + 3a(3 - \lambda^2 x^2) - 1 + \lambda^2 x^2 = 0 \tag{5.35}$$

$$\text{with solution [7]: } a = \frac{1}{2}\left(1 - \sqrt{1 + \lambda_r^2}\, sin\left[\frac{1}{3}arctan\left(\frac{1}{\lambda_r}\right)\right]\right) \text{ and} \tag{5.36}$$

$$\lambda_r := \lambda \cdot x \tag{5.37}$$

and the rest follows for a' from Eq. (5.34) for the flow angle from

$$tan(\varphi) = \frac{a'}{a} \cdot \lambda x \tag{5.38}$$

and finally dc_P from Eq. (5.28) (Fig. 5.5).

As a remark, it may be noted that Schmitz [49] formulated a comparable model. However, it was published in German only in an unknown journal behind the *iron curtain*. Nevertheless, a nice asymptotic formula $\lambda \to \infty$ results. To order λ^{-4}:

$$c_P^{Schmitz} = \frac{16}{27}\left(1 - 0.219\frac{1}{\lambda^2} - 0.106\frac{1}{\lambda^4} - \frac{4 \cdot ln(\lambda)}{9 \cdot \lambda^2}\right) \;. \tag{5.39}$$

It gives $x_P = 0.5852, 0.5703$ and 0.4 for $\lambda = 10, 5$ and 1. These values are in accurate agreement with values from Table 5.1 even for small values of λ.

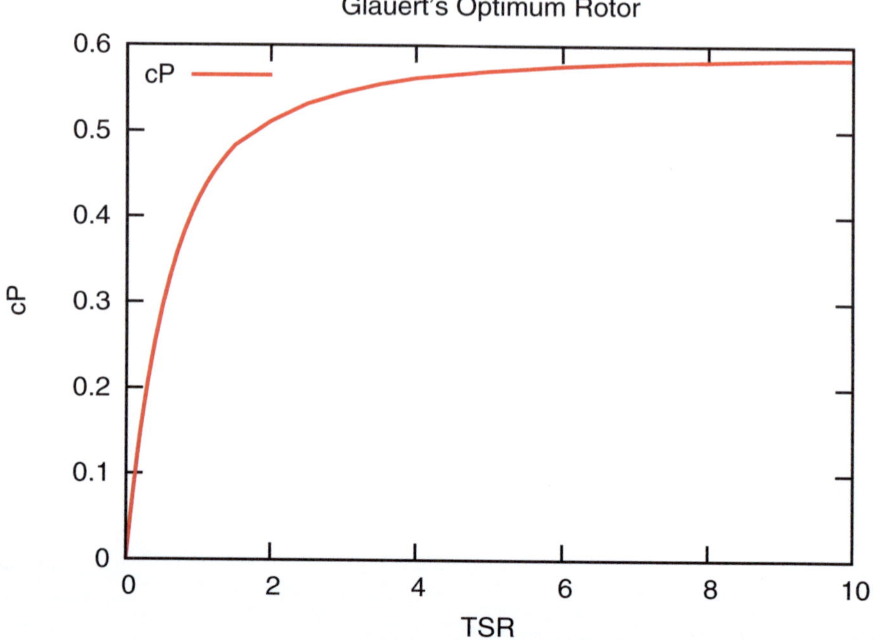

Fig. 5.5 cP versus TSR for a rotor according to Glauert's general momentum theory

Table 5.1 Numerical values for a locally optimized blade according to Glauert's general momentum theory

Local TSR	Axial induction	Rotational induction	Flow angle	Power	Solidity
λx	a	a'	φ	c_P	σ
0.25	0.280	1.364	50.6	0.176	0.3658
0.50	0.298	0.543	42.3	0.289	0.5205
1.0	0.316990	0.182981	30.0	0.4154	0.5355
2.0	0.327896	0.052349	17.7	0.5111	0.3799
3.0	0.330748	0.024016	12.2	0.5453	0.2749
4.0	0.331842	0.013669	9.3	0.5615	0.2128
5.0	0.332367	0.008797	7.5	0.5703	0.1728
6.0	0.332658	0.006127	6.3	0.5758	0.1452
7.0	0.332835	0.004510	5.4	0.5795	0.1250
8.0	0.332951	0.003457	4.9	0.5820	0.1098
9.0	0.333031	0.002733	4.7	0.5838	0.0977
10.0	0.333088	0.002215	3.8	0.5852	0.0881
11.0	0.333130	0.001831	3.4	0.5863	0.0802
12.0	0.333163	0.001540	3.2	0.5871	0.0736

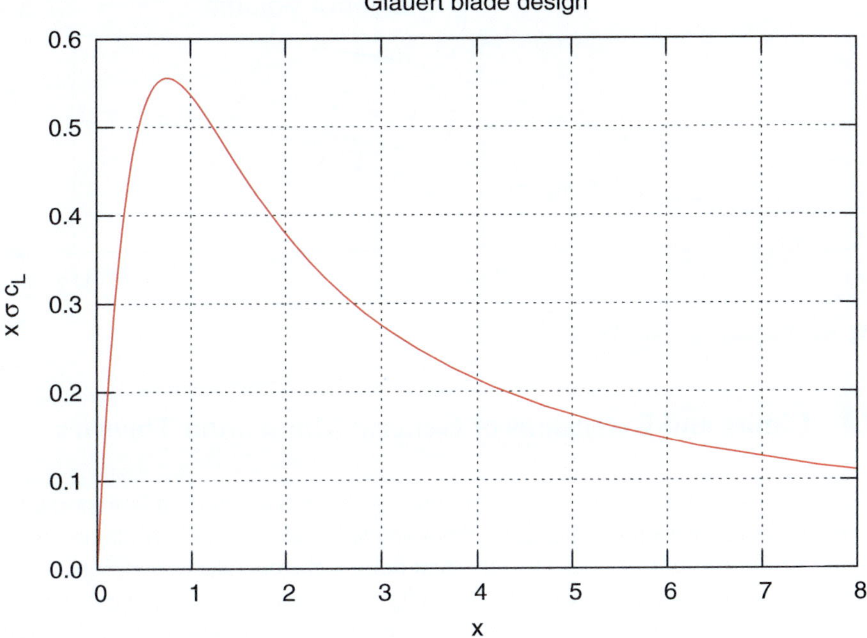

Fig. 5.6 Solidity of a blade from Glauert's theory

As a consequence, we now may derive a relation for the *chord* of the blade, if we assume that the forces come from lift only:

$$\sigma := \frac{Bcc_L}{8\pi \cdot r} = 4\,Ax\,\frac{sin^2(\varphi)}{cos(\varphi)} \tag{5.40}$$

$$A := \frac{a}{1-a} \tag{5.41}$$

which is shown in Fig. 5.6.

Recently, [60] Glauert's model has been re-investigated especially for low tip speed ratios for the case of constant circulation ([19], p. 193). The original singularity for $\lambda \to 0$ corresponding to $cP \to \infty$ could be eliminated by assuming an influence from the lateral pressure on the curved streamtube (compare to Fig. 5.7).

A CFD model (see Sect. 7.7 for details) [44, 56] gives a glimpse of how closely (an integral) analytical and (differential) numerical solutions may approach each other.

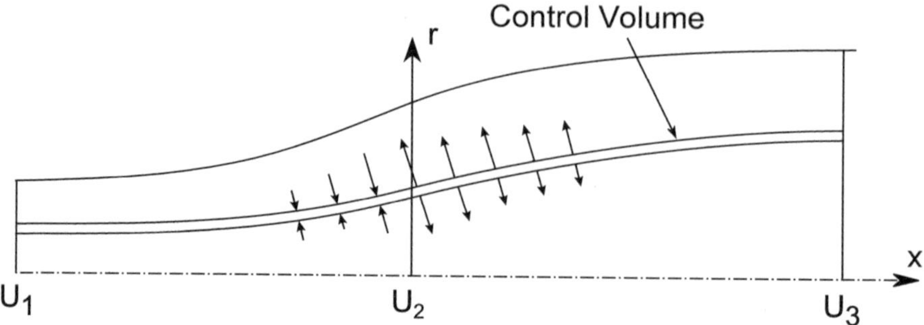

Fig. 5.7 Pressure on Control Volume

5.3 Limits and Extensions of General Momentum Theories

Momentum theory in itself is a very simple approach. Therefore, it is important to know the limits of validity and apply this knowledge to serve as a guide on how to improve this class of models. One important point is the boundary conditions for the (static) pressure far downstream. From a physicist's point of view, every quantity should be equally finite (best: 0) for $r \to \infty$. However, some authors [14, 28] argue that the static pressure inside the slipstream might have other values than outside. Clearly, this might be used as an energy source (Fig. 5.8). Sørensen [58] and van Kuik [55, 59] recently have published a more comprehensive discussion. The most general expression which can be deduced is

$$b(1 - a) = \frac{p_3 - p_1}{\rho u_1^2} \frac{1 - a}{1 - b} + \frac{\Delta X}{\rho u_1^2 \Delta A} + 2\lambda^2 x^2 a' \left(1 + a'\right) \tag{5.42}$$

in which $b =$ induction in the slipstream far downstream,

$$\Delta X = \oint_{ControlVolume} p(\mathbf{dA} \cdot \mathbf{e_x}) , \tag{5.43}$$

the force from the pressure on the slipstream.

With the following assumptions: $b \approx 2a$, $p_0 \approx p_3 \Rightarrow$, and $\Delta X = 0$ Glauert's expressions are recovered:

$$a(1 - a) = \lambda^2 x^2 a'(1 + a') , \tag{5.44}$$

$$\Delta T = \frac{1}{2}\rho(u_1^2 - u_3^2) \cdot \Delta A , \tag{5.45}$$

$$u_2 = u_{Disk} = \frac{1}{2}(u_1 + u_3) . \tag{5.46}$$

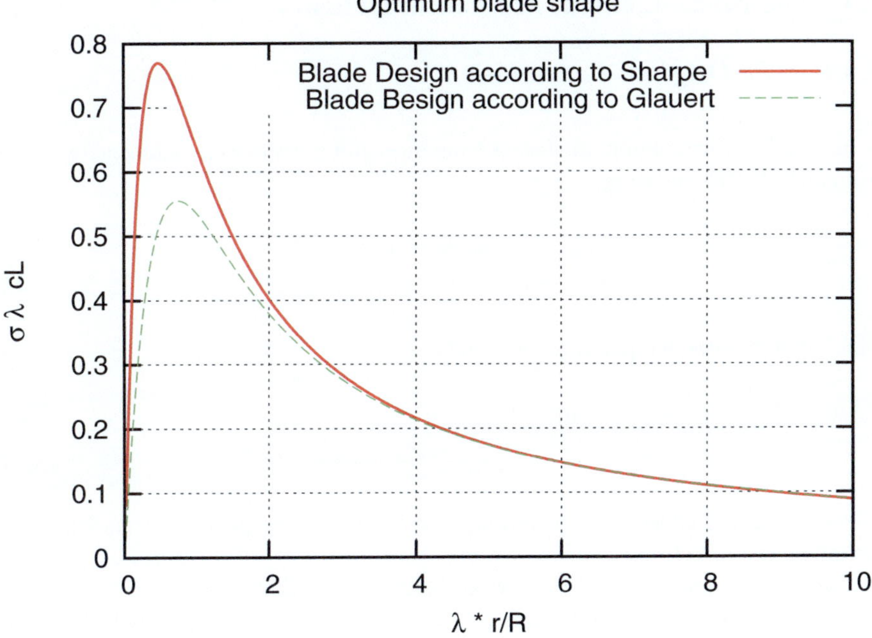

Fig. 5.8 Solidity of a blade from Sharpe's (constant Circulation) theory compared with Glauert's

The only way to extend momentum theory is to start with the DEQ approach Problem 3.11[68] and then to try to proceed as far as possible analytically.[4] Unfortunately, no—or almost no—purely analytical methods are available, but efficient numerical methods then must help. A group of scientists from Copenhagen have proceeded with the development of this line of theory. A recent summary and discussion of their findings may be found at [55, 59]. As may be expected, the deviations from simple momentum theory become particularly large when the rotor is highly loaded.

In a completely different approach, Conway ([12] of Chap. 6) (see Sect. 6.7) has found comparable results.

---

[4] van Kuik [66, 67] examined the type of singularity at the disk-edge in more detail and applied the consequences to real turbines. See also [61].

5.4 The Blade Element Momentum Theory

5.4.1 The Original Formulation

In its simplest formulation, the forces from momentum theory are balanced by forces from airfoils (lift and drag):

$$c_N = c_L cos(\phi) + c_D sin(\phi), \tag{5.47}$$

$$c_{tan} = c_L sin(\phi) - c_D cos(\phi) . \tag{5.48}$$

which then results in equations for a and a':

$$a = \frac{1}{4 sin^2(\phi)/(\sigma c_N)+1}, \tag{5.49}$$

$$a' = \frac{1}{4 sin(\phi)cos(\phi)/(\sigma c_{tan})-1} . \tag{5.50}$$

Here $\sigma = Bc/2\pi r$ again is the solidity and $N(ormal)$ means the normal while tan the tangential direction.

Clearly, this set of equations may be solved by iteration only.

For an ideal rotor, one may set $c_L = 2 \cdot \pi$ and $c_D = 0$, together with Eqs. 5.41 and 5.38, respectively.

5.4.2 Engineering Modifications

5.4.2.1 Real Airfoils

Airfoil data in the form of tables of c_L and c_D as a function of angle of attack (AOAα) have to be provided. In most cases, measurements are not available for AOAs greater than α_{stall}. In these cases, an empirical correction from Viterna and Corrigan (Chap. 2) [62], for $\alpha < \alpha_{Stall}$ and Aspect-Ratio μ:

$$c_L = \frac{c_{D,max}}{2} sin(2 \cdot \alpha) + K_L \cdot \frac{cos^2(\alpha)}{sin(\alpha)}, \tag{5.51}$$

$$c_D = c_{D,max} sin^2(\alpha) + K_D \cdot cos(\alpha), \tag{5.52}$$

$$K_L = \left(c_{L,S} - c_{D,max} sin(\alpha_S)\right) \cdot cos(\alpha_S)\, \frac{sin(\alpha_S)}{cos^2(\alpha_S)}, \tag{5.53}$$

$$K_D = \frac{c_{D,S} - c_{D,max} sin^2(\alpha_S)}{cos(\alpha_s)}, \tag{5.54}$$

$$if\ \mu \leq 50: \quad c_{D,max} = 1.11 + 0.018 \cdot \mu, \tag{5.55}$$

$$if\ \mu > 50: \quad c_{D,max} = 2.01 . \tag{5.56}$$

More about Post Stall may be found in [64].

Sometimes—especially in the inside parts(low r/R_{tip}) of a blade—the solution of Eq. 5.50 gives values outside $0 \leq a \leq 1/3$. Then empirical extensions are applied

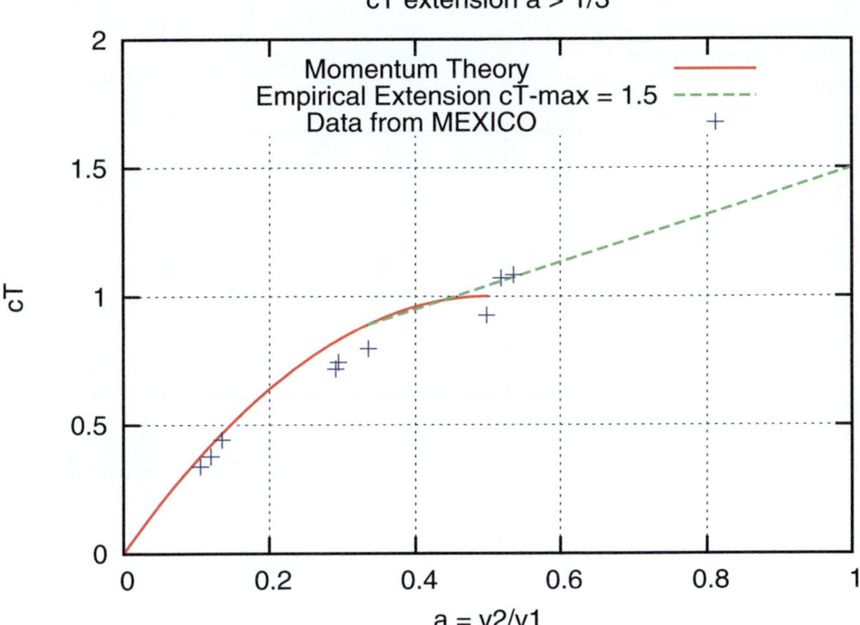

Fig. 5.9 Extrapolation of c_t values for $a > 1/3$ compared to measurements from [46]

which in their simplest form result in a linear interpolation to an assumed maximum thrust coefficient $c_{T_max} \approx 2$:

$$c_T(a) = 4a(1 - a), \ \ 0 \le a \le 1/3 , \tag{5.57}$$

$$c_T(a) = b + m \cdot a, \ \ 1/3 \le a \le 1 , \tag{5.58}$$

and $m = \frac{8}{9} \cdot c_{T,max} - 1, c_{T,max} > 8/9$. Figure 5.9 shows this for $c_{T,max} = 2.0$.

5.4.2.2   Finite Number of Blades

Prandtl [4] showed with an application of potential theory, Sect. 3.5, that close to the tip of a blade, the circulation must go down to zero, because of pressure compensation between the upper- and lower side at the tip of an airfoil. He compared that situation to a stack of flat plates, because this model is manageable using elementary methods, see Fig. 5.10.

Calculating using this model, he was able to show:

$$cos\left(\frac{\pi}{2}F\right) = exp\left(-\frac{B(1 - r/R_{Tip})}{(2r/R_{Tip})sin(\phi)}\right) , \tag{5.59}$$

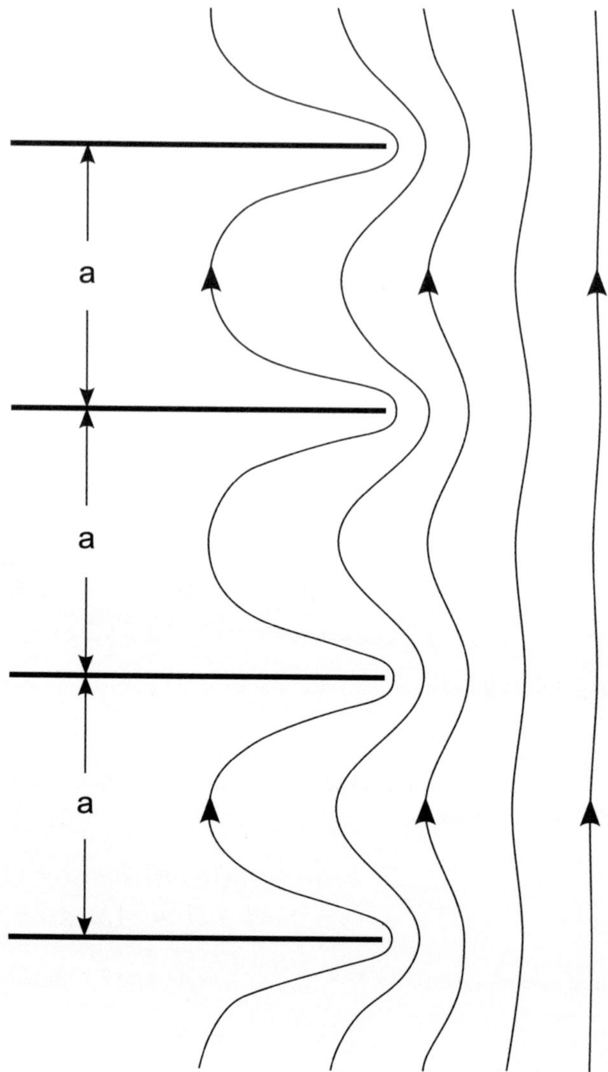

Fig. 5.10 Prandtl's flow model to derive a simple model for circulation loss at the tip

where F is the so-called *tip loss factor* (Fig. 5.11). Thereby we are able to discuss in approximate terms the effect of finite numbers of blades. It must be stated that, still today, these *tip loss models* are discussed because the influence on c_P is rather pronounced. A typical engineering expression states

$$\Delta c_P = const/(B \cdot \lambda) , \qquad (5.60)$$

with const ≈ 1.5. Main source of ambiguities is the question how to apply F to the induction factors, forces and/or airfoil data and how to calculate the flow angle.

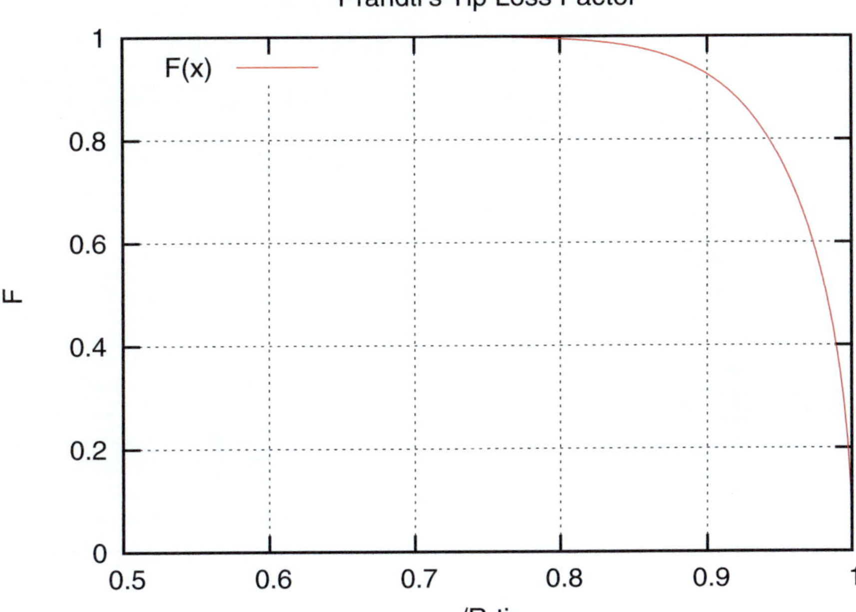

Fig. 5.11 $B = 3$, $a = 0.3$, $\lambda(\text{TSR}) = 6$

Ramdin [41] shows that a set of more than 70 combinations is possible. Equation 5.59 appears first in [19] and seems to be the most common implementation. However, Branlard [7] shows that the application of some of the proposed tip correction may result in changes of annual energy production of $1 \pm 2\%$ (Fig. 5.12). We will come back to this in more detail in Chap. 6. Ambitious readers may solve Problem 5.6. For the most often used blade number $B = 3$, we have in Fig. 5.13, a variation of airfoil data: If we compare the findings of Fig. 5.13 with Fig. 5.15, we see that a lift-to-drag ratio much larger than 100 is necessary to reach reasonable power production ($>0.5\ c_P$).

We may close this section by quoting a remark from van Kuik [68], p. 104, that there is

ample room for discussion and future work.

5.4.2.3 Behaviour Close to the Hub

In addition to the tip region, noticeable changes in aerodynamic forces are also seen in the inboard sections. In case of lift, this was already recognized by Himmelskamp in his Ph.D. thesis [22] in 1950. He used a Gö625 profile with 20% thickness. Compared

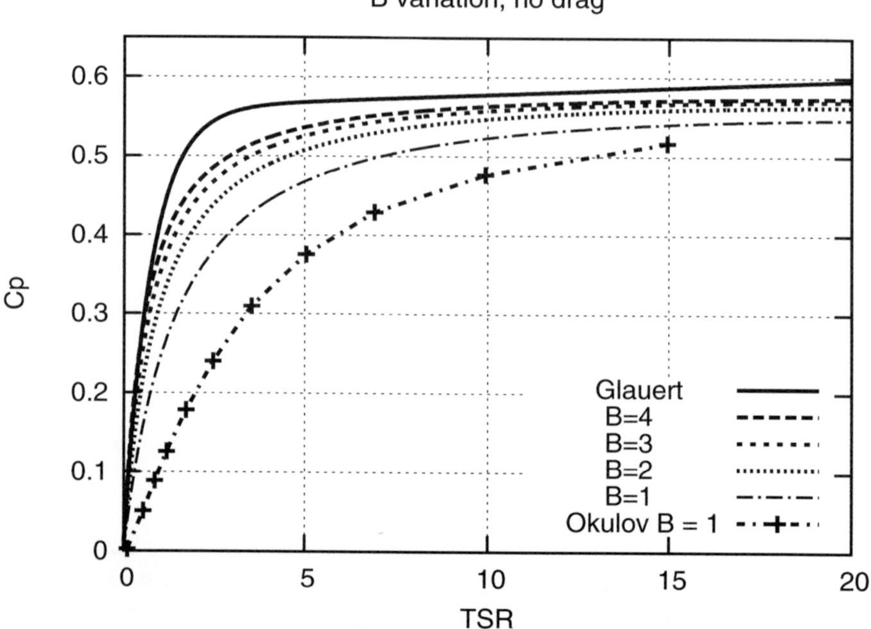

Fig. 5.12 c_P via λ, Number of Blades B varies

to the 2D wind tunnel measurements, he obtained almost a doubled $c_L \sim 3.0$ close to the hub from which he concluded that due to inertial forces, separation may be delayed to higher AOAs. Therefore, the so-called 3D correction using two parameters (c = chord of blade at radius r) and two constants (a and b) are most commonly written in the form of [55]:

$$c_{L,3D} = c_{L,2D} + a(c/r)^b \left[c_{L,inv} - c_{L,2D} \right] . \tag{5.61}$$

Bak, Johansen and Andersen [3] investigated three different rotors by CFD and compared to pure 2D and 3D corrections of other investigations. Since then, remarkable progress has been achieved.

On top of that, some authors [11] even argue for *blade root losses* with the same line of argument as for the top. As in most cases $c_l \to 0$ as $r/R_{tip} \to 0$ by use of a cylindrical section as a connection to the hub this might be regarded questionable.

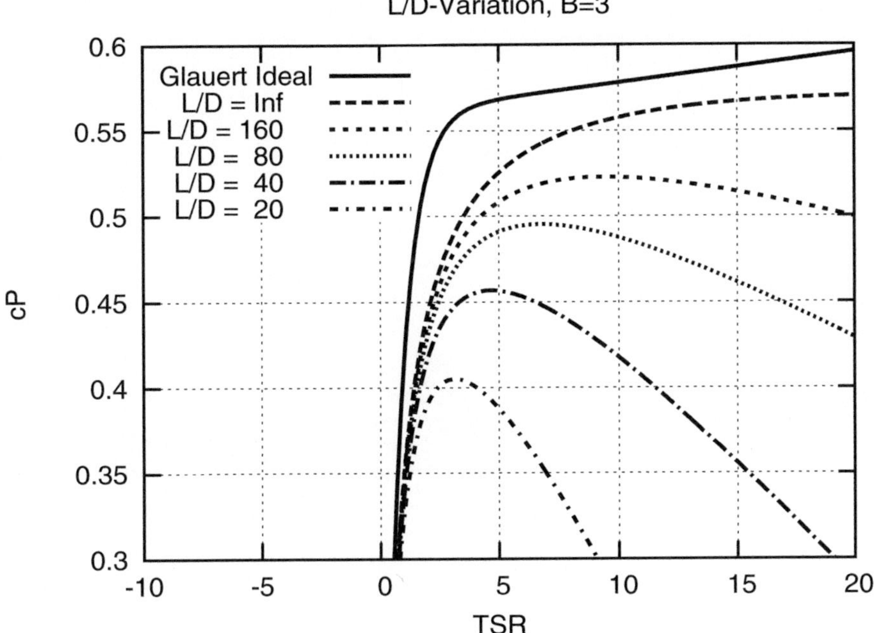

Fig. 5.13 c_P via λ. Number of blades fixed to B = 3. Lift to Drag ratio varies

5.4.3 Comparison to Actual Designs

Although this method is the most widely used in industrial applications for rotor performance and load prediction since it was published 80 years ago, it has to be noted that in general the agreement is not always the best. References [8, 9, 45] give a small insight into how much effort has been spent on the validation of these codes.

During the big NASA-Ames blind comparison (see Sect. 8.5, and Ref. [53] there) BEM-Codes have been compared also with the so-called *Wake Codes* and CFD (=RANS, see Sect. 7.8). As an example in Fig. 5.14, the comparison for the shaft torque Q ($P = Q \cdot \omega$) is shown. Because the rotor was designed somewhat unconventionally, BEM only gives reasonable results if the stall behavior of the S809 is modeled carefully [64].

In 2006, in Europe's largest wind tunnel, the DNW-LLF, a comparable experiment called *MEXICO* (=Model Experiments in Controlled Conditions) was conducted. Forces, pressures on the blades and the flow field were measured by PIV [48]. More about comparing modeling results with experimental results [47] will be presented in Sect. 8.6.

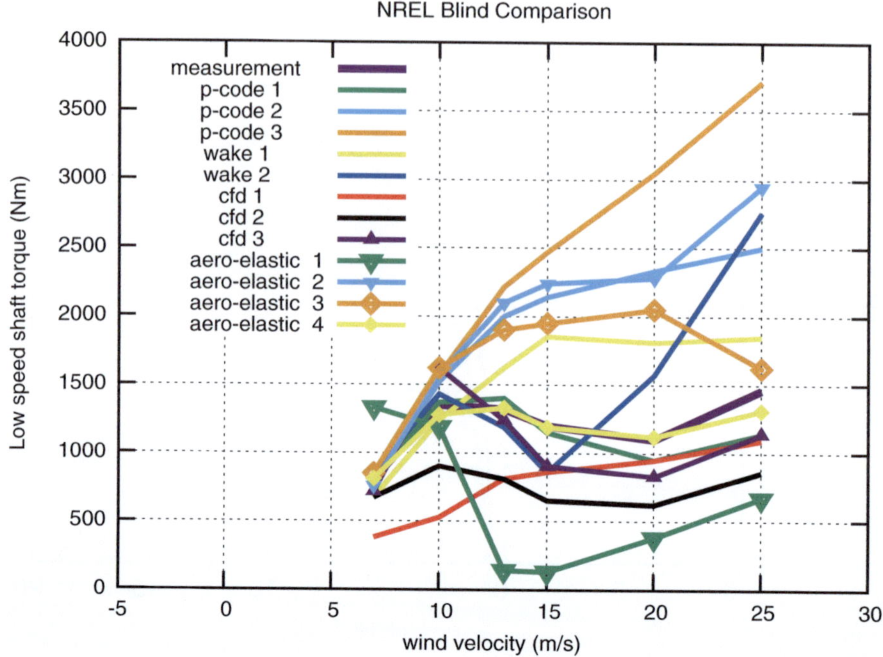

Fig. 5.14 Some results from NASA-Ames blind comparison

Madsen et al. [32] investigated the effect of pressure variation close to the hub due to wake rotation and non-uniformity of induction close to the tip, thereby considerably improving accuracy of c_P prediction for a high c_P (0.51–0.52) 5 MW rotor.

Much controversy was created in 2004 when ENERCON [38] claimed to have measured $c_P^{max} = 0.53$ (E2 of Fig. 5.15). Somewhat later this value were reduced to 0.51.

5.5 Optimum Rotors I

For a long time, many authors [15, 33, 70], to mention only a few of them, have tried to design an optimized (in the sense of reaching highest c_P^{max}) rotor which includes the presented findings of engineering BEM. One weak spot is of course that the model assumes an infinite number of blades and the Prandtl correction, Eq. 5.59 is valid only to a nominal accuracy.

It is therefore necessary to use a different approach, the vortex model of wind turbines, to look for an *optimum* rotor. This will be presented in the next chapter.

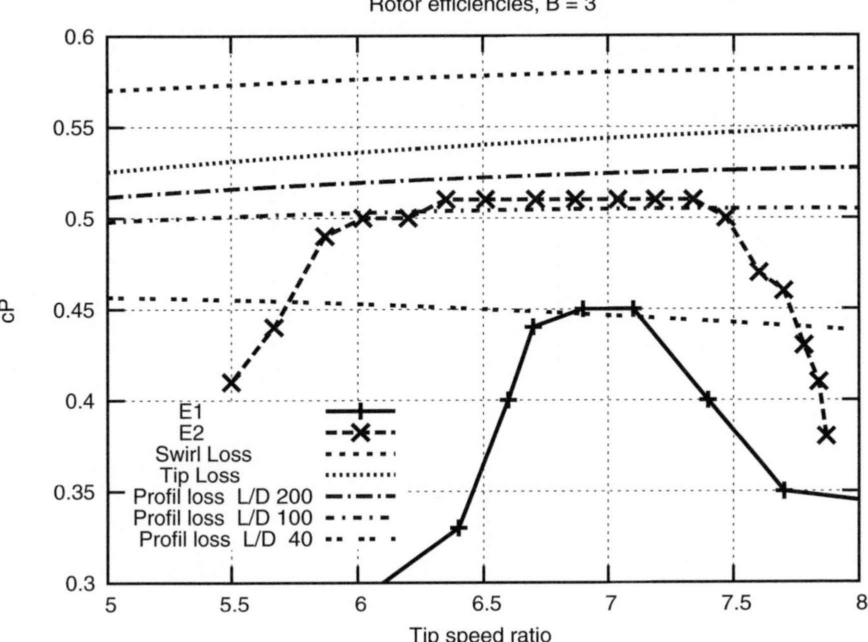

Fig. 5.15 Comparison of efficiency of 3-bladed wind turbines

5.6 Vertical Rotors

BEM may be adapted to Vertical Axis Rotors as well. As the motion is instationary and encompasses a whole volume (something between a cylinder and a sphere), the simplest possible model consists of two disks, see Fig. 5.16 and Chap. 2 [39]. It was discussed in detail by Chap. 2 [16] using vortex methods, see Sect. 6.3. In Figs. 5.18 and 5.19, we see a comparison of performance to measurements Chap. 2 [1] for a 34 m diameter 2-bladed *egg-beater*, see Fig. 5.17.

5.7 An Aerodynamics Analysis and Design Code

5.7.1 General Features

Here we summarize an independent FORTRAN 90/95 (F90 for short) implementation of a purely aerodynamic wind turbine blade design methodology. It is intended to be as detailed as possible although a fair amount of knowledge of wind turbine aerodynamics is necessary, see, for example, [50, 57]. Some—but not all—needs form a more practical point of view (maximum chord and twist, etc.) have been

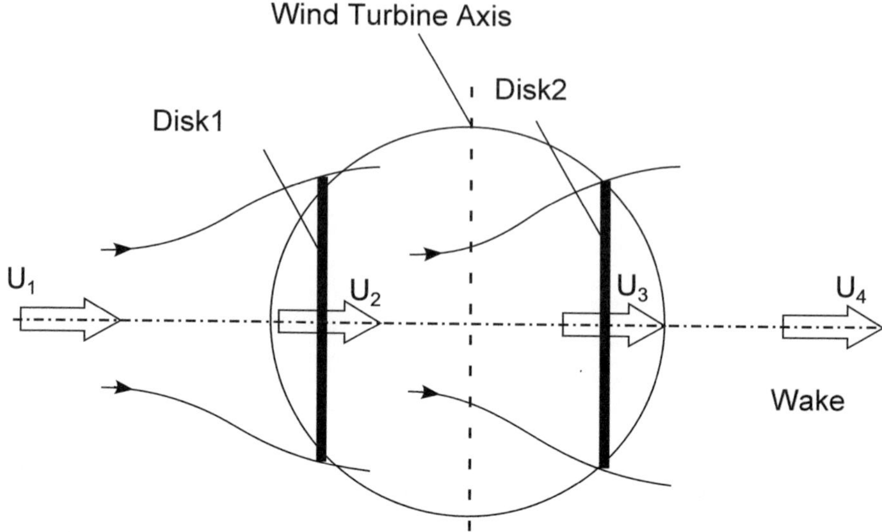

Fig. 5.16 Double actuator disk as a BEM model for vertical axis wind turbines

taken into account. This is in contrast to [28] who tried to find out what maximum c_P may be reached without any restrictions by *practicability*.

Our basis, a lecture of Korjahn [30], presents the so-called two-point design. In the first attempt, however, we only describe the implementation of the first part: to reach at a blade with as large parts with optimum $a(r)$[5] as possible. *Design point 2* consists of a certain safety against gusts corresponding to stalled parts on the blade. Therefore, AOA at design conditions have to have some safety margin against AOA stall.

The main objectives of our approach are:

- understand and formalize the approach and
- compare to other codes.

Further or additional sources from

1. Korjahn [30],
2. Bak [2],
3. Schmitz [50],
4. Johansen et al. [28].

are recommended to complement understanding.

Important: The code https://github.com/Schaffarczyk/KSS-Blade is by no means fool-proof:

---

[5] Usually, the term Betz optimum is related to $a(r) = 1/3 = \text{const}$. Very recently [27], it has been shown that, including wake rotation and tip loss, a somewhat different distribution results: Close to the hub $a \to 0.25$ and close to the tip $a \to 0.4$ follows from the Glauert approach [19].

Fig. 5.17 Sandia 34 m diameter turbine, reproduced with permission of Sandia National Laboratories, Albuquerque, New Mexico, USA

Bugs   may still be present.

Input   data, especially the aerodynamic one, has to be checked as far as possible before its use.

5.7.2 Multiple Solutions of the BEM-Equation

It was already recognized by Wilson et al. [42], p. 108 ff., that under some specific circumstances, the BEM-equations may not produce unique solutions. Maniaci [13] and Ning [43] further elaborate on this subject. To give an example, Fig. 5.20 (for a special section of a special commercial blade) shows the local thrust coefficient cT_{loc}

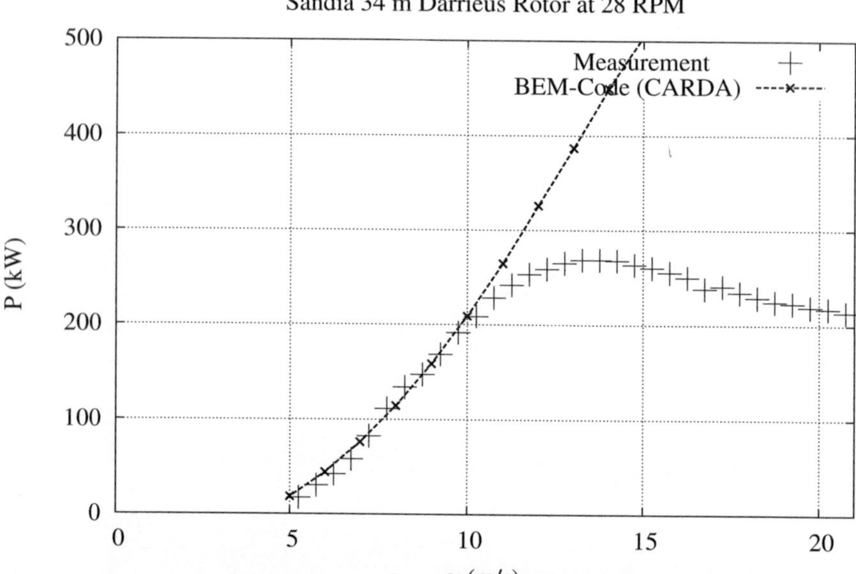

Fig. 5.18 Sandia 34 m diameter turbine: Power as function of wind speed

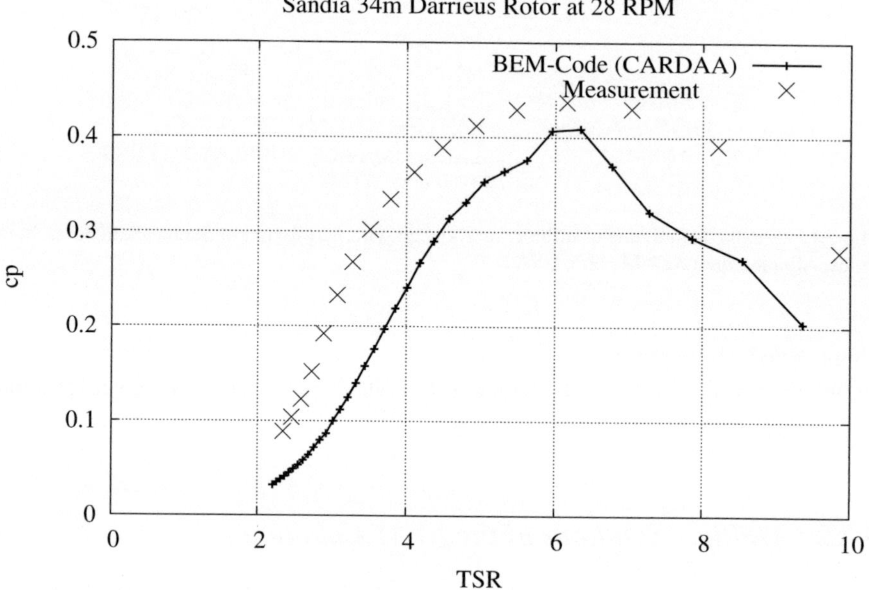

Fig. 5.19 Sandia 34 m diameter turbine: c_p as function of TSR

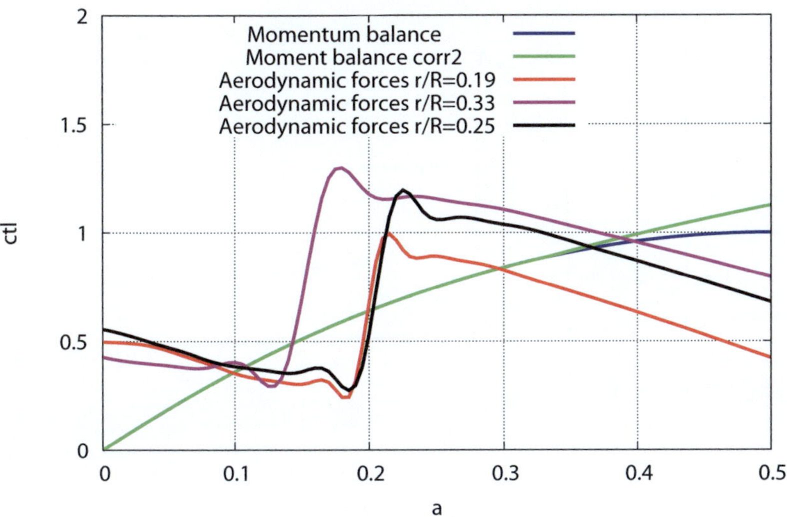

Fig. 5.20 Example of distributions of multiple solutions from balancing aerodynamic (lift) forces and local (axial) momentum

as function of a, the axial induction, defined by $u_1 := (1 - a) \cdot U_{-\infty}, 0 \le a \le 0.5$. Momentum balance gives

$$c_T = 4a \cdot (1 - a) \tag{5.62}$$

whereas the counterforce from the lift is its proper projection into the direction of the wind. Note that this is an approximate description only, as a' (circumferential induction) is neglected as well as tip correction (see Sect. 5.7.3) and a correction of c_T for larger a values of a $(a > a_c)$, see Sect. 5.7.4.

Our investigation for a 20 MW blade of 133 m length confirms this. It was recognized that the traditional fixed-point iteration did then not converge but simply jumps between two different values.

As a consequence, we improved our implementation and search algorithms as follows:

- Firstly, scan the whole a-region $0 \le a < 1$ in steps of 0.01. If there are solutions found by a binary search algorithm:
- start using standard fixed-point iteration until a maximum number is reached,
- switch to Newton(-Raphson) to to more iterations, if no convergence is reached,
- use a binary search algorithm *rtbis* from [40].

Figure 5.21 shows an example. Therefore, the argument of [42], p. 31 (Fig. 5.22).

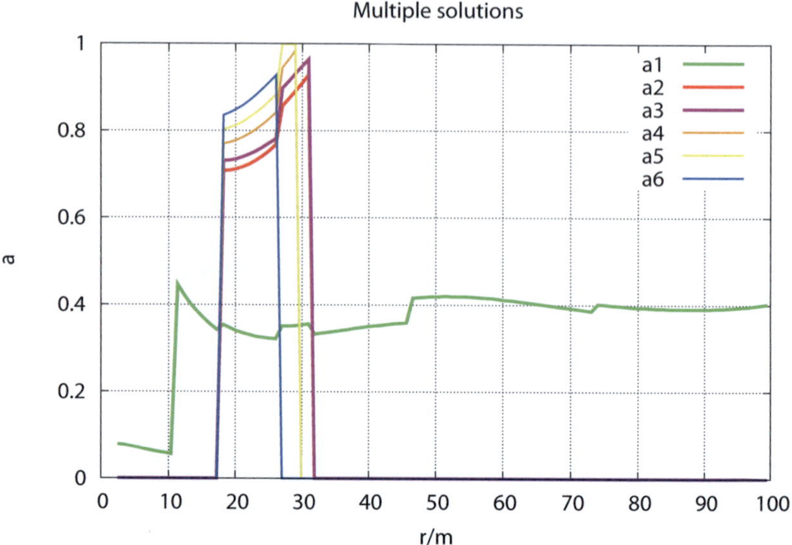

Fig. 5.21 Example of how a binary search algorithm finds multiple solutions along a whole blade (CIG10MW)

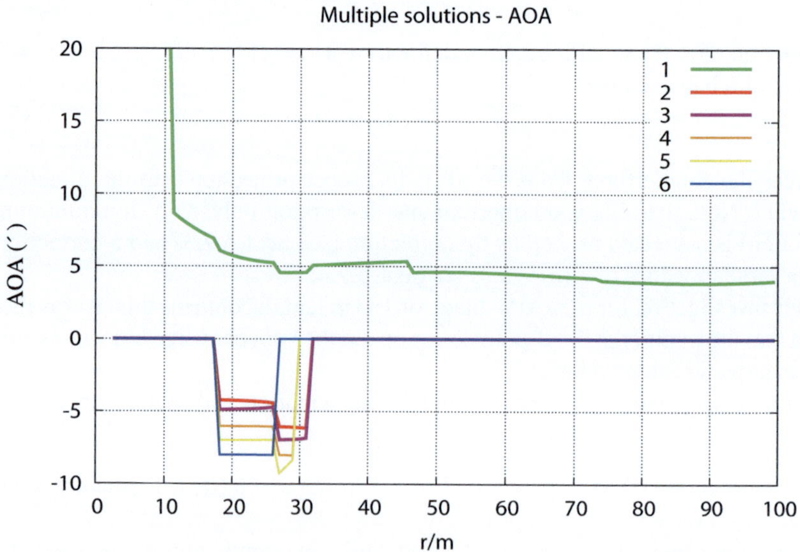

Fig. 5.22 Variation of angle of attack (CIG10MW)

> As a guiding principle, it is suggested that the solution to be chosen in the case of multiple solutions is the one that maintains continuity of angle of attack along the blade span.

can be confirmed.

5.7.3 Tip Correction

Momentum theory is valid only for an infinite number of blades $B \to \infty$. As in most designs, B is small ($B < 4$), this has to be taken into account. Indeed, as was shown very early, *tip losses* are as large as profile drag and must be accurately embedded into the code. Traditionally, the simplest (and oldest) version uses a factor F, which reduces circulation to zero when $r \to r_{tip}$:

$$F = \frac{2}{\pi} arccos \left(exp \left[-\frac{B/2(1-r)/R_{tip}}{r/R_{tip}sin(\varphi)} \right] \right) . \tag{5.63}$$

Here, φ is the all-important flow angle defined by

$$\tan (\varphi) = \frac{1-a}{\lambda(1+a')} , \tag{5.64}$$

which refers to $\lambda = \omega \cdot r / U_{-\infty}$ the local Tip Speed Ratio (TSR). $\omega = RPM \cdot \pi / 30$ is the angular velocity.

5.7.4 Glauert Correction $cT(a)$ for $a > a_c$

It is further well known that:

- Momentum theory is valid for $0 \le a \le 0.5$ only, and
- for some a_c, Eq. 5.62 becomes inaccurate.

To some extent, this can be cured by *empirical engineering* extensions which are listed, for example, in [23], Table 2.1. In principle, at some $a_c = 0.33 \ldots 0.4$, a more or less linear curve is attached to Eq. (5.62) so that $c_t (a = 1)$ approaches (somehow) 2 for $a \to 1$.

5.7.5 Sign Convention

Sign issues, especially for the pitch, are notorious. Following our basic interpretation

$$\alpha \text{ (AOA)} = \varphi \text{ (flow angle)} - \vartheta \text{ (twist]} , \tag{5.65}$$

we define **pitch** as being added to twist.

5.7.6   Code Description

5.7.6.1   A Short Note on FORTRAN

We will not defend ourselves why we use the *Grandfather of all (higher) programming languages* but only use it. It has been tried to write a code as simple as possible to make understanding and changes (**i.e., maintainability**) as simple as possible. A minimum set of algorithmic features is used. Especially, no tricks from Computer Science have been used. This code was(is) developed using *gfortran*, version 4.8.5 from 2015, see Sect. 5.7.6.7. A more detailed description is given in [35]. The code is available at https://github.com/Schaffarczyk/KSS-Blade.

5.7.6.2   Numerical Approach

Everybody knows from own experience the NEWTON-(Raphson) method for finding zeros of functions ($f(x_0) = 0$) or SIMPSON's rule for 1D-Integration. The ultimate manifestation of numerical algorithms may be found in the famous (series of) text book(s): *NUMERICAL RECIPES in Fortran 77* [40]. From here, we have copied

1. SPLINE and
2. SPLINT.

Unfortunately, some overshooting at sharp corners might still be possible. *Bezier* polynomials were coded independently from its definition, see [36].

5.7.6.3   Main Program

KKS.f contains the main code. It is merely a collection of routine calls for set-up and definition of the course of investigation:

1. (dynamic) memory allocation,
2. Read *machine.in*,
3. Read in spline thickness distribution from *ThickDis.in*,
4. Analyze a given set of profiles, L2D-max.
5. If design-mode is chosen:
6. An initial BEM calculation is performed and
7. chord and twist might be (Bezier-) smoothed or
8. Twist may changed to improve c_P, see Sect. 5.7.9.1.

5.7.6.4   Memory Management

mem.f defines all GLOBAL data. To use each module, one must start with a
   use mem
   command line.

5.7.6.5 Subroutines

1. BEM: Bem iteration as described,
2. ANALYSIS: Analyse a blade described in *BlaDes.in* using routine *BEM*,
3. DESIGN: Determines chord and twist by given profiles and Betz, optimum criteria. See Sect. 5.7.9.1 for more information about optimization.,
4. AIRFOILS: searches given α, c_L, c_D data for maximum of c_L/c_D, $c_{L,max}$ and slope at zero lift.
5. TipShapeEL: defines an elliptic (vertical tangent) tip shape geometry,
6. TipShapePara: defines a parabolic tip shape geometry determined by smooth slope at the beginning of the tip region and zero chord at the tip.
7. BEZIER: smoothing of a given data set according to the definition of Bezier polynomials,
8. smoothCHORD: smooths chord with subroutine BEZIER,
9. smoothTWIST: smooths twist with subroutine BEZIER,
10. improveTWIST: (1) tries to get locally as close as possible to 1/3 by adaption of twist, see Eq. 5.70. (2) Sørensen's approach, (3) a *constant circulation* approach.

5.7.6.6 Functions

1. FNOVERK: (*(Real) Function N Over K*) calculates Binomial coefficients (BC) needed in BEZIER, (special numerical care has to be taken because of the strong growth of BCs),
2. FPR: Prandtl's tip loss F, see ,
3. FBu: Burton's tip loss, see [10, 41],
4. chDES: optimum chord distribution from Betz (or Schmitz, or Glauert) [57],
5. chBe: Betz optimum chord, $a' \equiv 0$,
6. chSc: Schmitz optimum chord, a' is included,
7. TWIST: twist (in deg) from Betz, $a = 1/3$ and $a' = 0$, so only local TSR enters,
8. AHANSEN: gives Glauert's correction as stated in [20],
9. ANREL: gives Glauert's correction as stated in [31].

$$c_T(a) = \frac{8}{9} + \left(4\,F - \frac{40}{9}\right)a + \left(\frac{90}{9} - 4\,F-\right) \cdot a^2 \ . \tag{5.66}$$

10. thickpoly: cubic interpolation of thickness
11. twist bend: simple table for the inclusion of twist bend
12. apot: (1) Betz, (2) Sørensen and (3) constant circulation (very approximate).

5.7.6.7 Compilation

For compilation, we prefer to use (in a SUSE-LINUX environment) [35] with mandatory options (see **compDesCode.cmd**):

-fno-automatic helps to resolve notorious issues with local variables in SUBROU-
                TINES and FUNCTIONS,
-fbounds-check to avoid not allowed memory access,
          -O3 high optimization.

A typical job with 100 section and 100×100 wind speed and pitch variation takes a few minutes on a somewhat older *Intel(R) Core(TM) i3-4160 CPU @ 3.60GHz* CPU.

5.7.6.8   Windows

minGW (=minimum gcc for Windows) was used to compile a windows executable file. It is also available at git.

5.7.6.9   Modularization

Apart from mem and KKS, the following modules

- mem.f, declares global variables,
- Sub.f, a set of used subroutines and functions,
- SubNum.f, numerical functions: SPLINE and SPLINT from [40],

help to put some structure into the code.

5.7.6.10   Work Flow

All input has to be provided as ASCII files, as well as all output will be provided in the same format. Post-processing then has to be performed by other tools. We prefer GNUPLOT [26, 37], see Sect. 5.7.7.2.

5.7.7   Output Files and Tools for Visualization

5.7.7.1   Description of Output Files

- *Bem.out*: sectional output from BEM, see Table 5.6,
- *des.out*: sectional output from design: rsec, twist, chird, thick, c_L^{des} and AOA_{des},
- *ProProp.out*: from routine AIRFOILS: design properties of given airfoil data.

5.7.7.2   Description of GNUPLOT Scripts

- *a.gpl*: plots a(r). data from Bem.out
- *chord.gpl*: plots c(r). data from Bem.out an BlaDes.in and des.out
- *cL-aoa.gpl* and *cL-cD.gpl*: plots polars. data from **.aer*
- *cL-des.gpl* plots c_L^{des}(r). data from *PrOpt/Bem.out*
- *twist.gpl*: plots ϑ(r). data from Bem.out an BlaDes.in and des.out
- and some more.

5.7.8   *Mode: Analysis*

5.7.8.1   Input Data

The code is able to operate in two modes:

1. Design mode, set bit *DesMode* to .T. in file *Machine.in*,
2. if not, then the code will operate to analyze a given blade.

Then, in addition, other files:

- *BlaDes.in* (r, twist, chord, attached prfile name),
- *ProThick.in* which maps airfoil names to their (relative) thickness,
  Important: This file has to have a header line
- *ThickDis.in* which gives with few entries r/R_{tip} and corresponding t/c,
- **.aer*: files with airfoil data $AOA(\alpha)$, c_L, c_D, c_M (not used).

are needed. Most of the input (more or less) should be self-explaining. Lines with 3 numbers consist of first, last and number of values (wind speed and pitch, for example).

Below examples are given (Tables 5.2, 5.3, 5.4 and 5.5).

Table 5.2 Part of a sample job description file *Machine.in* based on NREL's 5MW reference wind turbine

```
DesMode     .F.
BladeNo     3.0
dens        1.225
Rhub        0.3
Rtip        63.0
windrage    11.4  11.4  1
rpm         12.1
Pitchr      0.0  0.0  1
Nsec          70

 .   .   .   .
```

Table 5.3 Sample blade description file *BlaDes.in* for NREL/IEAwind 15MW reference wind turbine

r	Twist	chord	Profile Name
0.0000	15.5946	5.2000	CYL1
2.4000	15.5911	5.2083	CYL2
4.8000	15.4288	5.2357	FF50
7.2000	14.9891	5.2889	FF50
9.6000	14.3263	5.3606	FF50
12.0000	13.4947	5.4433	FF50
15.0000	12.3001	5.5510	FF50
18.0000	11.0323	5.6501	FF50
20.2841	10.0834	5.7110	FF36
. . . .			
118.7179	−1.3914	1.7239	FF21
118.9744	−1.3628	1.6717	FF21
119.2308	−1.3336	1.5813	FF21
119.4872	−1.3038	1.4157	FF21
119.7436	−1.2734	1.1027	FF21
120.0000	−1.2424	0.5000	FF21

Table 5.4 Sample Thickness assignment file *ProThick.in* for NREL/IEAwind 15MW reference wind turbine

NAME	rel thickness
CYL1	1.0
CYL2	0.95
FF50	0.5
FF36	0.36
FF33	0.33
FF30	0.3
FF27	0.27
FF24	0.24
FF21	0.21

Table 5.5 Sample Thickness distribution file *ThickDis.in* for NREL/IEAwind 15MW reference wind turbine

0.0000	1.0
0.0200	0.98
0.1500	0.500
0.2452	0.360
0.3288	0.330
0.4392	0.301
0.5377	0.270
0.6382	0.241
0.7717	0.211
1.0000	0.211

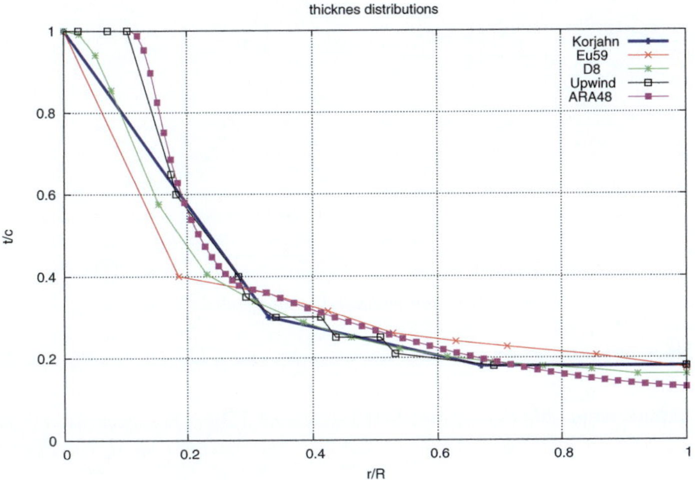

Fig. 5.23 Sample thickness distributions

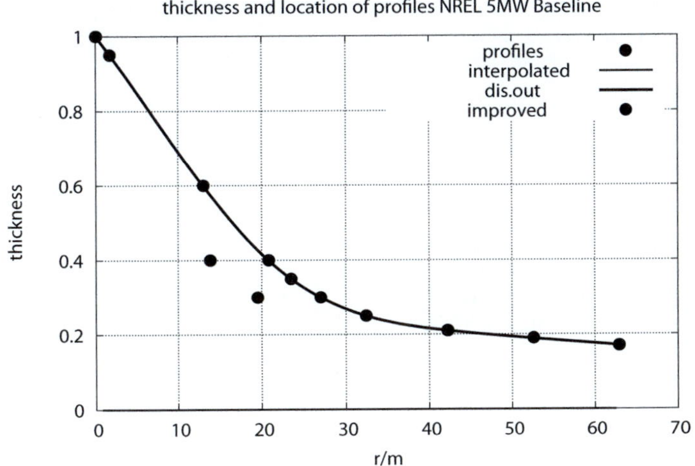

Fig. 5.24 Refined thickness distribution from NREL 5 MW

Figures 5.24 and 5.23 give an impression of what kind of thickness distribution has been and will be used.

Recently, [51] investigated a 15 MW reference blade in more detail. For reducing the mass from 71 to 63, a considerable change of thickness (see Fig. 5.25) was necessary. This underlines again the special importance of thickness (profile) distribution.

Main output file is *Bem.out* (Tables 5.6 and 5.7).

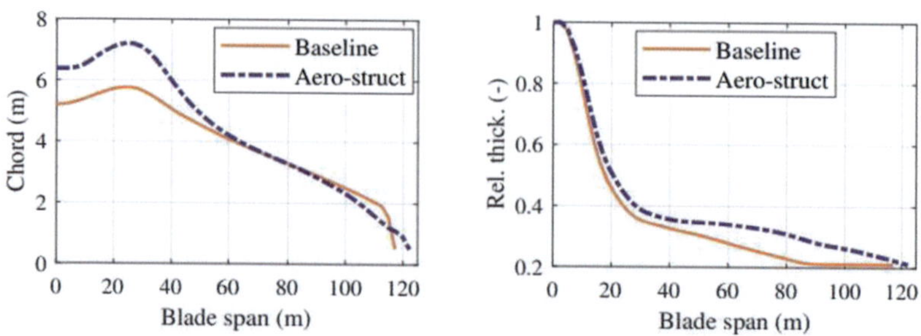

Figure 4: Chord and relative thickness distributions.

Fig. 5.25 Thickness distribution change to decrease mass for a 15 MW reference blade [51]

Table 5.6 Sample output file *Bem.out* for NREL/IEAwind 15MW reference wind turbine

r	a	ap	F	w	chord	twist	phi	aoa	cL	cD	L2D	cNo	cTa	dT	dQ	iter	err	th	Prof
0.598	0.334	−0.334	1.0	6.004	5.202	15.596	87.4	71.8	0.000	0.481	0.000	0.481	−0.022	0.20	−0.01	3	0.0E+00	0.99	CYL2
1.794	0.134	−0.134	1.0	7.861	5.205	15.598	82.3	66.7	0.000	0.444	0.000	0.440	−0.059	0.31	−0.04	3	0.0E+00	0.98	CYL2
2.990	0.076	−0.076	1.0	8.519	5.213	15.573	77.3	61.7	0.000	0.388	0.000	0.379	−0.085	0.31	−0.07	3	−0.2E−06	0.97	CYL2
4.186	0.045	−0.045	1.0	9.009	5.226	15.495	72.5	57.0	0.000	0.303	0.000	0.289	−0.091	0.27	−0.08	3	−0.2E−06	0.95	CYL2
5.382	0.040	−0.020	1.0	9.353	5.246	15.347	67.5	52.2	0.060	0.303	0.197	0.303	−0.061	0.31	−0.06	4	−0.1E−04	0.92	FF50
6.578	0.038	−0.002	1.0	9.730	5.273	15.128	62.8	47.7	0.144	0.297	0.484	0.330	−0.008	0.36	−0.01	4	−0.3E−04	0.89	FF50
7.774	0.040	0.011	1.0	10.149	5.305	14.848	58.4	43.5	0.245	0.283	0.864	0.370	0.060	0.44	0.07	5	−0.9E−07	0.85	FF50
8.970	0.043	0.021	1.0	10.608	5.340	14.519	54.2	39.7	0.356	0.263	1.354	0.421	0.135	0.56	0.18	6	0.0E+00	0.81	FF50
10.16	0.049	0.028	1.0	11.106	5.379	14.144	50.4	36.3	0.469	0.238	1.966	0.482	0.210	0.70	0.31	6	0.0E+00	0.76	FF50

Table 5.7 Sample aerodynamic data *FF21.aer* for NREL/IEAwind 15MW reference wind turbine

```
.   .   .   .
 10.00000  −0.88209  0.01386  −0.04397
 −8.00000  −0.62981  0.01075  −0.05756
 −6.00000  −0.37670  0.00882  −0.06747
 −4.00000  −0.12177  0.00702  −0.07680
 −2.00000   0.12810  0.00663  −0.08283
 −1.00000   0.25192  0.00664  −0.08534
  0.00000   0.37535  0.00670  −0.08777
  1.00000   0.49828  0.00681  −0.09011
  2.00000   0.62052  0.00698  −0.09234
  3.00000   0.74200  0.00720  −0.09447
  4.00000   0.86238  0.00751  −0.09646
  5.00000   0.98114  0.00796  −0.09828
  6.00000   1.09662  0.00872  −0.09977
  7.00000   1.20904  0.00968  −0.10095
  8.00000   1.31680  0.01097  −0.10163
  9.00000   1.42209  0.01227  −0.10207
.   .   .   .
```

5.7.9 Mode: Design

If the design flag is switched on, the lower part of *Machine.in* is important and will be used (Table 5.8). As chord from Betz and/or Schmitz usually will give non-zero chord at any sections, a tip-length must be given within its range chord goes down to zero. Two shapes can be chosen:

- Parabolic "P" and
- Elliptic "E".

The last one gives a vertical tangent at the tip, and the first one is a simple quadratic parabola defined by two points (begin and end (=rtip)) and slope of c(r) at the begin of the tip region.

   Note: *ImpChord* flag is not used at the moment. Improvement of chord (to proceed to more monotonic-like behavior, see [30] sheet 11) is achieved by smoothing with *Bezier* polynomials.

   In addition to Korjahn's $a = 1/3$ approach *DesSchema = 1*, we use *DeSchema = (2, 3)* for other Schemas (with very low differences), as described in Sect. 5.7.9.1.

5.7.9.1   Aerodynamic Optimization

5.7.9.2   Betz' and Other Criteria

We will not give a complete list or discuss in much detail the various interpretations of what an optimization might be and how this may be reached. Only the following three are implemented:

- Betz: $a(r) = 1/3 = const.$,
- Sørensen's (2022), modified *Glauert's* approach [27], see Fig. 5.26 and
- Joukovsky's: $\Gamma(r) = const.$

   (1) Betz: no further explanation is needed.
   (2) Sørensen [27] argued that close to tip $a \rightarrow 2/5$, see Fig. 5.26.

Table 5.8 Parameters describing a design case

```
.  .  .
Tiplen        3.0
twmax         20.
chmax          5.5
ImpChord      .F.
ImpTwist      .F.  1
ImpThick      .F.
TwistB        .F.
DesSchema   2
```

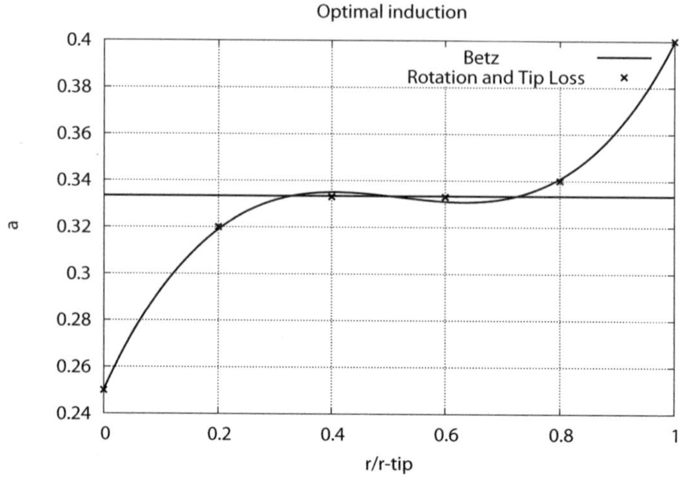

Fig. 5.26 Betz' and Sørensen' optimum a-distributions

(3) The last one is more subtle; to first order (in a), it is equivalent to Betz', but to second order:

$$a(1 - a) = \frac{\lambda B \Gamma (1 + a')}{4\pi} \, .$$

(5.67)

To complete, a model for a' has to be added, for example

$$a' = \frac{1 - 3a}{4a - 1} \ \text{(Glauert) or}$$

(5.68)

$$a' = \frac{2}{9} \frac{1}{\lambda_r^2} \ \text{(Simple approximation)} \, .$$

(5.69)

DesSchema = 1 refers to Betz' aerodynamic optimized blade $a = 1/3$ for all spans.

DesSchema = 2 refers to Sørensen's different a(r) distribution, see [27] and Fig. 5.26. See [28] for further discussion about aerodynamic optimization and textbooks, especially [50].

DesSchema = 3 is implemented but seems not to give any significant changes.

In any case, only a change of twist is used for optimization.

5.7.9.3 Change of Twist

Our approach to relate twist and a runs via the flow angle and reads as:

If $da := a_{opt}(r) - a_{loc} \neq 0$, then locally change twist by

$$d\varphi = -\frac{1}{\lambda_r(1 + a')}da \ . \tag{5.70}$$

5.7.9.4 Change of Thickness—Arrangement of Profiles

As stated in [30] (sheet 19, first figure), a further means to improve performance is to reduce thickness in the inner part. This is provided in our code by setting the *ImpThick*-bit and providing a changed thickness distribution *ThickDis.new*.

At the very end of *Machin.in*, a flag *TwistB* (binary) can be set. It is a very simple way to take *pre-twist* into account. If set to T, a table of additional twist deformation (along the blade) as a function of wind speed has to be given.

5.7.10 Sample Cases

5.7.10.1 Case 1: The NREL 5MW Baseline Blade

As this turbine [29] was also used by Korjahn [30] as a sample case, we here report first on output from KSS as well (Table 5.9).

Except of (out-dated) *wt-perf* (and some other codes) accuracy is within a 1% range. Difference of KSS to TP38060 is only 0.4% This may serve as a hint that our BEM implementation is not totally wrong.

Table 5.9 Rated rotor power of NREL 5MW baseline from various sources. HAWC data from Mr. Hinrichs, aerodyn engineering GmbH, bladed from Mr. Manjock, DNV. If all values are equally weighted, P = 5356 ± 89 kW results

Source	Value/kW	Deviation (%)
TP 38060	5267	0.0
KSS	5248	−0.4
FAST	5293	0.5
bladed	5322	1.0
HAWC2	5323	1.1
q-blade	5364	1.8
Xturb	5418	2.9
WZX	5478	4.0
wt-perf	5494	4.3

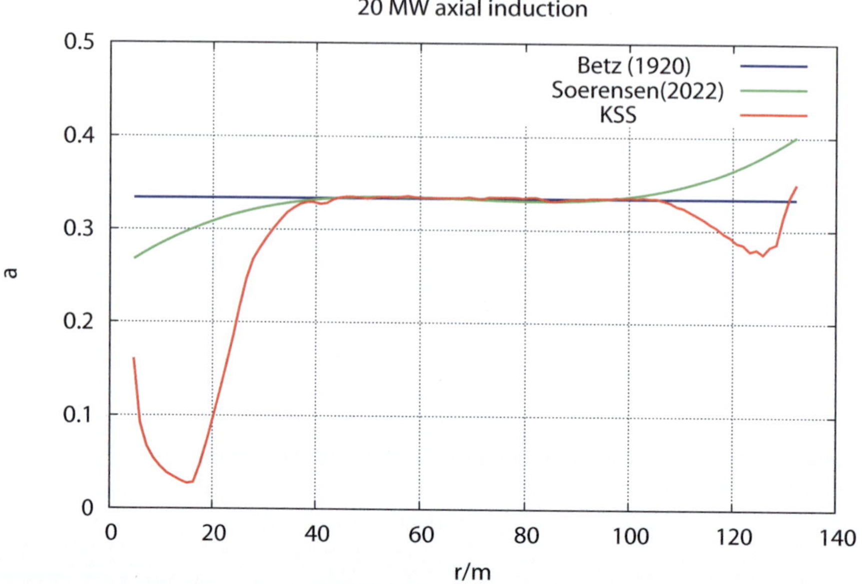

Fig. 5.27 KSS optimisation (Own 20MW-balde, $r_{tip} = 133m$) after 5 twist loops: a(r)

To have a glimpse of the design capability, Fig. 5.27 shows a(r) after 5 loops of twist optimization. cP improves from 0.4682 to 0.4685. We agree with [30], sheet 16, that

> For radius < 15m optimum induction is not possible.

In addition, Fig. 5.28 shows some angles for further assessment.

5.7.10.2   Case 2: A 15MW WT Reference Blade

The blade described in [17] was also analyzed with KSS, see Figs. 5.29 and 5.30. Some differences are visible, but agreement at rated conditions, see Table 5.10 (comparing c_P and c_T), again is within the 1% range.

5.7.10.3   Case 3: A 20 MW Blade

Reference [63] investigated the aerodynamic design of a 20MW blade in more detail. A maximum $cP = 0.4826$ was reached by Betz-mode optimization after 6 loops starting with an initial design of $cP = 0.4821$. This is less than 1% improvement. It seems that restrictions are so strong that a noticeable improvement is difficult to achieve.

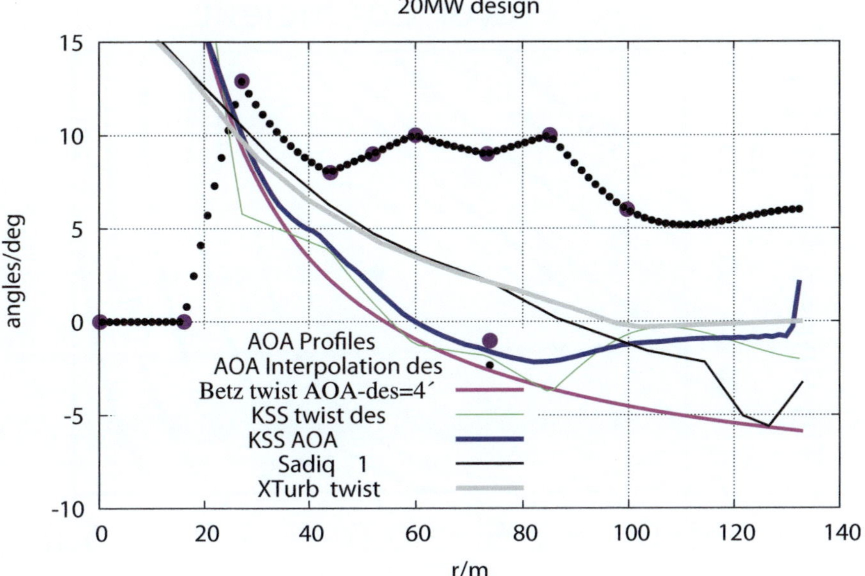

Fig. 5.28 KSS optimisation after 5 twist loops: some angles

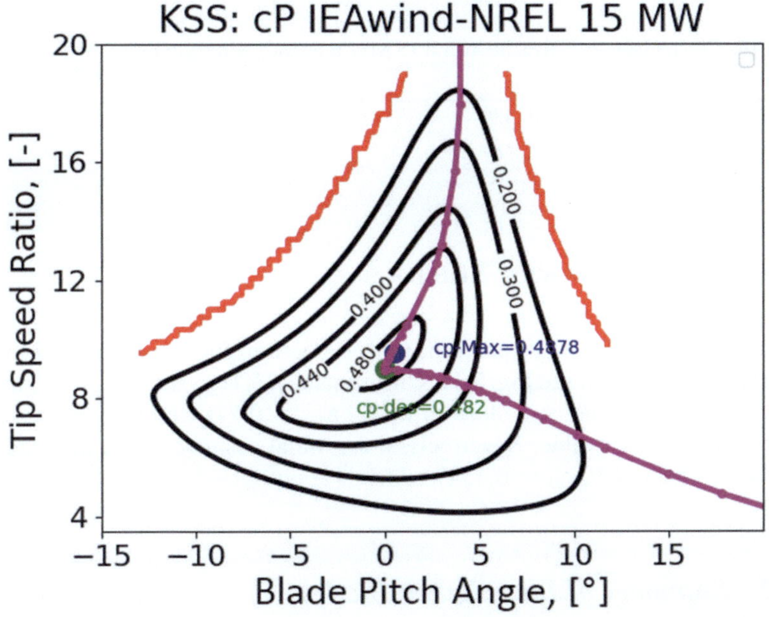

Fig. 5.29 cP map from KSS, including control path (magenta) and an extended pitch variation $-15° < \vartheta < 20°$

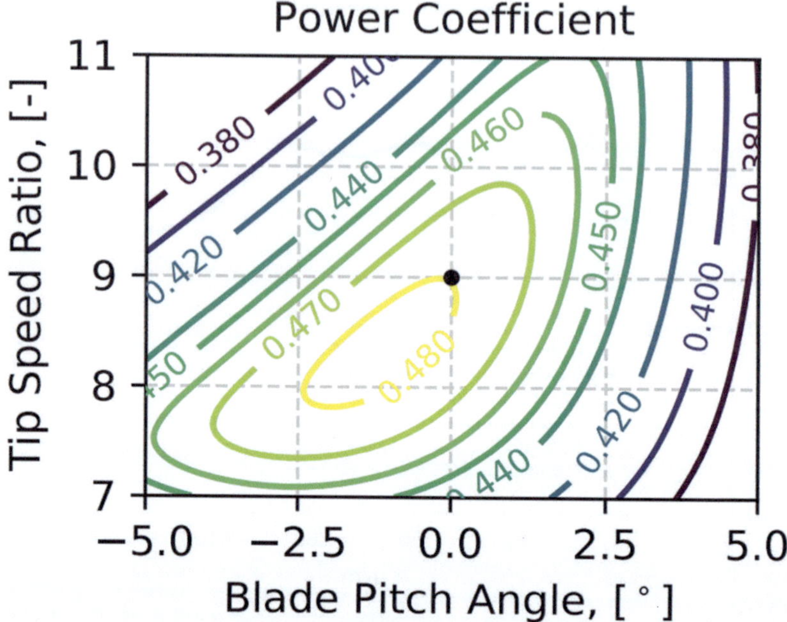

Fig. 5.30 cP map from NREL [17]

Table 5.10 Comparison for NREL/IEAwind 15 MW reference wind turbine

Source	c_P	c_T
TP 76698	0.489	0.804
KSS	0.484	0.768

5.7.11   Comparison with Xturb

Xturb by Sven Schmitz [50] offers some tools for optimization as well. We compared our 20 MW design [63], see Fig. 5.31. Unfortunately, XTurb showed an unexpected and non-explainable decrease close to tip ($r/R_{tip} > 0.77$) which makes a reasonable comparison impossible. Nevertheless, the initial design can be improved from $cP_{xT,A} = 0.4299$ to $cP_{xT,D} = 0.4347$ which is about 1.1% higher.

5.7.12   Summary of Investigated Blades

Table 5.11 shows a summary of investigated blades so far.

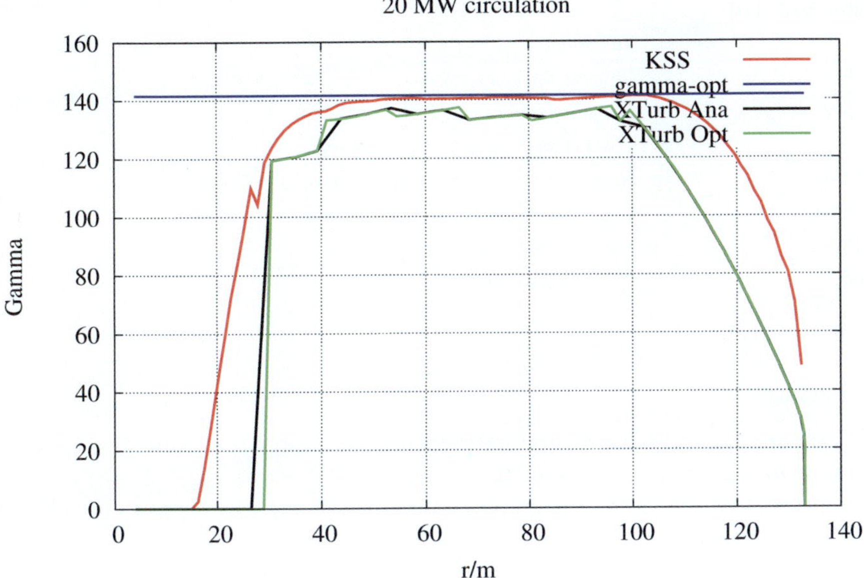

Fig. 5.31 Comparison of circulation for an own 20 MW blade

Table 5.11 Summary of investigated blades

Name	Rated power	Rotor diameter/m
Baltic Thunder	$\sim$ 1–5 kW	1
SWT(WINDFLOH)	14 kW	10.3
NecMicon 80	2 MW	80
NREL-Baseline	5 MW	126
DTU	10 MW	178
CIG10MW	10 MW	200
		220
NREL	15 MW	239
		248
KUAS20	20 MW	266

5.7.13 Summary

Based on the lecture notes from [30], a F90 blade analysis and design tool has been
implemented using F90. It was applied and tested to a wide range of WT-blades, see
Table 5.11. Agreement with other sources/codes seems to be within a 1% range or
smaller seems to be possible.

Table 5.12 List

	Meaning
CIG	Chinese, India, Germany
DNV	Det Norske Veritas
DTU	(The) Danish Technical University
HAWC	Horizontal Axis Wind Turbine simulation Code
KUAS	Kiel University of Applied Sciences
NREL	National Renewable Energy Laboratory (USA)
SWT	Small Wind Turbine

Findings:

- Choice of thickness distribution and polar data has probably the strongest influence on performance data for a wind turbine blade.
- Improvement (of c_P) by the proposed one-point algorithm seems to be within a 1%-range only.

5.7.14   List of Abbreviations

See Table 5.12.

5.8   Problems

Problem 5.1 Strange Farmer?[6]

(a) A wind turbine with D = 33 m and a rated power of P = 330 kW may serve as a fan. Show by simple momentum theory:

$$T = \sqrt[3]{2\rho A_r \cdot P^2} \tag{5.71}$$

$$v_2 = P/T \tag{5.72}$$

and finally estimate the maximum possible blowing velocity $v_3 == 2 \cdot v_2$.

(b) Now make a more realistic BEM model by using reasonable blade geometry data and Cl-Cd-data and vary pitch until either maximum **output** power or pitching angle is reached. Make sure that your BEM model may switch properly from turbine ($a > 0$) to fan mode ($a < 0$).

---

[6] In fact this problem was real. In 1993, a farmer from the German West Coast (close to the North Sea) asked if an ENERCON33 wind turbine could serve as a fan to blow warmer air from above to his blooming apple trees to protect them against night frost.

Problem 5.2 Multiple Actuator Disks.

A Darrieus-type VAWT serves as a simple model of a counter-rotating (see Sect. 2.7) *double actuator disk*. The model assumes two fully developed slipstreams, and the far downstream outlet of the first disk is regarded as the inlet of the second one. So we have five special locations and two induction factors a and b:

- 1: far upstream whole assembly, u_1,
- 2: disk 1, $u_2 = u_1(1 - a)$,
- 3: far downstream disk 1 = far upstream disk 2,

$$u_3 = u_1(1 - 2a), \tag{5.73}$$

- 4: disk 2, $u_4 = \mathbf{u_3}(1 - b)$ and
- 5: far downstream whole assembly, $u_5 u_3(1 - 2b)$.

(a) show that

$$c_{P1} = 4a(1 - a)^2 \tag{5.74}$$

$$c_{P2} = \frac{u_3^3}{u_1^3} 4b(1 - b)^2, \tag{5.75}$$

(b) maximize total power $c_P^{total} = c_{P1} + c_{P2}$ and show

$$b_{max} = \frac{1}{3} \tag{5.76}$$

$$c_{P2} = 4a(1 - a)^2 + (1 - 2a)^3 \cdot \frac{16}{27} \rightarrow max \tag{5.77}$$

$$a_{max} = 0.2(!) \tag{5.78}$$

$$c_P^{max,total} = \frac{64}{125} + \frac{16}{125} = 0.64 > 0.59, \tag{5.79}$$

(** c) show that for $N \in \mathbb{N}$ disks:

$$c_p^{max,Ndisks} = \frac{8}{3} \cdot \frac{1 + (1/N)}{(2 + (1/N)^2} \tag{5.80}$$

$$(N \rightarrow \infty) \rightarrow \frac{2}{3} = 0.67 . \tag{5.81}$$

Problem 5.3 Use the code from Problem 5.1 to deduce the power curve NTK for 500/41 from Chap. 3 [21]

Problem 5.4 ((Chap. 2) [62]) Uses a kind of a Rankine vortex model (see Sect. 3.6, Fig. 3.11) to include wake rotation. Try to show and understand them in finding:

$$c_P = \frac{b^2(1-a)^2}{b-a} \tag{5.82}$$

$$= \frac{b^2(1-a)^2}{b-a}\,[b+(2a-b)\Omega/\omega_{max}]\,. \tag{5.83}$$

Here ω_{max} is the maximum velocity at the edge of the core.

Problem 5.5 BEM model of a wind-driven vehicle using a turbine:

If a wind turbine is used to drive a vehicle (with velocity w), thrust has to be transported, the extracted power now being

$$P = (T+D)w \tag{5.84}$$

$$\text{with } D = \frac{1}{2}\rho A_{vehicle}(u_1 + w)c_D\,. \tag{5.85}$$

(a) Estimate Force and Power for a wind car with $4\,\mathrm{m}^2$ rotor area. Assume $v_{wind} = 8\,\mathrm{m/s}$ and $v_{car} = 4\,\mathrm{m/s}$.

(b) Show [54] that for maximum car speed $w/u_1 \rightarrow max$ the (usually defined) induction a depends only on one parameter K:

$$a^3 - a^2 - 3\,Ka + K = 0\,, \tag{5.86}$$

with

$$K := \frac{1}{4}\left(\frac{A_{vehicle}}{A_{rotor}}\right)\cdot c_d\,. \tag{5.87}$$

(c) Solve this equation for a(K) and prepare a table with columns:
a, w, c_P, c_T and c_P/c_T of K with $0.001 \leq K \leq 10$.
(d) Prepare a (double logarithmic) graph and discuss the results.
(** e) now include drive train efficiencies [18].

Problem 5.6 Show by conformal mapping how Eq. 5.59 may be deduced.

Problem 5.7 Show by using a BEM code and empirical equations for ideal chord, twist and lift

$$c(r/R_{Tip}) = \frac{16\pi}{9}\left((\lambda r/R_{Tip})^2 + 4/9\right)^{-1/2} \tag{5.88}$$

$$tan(\theta) = \frac{1}{\lambda \cdot r/R_{Tip}} \tag{5.89}$$

$$c_L = 2\pi\alpha \tag{5.90}$$

how Figs. 5.12 and 5.13 may be deduced.

References

1. Ashwill TD (1992) Measured data for the sandia 34-meter vertical axis wind turbine. Technical report, SAND91-228, Albuquerque, New Mexico, USA
2. Bak C (2013) Aerodynamic design of wind turbine rotors. In: Advances in wind turbine blade desing and materials, pp 59–108
3. Bak C, Johansen J, Andersen PB (2006) 3d-corrections of airfoil characteristics based on pressure distributions. In: Athens G (ed) Proceedings of EWEC
4. Betz A (1919) Schraubenpropeller mit geringstem energieverlust mit einem zusatz von l.prandtl. In: Nachr. d. Königl. Gesell. Wiss. zu Göttingen, Math.-phys. Klasse, pp 198–217. Königl. Gesell. Wiss. zu Göttingen
5. Betz A (1926) Wind-Energie und ihre Ausnutzung durch Windmühlen. Vandenhoeck und Ruprecht, Göttingen
6. Betz A (2013) Das maximum der theoretisch möglichen ausnützung des windes durch windmotoren. Wind Eng 32(4):307–309
7. Branlard E (2017) Wind turbine aerodynamics and vorticity-based methods. Springer, International Publishing
8. Buhl ML, Manjock A (2006) A comparison of wind turbine aero elastic codes used for certification. Technical report, NREL/CP-500-39113, NREL, CO, USA
9. Buhl ML, Wright AD, Tangler JL (1997) Wind turbine design codes: a preliminary comparison of the aerodynamics. Technical report, NREL/CP-500-23975, NREL, CO, USA
10. Burton T, Jenkins N, Bossanyi E, Sharpe D, Graham M (2020) Wind energy handbook, 3rd edn. Wiley, Hoboken, NJ, USA
11. Burton T, Sharpe D, Jenkins N, Bossanyi E (2011) Wind energy handbook, 2nd edn. Wiley, Chichester
12. Conway JT (2002) Application of an exact nonlinear actuator disk theory to wind turbines. In: Proceedings ICNPAA, Florida, USA. Melbourne
13. Maniaci DC (2011) An investigation of WT-perf convergence issues. In: 49th AIAA aerospace sciences meeting
14. de Vries O (1979) Fluid dynamic aspects of wind energy conversion. Technical report, AGAR-Dograph, No. 243, Neuilly sur Seine, France
15. de Vries O (1979) Fluid dynamic aspects of wind energy conversion. Technical Report AGAR-Dograph, 242, NATO, Neuilly sur Seine, France
16. Ferreira CS (2009) The near wake of the VAWT. PhD thesis, TU Delft, The Netherlands
17. Gaertner E, Rinker J, Sethuraman L, Zahle F, Anderson B, Barter G, Abbas N, Meng F, Bortolotti P, Skrzypinksi W, Scvott G, Feil R, Bredmose H, Dykes K, Shileds M, Alen Chr, Vislli A (2020) Definition of the IEA 15-megawatt offshore reference wind turbine. NREL/TP-5000-75698, NREL
18. Gaunå M, Øye S, Mikkelsen R (2009) Theory and design of flow driven vehicles using rotors for energy conversion. In: Proceedings of EWEC, Marseille, France
19. Glauert H (1936) Aerodynamics of propellers. J. Springer, Berlin
20. Hansen M (2015) Aerodynamics of wind turbines, 3rd edn. Routledge, London, UK
21. Hansen MOL (2008) Aerodynamics of wind turbines, 2nd edn. Earthscan, London, UK
22. Himmelskamp H (1950) Profiluntersuchungen an einem umlaufenden propeller. Technical report, Mitt. Max-Planck-Inst. f. Strömungsforschung, Nr. 2, Göttingen, Germany (in German)
23. Ledoux J, Riffo S, Salomon J (2021) Analysis of the blade element momentum theory. SIAM J Appl Math 81(6):2596–2621
24. Jamieson P (2008) Generalized limits for energy extraction in a linear constant velocity flow field. Wind Energy 11(5):445–457
25. Jamieson P (2018) Innovation in wind energy, 2nd edn. Wiley, Chichester, UK
26. Jannert PK (2016) gnuplot IN ACTION, 2nd edn. Manning Publications
27. Sørensen JN (2022) Glauert revisited. In: Proceedings of EAWE conference the science of making torque from the wind

28. Johansen J, Madsen HA, Gaunaa M, Bak C (2009) Design of a wind turbine rotor for maximum aerodynamic efficiency. Wind Energy 12:261–273
29. Jonkman J, Butterfield S, Musial W, Scott G (2009) Definition of a 5-mw reference wind turbine for offshore system development. Technical Report NREL/TP-500-38060, NREL, Golden, Co, USA
30. Korjan M (2021) Predesign rotor blade (full example). Technical report, bewind GmbH, Rendsburg, Germany
31. Buhl ML (2005) A new empirical relationship between thrust coefficient and induction factor for the turbulent windmill state. NREL/TP-500-36834, NREL
32. Madsen HÅ et al (2010) Validation and modification of the blade element momentum theory based on comparisons with actuator disc simulations. Wind Energy 13:373–389
33. Mikkelsen R, Øye S, Sørensen JN (2006) Towards the optimally loaded actuator disc. In: Proceedings of IAEwind Annex XI (Basic Information exchange). Kiel University of Applied Sciences, Kiel, Germany
34. Munk M (1920) über vom wind getriebene luftschrauben. Z f Flugtechnik u Motorluftschiffahrt X I(15):200–223
35. NN. https://gcc.gnu.org/fortran/
36. NN. https://de.wikipedia.org/wiki/Bezierkurve
37. NN. www.gnuplot.info
38. NN (2004) Considerably higher yields—revolutionary rotor blade design. Technical Report 3, Wind Blatt, Aurich, Germany
39. Paraschivoiu I (2002) Wind turbine design. With emphasis on darrieus concept. Polytechnic International Press, Montreal, Canada
40. Press W, Teukolsky A, Vetterling W, Flannery B (1992) Numerical recipes in fortran, 2nd edn. The art of scientific computing. Cambridge University Press, Cambridge, UK
41. Ramdin SF (2017) Prandtl tip loss factor assesed. Master's thesis, TU Delft, Delft, The Netherlands
42. Wilson RE, Lissaman PBS, Walker SN (1976) Aerodynamic performance of wind turbines. Technical report, Oregon State University
43. Ning SA (2014) A simple solution method for the blade element momentum equations with guaranteed convergence. Wind Energy 17:1327–1345
44. Schaffarczyk AP (2012) Ad-cfd implementation of the glauert-rotor. Unpublished internal report, p 89
45. Schepers JG (2002) Verification of European wind turbine design codes, vewtdc; final report. Technical report, ECN-C-01-055, ECN, The Netherlands
46. Schepers JG, Boorsma K, Munduate X (2012) Final results from mexnext-i: analysis of detailed aerodynamic measurements on a 4.5 diameter rotor places in the large german dutch wind tunnel dnw. Technical report. In: Proceedings of 4th conference of the science of making torque from wind, Oldenburg, Germany
47. Schepers JG et al (2012) Final report of iea task 29, mexnex (phase 1): analysis of mexico wind tunnel measurements. Technical report, ECN-E-01–12-004, ECN, The Netherlands
48. Schepers JG, Snel H (2007) Model experiments in controlled conditions, final report. Technical report, ECN-E-07-042, ECN, The Netherlands
49. Schmitz G (1955) Theorie und entwurf von windrädern optimaler leistung. Wiss. Zeitschr, Uni Rostock
50. Schmitz S (2020) Aerodynamics of wind turbines. Wiley, Hoboken, USA
51. Scott S, Greaves P, Macquart T, Pirrera A (2022) Comparison of blade optimization strategies for the IEA 15 MW reference wind turbine. In: Proceedings of EAWE conference the science of making torque from the wind
52. Sharpe DJ (2004) A general momentum theory applied to an energy-extracting actuator disk. Wind Energy 7:177–188
53. Simms D, Schreck S, Hand M, Fingersh LJ (2001) Nrel unsteady aerodynamics experiment in the nasa-ames wind tunnel: a comparison of predictions to measurements. Technical Report NREL/TP-500-29494, NREL, Golden, CO, USA

54. Sørensen JN, Aero-mekanisk model for vindmølledrevet køretøj. unpublished report, Copenhagen, Denmark
55. Sørensen JN (2011) Aerodynamic aspects of wind energy conversion. Ann Rev Fl Mech 43:427–448
56. Sørensen JN (2012) priv comm. private communication
57. Sørensen JN (2016) General momentum theory for horizintal axis wind turbines. Springer International Publishing Switzerland
58. Sørensen JN (2016) General momentum theory for horizontal axis wind turbines. Springer, Cham, Switzerland
59. Sørensen JN, Mikkelsen R (2012) A critical view on the momentum theory. Technical report, In: Proceedings 4th conference on the science of making torque from wind, Oldenburg, Germany
60. Sørensen JN, van Kuik GAM (2010) General momentum theory for wind turbines at low tip speed ratios. Wind Energy 43:427–448
61. Spalart P (2003) On the simple actuator disk. J Fluid Mech 494:399–405
62. Spera D (ed) (2009) Wind turbine technology, 2nd edn. ASME Press, New York, USA
63. Syed S (2022) Aerodynamic design of a 20MW blade. Master's thesis, Kiel University of Applied Sciences
64. Tangler JL, Kocurek JD (2005) Wind turbine post-stall airfoil performance characteristics guidelines for blade-element momentum methods. Technical report, NREL/CP-500-36900 and 43rd AIAA aerospace meeting and exhibition, Reno, Nevada, USA
65. van Bussel GJW (2007) The science of making more torque from wind: diffuser experiments and theory revisited. J Phys: Conf Ser 75:012010
66. van Kuik GAM (1991) On the revision of the actuator disc momentum theory. Wind Eng 15:276–289
67. van Kuik GAM (2003) The edge singularity of an actuator disk with a constant normal load. AIAA paper, pp 2003–356
68. van Kuik GAM (2018) The Fluid Dynamic Basis for Actuator Disc and Rotor Theories. IOS Press, TU Delft, The Netherlands
69. van Kuik GAM (2022) The actuator disc concept. Springer International Publishing, Cham, pp 47–94
70. Wilson RE, Lissaman PBS, Walker SN (1976) Aerodynamic performance of wind turbines. Technical report, Corvallis, Oregon, USA
71. Wood DH, Hammam MM (2002) Optimal performance of actuator disc models for horizontal-axis turbines. Front Energy Res 10

Chapter 6
Application of Vortex Theory

6.1 Vortices in Wind Turbine Flow

It was already mentioned in Chap. 3 that Saffman [40] quoted Küchemann, who said that

vortices are the sinews and muscles of fluid motion.

It has to be noted that, based on some progress in the 2D case, some authors [10] even argue for a vorticity-based theory of turbulence.

As shown in Chap. 3, this metaphor can be expressed with equations. Remember that local vortex strength, or vorticity was defined via Eq. 3.39

$$\omega := \nabla \times \mathbf{v} \,.$$

Now if a velocity field $\mathbf{v}$ satisfies $\nabla \cdot \mathbf{v} = 0$ (is *solenoidal*) , then

$$\mathbf{u} = \nabla \times \mathbf{A} \,.$$

with $\mathbf{A}$ the vector potential Eq. 3.44. Now ω being given, $\mathbf{A}$ and therefore $\mathbf{u}$ is given by *Poisson's equation* (inhomogeneous Laplace equation):

$$\nabla(\nabla \cdot \mathbf{A}) - \nabla^2 \mathbf{A} = \omega \tag{6.1}$$

Now, if a flow has (large) regions of vanishing vorticity, we may introduce the concept of *vortex lines* and *vortex sheets* . As can be immediately seen, this is in conflict with the conventional smoothness assumption for the flow and vorticity fields.

© The Author(s), under exclusive license to Springer Nature Switzerland AG 2024
A. P. Schaffarczyk, *Introduction to Wind Turbine Aerodynamics*, Green Energy
and Technology, https://doi.org/10.1007/978-3-031-56924-1_6

This is the place where δ functions[1] are introduced. We refer the interested reader to [20] of Chap. 3 for more mathematical rigor. In the simplest case for a vortex line we may write formally [40]:

$$\omega = \Gamma\delta(n)\,\delta(b)\,\mathbf{t} \tag{6.3}$$

where Γ (Eq. 3.40) was the circulation and $\mathbf{t}$, $\mathbf{n}$ and $\mathbf{b}$ are the tangent, normal and b-normal of the space curve. A closed loop is called *vortex ring*. In case of an irregular shape, we talk about *vortex filaments*.

The properties of vortex elements are governed by **Helmholtz' Theorem**, which will be presented in a more modern nomenclature:

Theorem 1 *In inviscid flow, in which only conservative forces act*[2]

- *fluid particles free of vorticity remain so,*
- *vortex lines move with the fluid,*
- *circulation is conserved over time.*

See [53] for a rigorous derivation of these conservation laws and a possible connection to turbulence, even in the inviscid case.

To summarize: ordered by dimension, the following forms of sustained vorticity ay be important to wind turbine flow and will be discussed in more detail in the text that follows:

P 0D patches: 2D transient VAWT,
L 1D lines of finite length: Diffuser design,
H 1D helical lines of infinite length: Joukovski,
S 2D vortex sheets: Betz/Goldstein Rotor,
V 3D vortical wake volume: Conway.

But before we start, we should introduce some well-known quantifiable examples.

6.2   Analytical Examples

As an initial example, we now generalize from Sect. 3.6, Fig. 3.11, the *Rankine vortex* which has a non-smooth slope at the core's border. The so-called $n = 2$ *Scully vortex* (see [26] of Chap. 1):

---

[1] In fact, this is a distribution (sometimes also called a generalized function) with the property that

$$\forall f \in \mathcal{S} : \int \delta(x)f(x)\,dx = f(0) \tag{6.2}$$

$f(x)$ is called a *test function*. $\mathcal{S}$ is the set of test functions named after *Laurent Schwartz*, Field's medal winner of 1950.

[2] van Kuik [57, 59, 62] recently pointed out how to combine non-conservative forces with the conventional lifting-line Helmholtz-Kelvin approach.

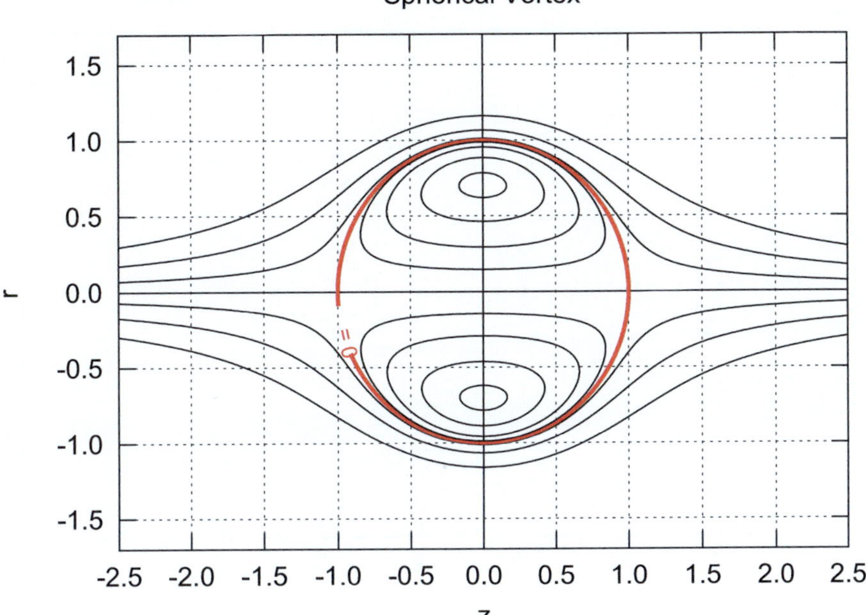

Fig. 6.1 Hill's spherical vortex

$$v_\phi(r) = \left(\frac{\Gamma}{2\pi\, r_c} \frac{r}{\sqrt[n]{1 + r^{2n}}} \right) \tag{6.4}$$

gives a somewhat smoother shape. [27] gives pictures of recent measurements of *tip vortices* in water. During the MexNext experiment, the shape and movement of the vortex was measured in some detail (Chap. 5) [45, 46].

A well-known (second) example is *Hill's Spherical Vortex* with vorticity in cylindrical coordinates $\omega = (\omega_r = 0, \omega_\phi, \omega_z = 0)$ (Fig. 6.1):

$$\omega_\phi = \begin{cases} A \cdot r & \text{if } r^2 + z^2 < a^2 , \\ 0 & \text{if } r^2 + z^2 > a^2 . \end{cases} \tag{6.5}$$

The stream function is:

$$\Psi = \begin{cases} \frac{-A}{10}\left(r^4 + r^2 z^2 + \frac{5}{3} r^2 a^2\right) & \text{if } r^2 + z^2 < a^2 , \\ \frac{A r^2 a^5}{15(r^2 + z^2)^{3/2}} & \text{if } r^2 + z^2 > a^2 . \end{cases} \tag{6.6}$$

It has the remarkable property, that it moves with a self-induced velocity of

$$u = 2 A a^2 / 15 . \tag{6.7}$$

through the fluid.

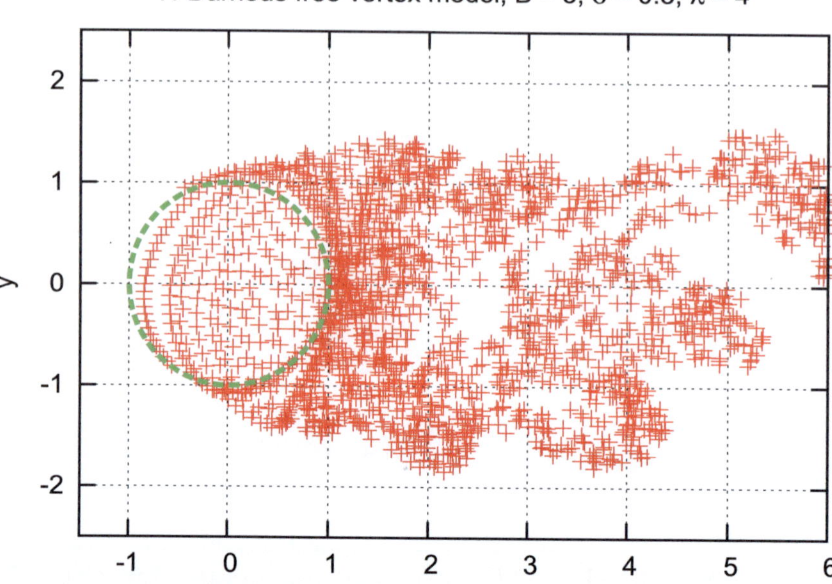

Fig. 6.2 Shedded vorticity patches

6.3 Vortex Patches

Vortex patches are small finite volumes (in 3D) or areas (in 2D) with a finite amount of vorticity. A very simple way and easy to calculate example is the H-shaped VAWT where due to conservation of circulation, vorticity is continuously shedded (compare Fig. 6.2), because AOA and therefore lift changes. The motion of the patches is calculated by Helmholtz's 2nd law. As an example, a 2D H-Darrieus (from Chap. 2 [18]) was coded (in FORTRAN, see listing in Problem 6.5) with the method from [52]. As a result, we see in Fig. 6.3 the evolution of the normal and tangential forces as well as the power coefficient.

6.4 Vortex Lines

6.4.1 Vortex Lines of Finite Length

Extending the almost point-wise *patches* to a line of finite length gives a good example of investigations of the action of a diffuser (see Sect. 2.5 and Fig. 2.11) if regarded as a *ring wing*. We follow Appendix A of [17]. If we associate the bound vorticity Γ_{bound} with a simple circular ring vortex of zero core size, we may find [40] arbitrary large

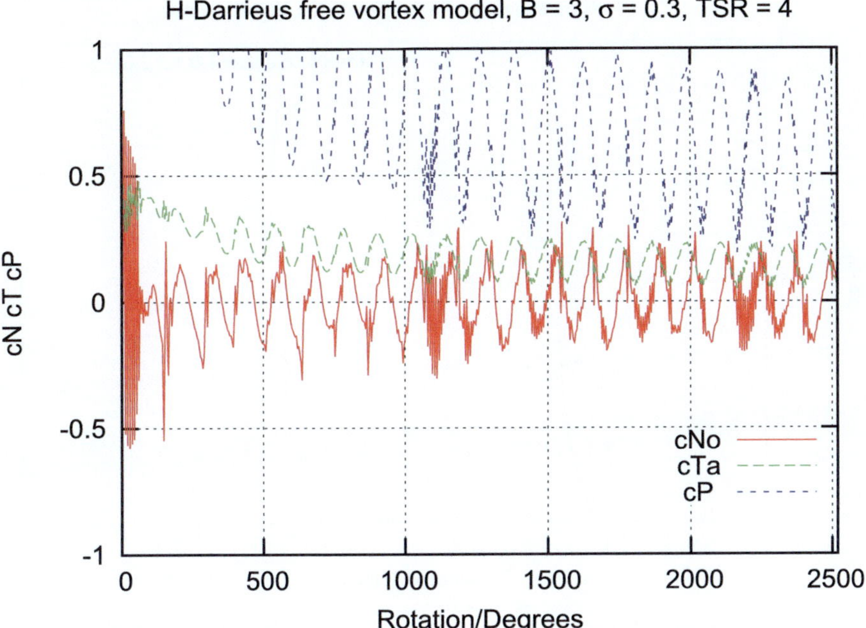

Fig. 6.3 Evolution of forces after seven revolutions

values close to the ring. de Vries found an elegant way out—apparently inspired by thin airfoil theory—and assumed distributed ($0 \leq x \leq c$) vorticity along the whole chord. After some algebra on elliptic integrals [1]:

$$E(k) := \int_0^{\pi/2} d\phi \sqrt{1 - k^2 \, sin^2(\phi)} \tag{6.8}$$

$$K(k) := \int_0^{\pi/2} \frac{d\phi}{\sqrt{1 - k^2 \, sin^2(\phi)}} \tag{6.9}$$

he was able to provide an approximate expression for the averaged induced velocity, see Fig. 6.4. This concept was successfully applied to diffuser designs for small wind turbines up to 5 kW by Schaffarczyk [42].

6.4.2   Preliminary Observations on Vortex Core Sizes

As already noted, it may be shown easily that a straight vortex line with concentrated vorticity $\omega = \Gamma \delta(x)$ induces a velocity field with specific energy (J/m) of

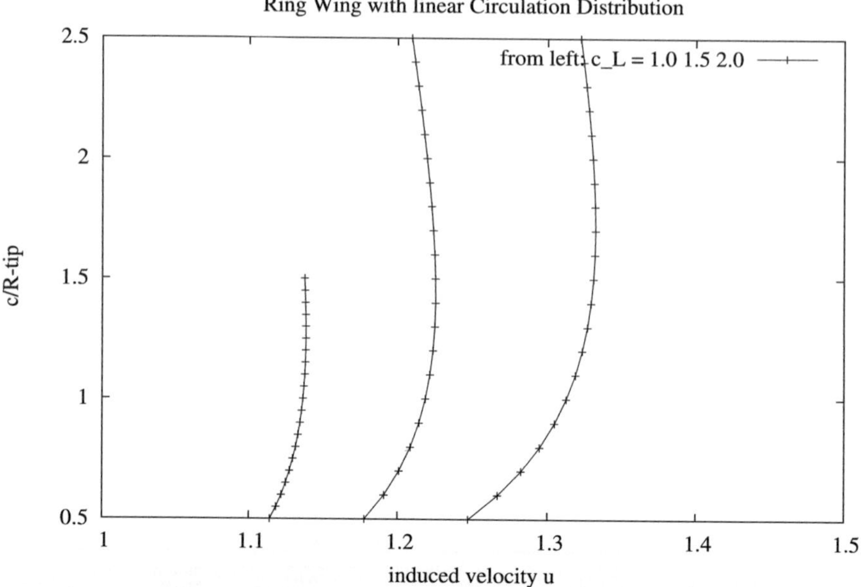

Fig. 6.4 Averaged induced velocity of a ring wing with diameter R_{tip}, chord c, and lift coefficient c_L

$$E \sim \log \left(R/r_{core} \right) .$$
(6.10)

which is singular for $r_{core} \to 0$ and $R \to \infty$ as well.

Fortunately, due to viscosity, we always have finite core sizes. For example, [67] provides for tip vortices of helicopter blades an empirical formula

$$ln \left(r_{core}/R_{tip} \right) \sim c_T^{-0.5}$$
(6.11)

Note that for a propeller c_T is defined as

$$c_T = T/\rho A_R (\omega \cdot R_{tip})^2 .$$
(6.12)

During the European MEXICO experiment (see Sect. 8.6), velocity fields were measured by *PIV*, so that the tip vortices could be observed. Figure 6.5 gives some results [46] from Chap. 5.

6.4.3 Helical Vortex Lines

From early flow visualization for ship propellers Chap. 1 [6], Joukovski [24] (see Fig. 6.6) deduced a very tidy and simple picture in which the **constant** bound circu-

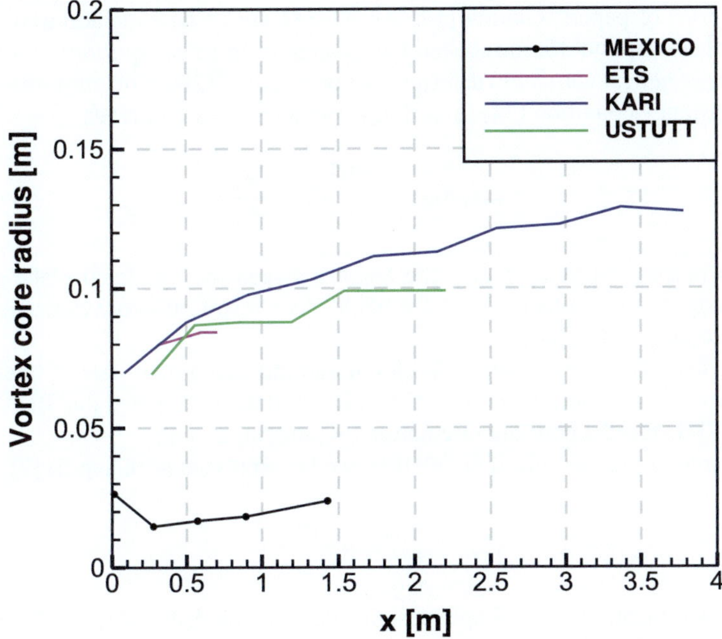

Fig. 6.5 Core radius of the Mexico-Rotor-Vortex at 10 m/s inflow condition. With permission of University of Stuttgart, Germany

Fig. 6.6 Vortex system from Joukovski [24] for a two-bladed propeller

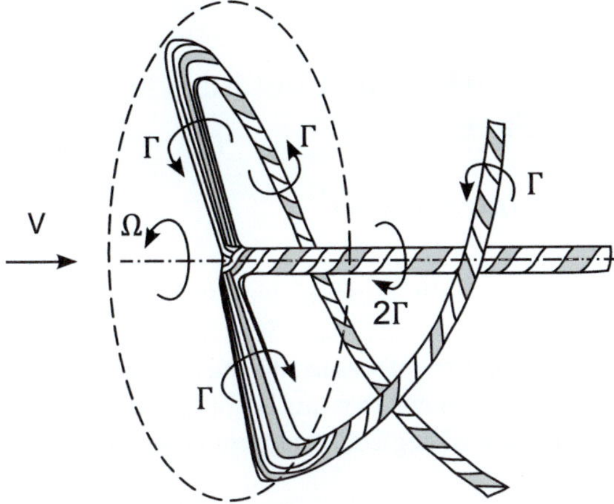

lation Γ of each blade is convected on helical lines by the flow. On the axis of rotation, they combine with the so-called *hub vortex*. Unfortunately, this picture could not be used for numerical predictions due to the calculation complications because of the singular behavior close to the vortex line.

In a series of papers, Okulov [19, 30, 32–34] used a new method to develop a consistent description. He introduced the concept of *induced equilibrium motion of a multiple of helical vortices* to determine a vortex core of size ϵ of constant vorticity. The consistency condition [34] (in a slightly adapted notation) reads

$$\bar{u}_\chi(\epsilon/R_{tip} = -\frac{B^2}{\tau}\,(1 + \frac{\epsilon}{R_{tip}})$$
(6.13)

with $\tau = tan(\varphi) = \ell/R_{tip}$, and φ the angle between the rotational plane and the direction of the total velocity V_0 at the blade. As a result, this core radius is on the order of $B\epsilon/R_{tip} \approx 0.4$ [34].

Numerical values are shown in Table 6.4 and indicate a flow state of higher efficiency than the so-called *Betz rotor* (Chap. 5) [2]. It has to be noted that Burton et al. Chap. 5 [7] deduced a constant circulation by other arguments.

The value of this optimized circulation can be expressed as (Chap. 5) [7] :

$$\Gamma_{opt,constant} = 4\pi\,\frac{u_1^2}{\omega}\cdot\frac{2}{9}\,.$$
(6.14)

From this equation, very simple design rules may be deduced ($\sigma = Bc/R_{tip}$, the shape parameter) as was shown in Figs. 5.6 and 5.8.

Recent experiments by Leweke et al. [27] investigated the dynamic stability of the vortical structure from a constant-circulation rotor in a water tunnel.

6.5 Modeling the Contant Loaded Actuator Disk by Vortex Lines

6.5.1 Introduction

Rotors, especially those of wind turbines, have been modeled by an axisymmetric actuator disk for more than 150 years, see for example [61]. Originally, only the axial velocity component was used in an integral momentum theory description [50], but it can easily be shown that this is not consistent with the differential equations for fluid flow. In case of neglecting any influence of viscosity, Euler's Equation becomes:

$$\nabla \cdot \mathbf{u} = 0\,,$$
(6.15)

$$\rho\frac{D\mathbf{u}}{Dt} = -\nabla p + \mathbf{f}.$$
(6.16)

Therefore, to include the radial velocity component, the vortex rings method [5, 58] seems to be the simplest way to overcome this shortcoming if one does not want to solve the differential equations itself numerically.

Beginning with Stig Øye's paper form 1990 [29] this approach has been further developed and applied, see [3, 4, 60]. It was generalized to non-uniform loading by Conway [13, 14].

The main objectives of our approach are:

- scrutinize the accuracy of the model,
- try to find hints of the properties of a possible edge singularity [9, 56].

6.5.2 Implementation

6.5.2.1 Model Description

We use 2D real axisymmetric coordinates z, r with velocity $\mathbf{u} = (u_z, u_r)$. The Actuator Disk is located at $z = 0$ and $0 \leq r < 1$. The edge ($\mathbf{r_E} = (0, 1)$) may carry a singularity, a point where $\mathbf{u}$ is not defined.

The model itself is defined by a piece-wise linear vortex sheet which starts at $\mathbf{r_{edge}} = (0, 1)$ and ends at $\mathbf{r_{Cyl}} = (zsivc, rsivc)$, the location where a semi-infinite vortex cylinder is appended. For $0 < z < zsivc$ NP vortex rings with strength γ are placed with non-uniform distance at locations zsl, rsl. This part of our approach is identical to that from [3, 4]. The locations of the z-positions of the rings are distributed according to

$$z(i) = z_{SIVC} * (1. - cos(\varphi)) \tag{6.17}$$

$$\varphi = 0.5 \cdot \pi \cdot i/np \tag{6.18}$$

$$1 \leq i \leq np \tag{6.19}$$

To circumvent singularities on the slipstream (i.e. infinite velocities), all velocities are evaluated just in the (arithmetic) middle of two slipstream (singularity carrying) points. This approach circumvents introducing an artificial *vortex core*.

The basic model parameter is the thrust coefficient c_T, where $c_T > 0$ refers to the propeller case (induced velocities > 0) and $c_T < 0$ to the turbine case. Momentum theory is valid for $-1 \leq c_T < \infty$. Many parameters can be deduced from momentum theory, for example, the asymptotic radial extension of the slipstream, which is given by

$$r^{sl}_{z \to \infty} \equiv RSIVC = \sqrt{\frac{1 + \sqrt{1 + c_T}}{2 \cdot \sqrt{1 + c_T}}} \, , \tag{6.20}$$

$$\text{For } c_T = 1 \text{ we have: } r_\infty = 0.9239 \text{ and} \tag{6.21}$$

$$\text{for } c_T = -8/9 : r_\infty = \sqrt{2} = 1.4142 \, . \tag{6.22}$$

Table 6.1 Table of major and minor influencing parameters

Name	Meaning
Major	
c_T	Thrust coefficient
Restart	LOGICAL. if .T. code will restart from existent sls.DAT
NP	Number of rings
ZSIVC	Axial location of semi-infinite cylinder
Minor	
Selfind	(0,1), switch on/off self-induction
Update	(0,1), def = 1, debug only
Maxiter	Maximum number ($<$100) of slipstream iterations
Under	Under relaxation (0.1 ... 1.)
epscP	
Wakety	Initial wake type (3 models available)
Wakety = 1	rsl(i) = rinf+(1.-rinf)*exp(-al*zsl(i))
Wakety = 2	gnuplot fit of vK data cT = 1 and cT = −8/9 only
Wakety = 3	van Kuik's Eq. (D.10)
Wake-a	Parameter al for wake-ty = 1
Iter-sc	Slipstream iteration (3 = psi)

Value for the stream function at infinity:

$$\psi_{wake} = \frac{c_T}{4 \cdot \sqrt{1 + c_T}} \, . \tag{6.23}$$

Our model defining parameters are summarized in Table 6.1. Inflow velocity is set to $u(r)_{z \to -\infty} = 1.0$.

6.5.2.2 Induced Velocities

Accurate calculation of induced axial and radial velocities is the key for any arbitrary accurate (= successful) model. Calculation is divided into three parts:

1. Induced velocities from the NP-1 other vortex rings to the regarded one for $0 \leq z \leq ZSIVC$,
2. Self-induced velocities of a regarded ring:
   We only have implemented Bontempo's [4] Eqs. (6) and (7) which surprisingly do not contain any core-parameter:

$$v_{z,m \to m} = \gamma_m \left\{ -\frac{\beta_{m+1} - \beta_m}{8\pi} \cos \beta_m - \frac{ds_m}{4\pi r_m} \left[\log\left(\frac{8\pi r_m}{ds_n} \right) - 0.25 \right] \right\} \tag{6.24}$$

$$v_{r,m \to m} = \gamma_m \left\{ \sin(\beta_m) - \frac{\beta_{m+1} - \beta_m}{8\pi} \right\} \tag{6.25}$$

with β_m being the slope at ring m

with r_m being radial distance of ring m

with s_m being the distance of the two rings nearby

$$\tag{6.26}$$

For more details on self-induction on vortex rings, see, for example, [41].

3. Induced velocities from a semi-infinite vortex cylinder with tangential vorticity located at ZSIVC [5], Chap. 36.

To avoid errors in implementation three different *models* (i.e. set of equations) can be chosen:

1. Eqs. (1) to (7) from [3, 4] called model *Bontempo*,
2. Eqs. (D1) from [58] called model *van Kuik*,
3. Eqs. from Chaps. (33) and (36) from [5] called model *Branlard*.

To be as consistent in nomenclature as far as possible, we prefer (and use) equations from Branlard [5].

6.5.2.3 Slipstream Iteration

The shape of the slipstream is determined by the induced velocities and vice versa by constraints described in [3, 4, 58]:

- Force free condition implemented as $\gamma = c_t/(2 \cdot u_s)$ with $u_s^2 = (1 + u_z)^2 + u_r^2$,
- Slipstream aligned with flow, implemented as $\psi_w(z) = \psi_w(\infty)$.

An initial slipstream shape may be helpful to reduce the number of iterations. The following models are implemented:

1. A cylinder with RSIVC = 1, or
2. A prescribed exponential shape described by

$$r_{sl}(z) = RSIVC + (r_0 - RSIVC \cdot \exp(-a \cdot z)) \ . \tag{6.27}$$

3. A prescribed algebraic shape deduced from the axial velocity at r = 0 for a vortex cylinder (attributed by van Kuik [58] to Snel and Schepers, which RSIVC is probably

$$\text{or from } u_z(r = 0) \ \frac{\gamma}{2} \left(1 + \frac{z}{\sqrt{(z^2 + RSIVC^2)}} \right) \tag{6.28}$$

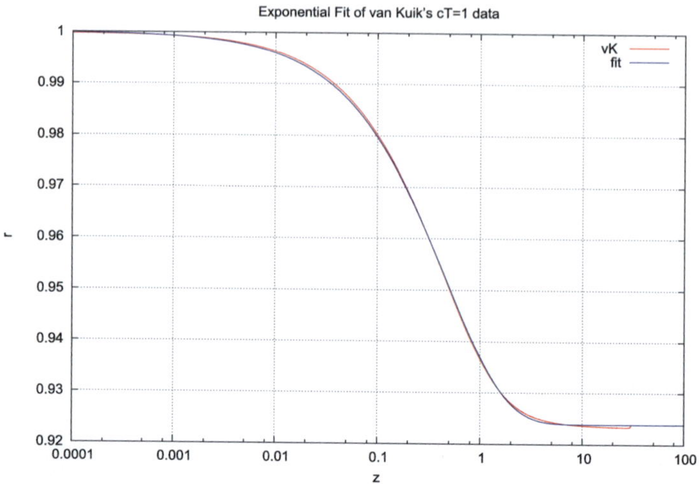

Fig. 6.7 Two parameter a_2, k fit of van Kuik's $c_T = 1$ slipstream according to $r(z) = r_{inf} + (1. - r_{inf}) * \exp{(-a_2 * z^k)}$. Here: $a_2 = 1.7743$ and $k = 0.7570$

Given by Eq. (6.22) but a (not to be confused with axial induction) has to be guessed $\approx 1 \dots 2$. Figure 6.7 shows a GNUPLOT fit $r(z) = r_\infty + (1 - r_\infty) \cdot \exp{(-a \cdot z^k)}$ from van Kuik's $cT = 1$ data. As he sets $RSIVC_{vk} := 1$, a fit results in:

1. $r_0 = 1.03498$,
2. $RSIVC = 1.00006$ (as it should) and
3. $a = 1.01452$.

We have implemented three different algorithms for the slipstream iteration:

1. Equations (13) and (14), Fig. 2 from [4],
2. Shift points normal to actual slipstream direction, or
3. Use van Kuik's [58] Eq. (D.12).

$$\partial r(z) = \frac{\partial \psi}{r \cdot u_z}$$

with $\partial \psi = \psi(slipstream, z) - \psi_{wake}$ from Eq. 6.23.

Method No. (3) is to be preferred All algorithms may be under relaxed. Corresponding factors have to be guessed:

1. $0.001 \cdots 0.01$,
2. < 0.01,
3. $0.1 \cdots 0.9$.

Slipstream iteration ends if

1. number of iterations MAXITER is exceeded or
2. errcP := $(c_P$ from code - $c_{P,mom}) \leq EPSCP$.

The accuracy of results can be considerably influenced by a careful choice of under-relaxation parameters and by increasing the number of vortex rings.

6.5.3 *Numerics*

6.5.3.1 The Goal of Numerical Computations

Everybody knows from own experience the NEWTON-(Raphson) method for finding zeros of functions $(f(x_0) = 0)$ or SIMPSON's rule for 1D-Integration. The ultimate manifestation of numerical algorithms may be found in the famous (series of) text book(s): *NUMERICAL RECIPES in Fortran 77* [38].

Let us state

Theorem 6.1 *A numerical schema is meaningful if and only if,*

$$\forall \epsilon > 0 : \ \exists n \in \mathbb{N}$$

that after n iterations the result is accurate by ϵ *.*

In a new version we have some reasonable hope that the proposed code is able to obey this theorem.

6.5.3.2 A Short Note on FORTRAN

We will not defend ourselves why we use the *Grandfather of all (higher) programming languages* but only use it. It has been tried to write a code as simple as possible to make understanding and changes (**i.e. maintainability**) as simple as possible. A minimum set of algorithmic features is used. Especially, no tricks from Computer Science have been used.

6.5.3.3 Compilation

For compilation, we use (in a LINUX environment) *GNU Fortran* (in short: gfortran) [28] with mandatory options (see **compVortexCode.cmd**):

-fno-automatic helps to resolve notorious issues with local variables in SUBROU-
            TINES and FUNCTIONS

-fdefault-real-8 set all floating numbers to REAL*8 (= 64 bit). This then should ensure an internal accuracy of 52 bits (more than 30 decimal digits) and a computational range of $\pm 10^{\pm 302}$

6.5.3.4   Modularization

The code is divided into 3 pieces only:

- mem.f, declares global variables,
- VortexCode.f, main slipstream iteration loop,
- SubA.f, a set of used functions, including evaluation of Elliptic Integrals (from an external source).

6.5.3.5   Elliptic Integrals

Calculation of Elliptic Integrals, see chapter (C.4) of [5], is performed via an open-source subroutines **RFL** by [8]

```
c***begin prologue   rfl
c***date written    790801     (yymmdd)
c***revision date   831228     (yymmdd)
c***category          no.  c14
c***keywords   duplication theorem,elliptic  integral ,incomplete,complete,
c                    integral  of  the  first  kind,taylor  series
c***author   carlson, b.c., ames laboratory–doe
c                  iowa  state  university , ames, iowa  50011
c                notis, e.m., ames laboratory–doe
c                  iowa  state  university , ames, iowa  50011
c              pexton, r.l., lawrence livermore national laboratory
c                  livermore, california  94550
```

It was tested and compared with MATHEMATICA to 10 digits.

6.5.3.6   Evaluation of Induced Velocities

Testing the accuracy of the induced velocities was done separately in the following manner: using all cases from Table 6.3:

- check $v_{z,r}$ for the ring part $0 < z < zsivc$
- check $v_{z,r}$ for the cylinder part $z \leq zsivc$
- compare to data from van Kuik [21].

Table 6.2 Sample input (configuration) file

```
Restart    .false.
N-Ring     8000
ct         −0.888888
zsivc      200
selfind    .true.
update     .true.
maxiter    10
under      0.1
epscp      1.e−5
IndVel     3
wakety     3
wake-a     1.0
itersc     3
```

6.5.3.7 Verification and Evaluation

Power is determined by using Bontempo's Eqs. (15) to (20) [4]. Radial integration of $u_z(z = 0, r)$ from $r = 0$ to $r = 1$ is performed by Simpson's rule. This schema is accurate to 2nd order. Thus, for an accuracy of $< 10^{-6}$, at least 10^3 evaluation points are needed. We use 2000 points.

A computational case is defined in **inpa.dat**. See Table 6.2 for set-up.

Before testing the iteration schema, two types of a fixed shape for the slipstream can be examined

1. A fixed cylinder with $r = 1$ or , $r = r_{SIVC}$
2. prescribed shape from Eq.(6.27), see Fig. 6.7 but for ($c_T = 1$ and $c_T = -8/9$ only).
3. An algebraic shape according to Eq. 6.28.

6.5.4 Output Files, Tools for Vizualation and Testing

6.5.4.1 Description of Output Files

Following (ASCII) output is provided (list not complete):

1. sls.DAT: geometry, velocity, vortex strength and stream function on vortex sheet,
2. vzd.DAT, vrd.DAT: velocities at the disk $(z = 0)$,
3. vza.DAT: axial velocity at the axis $(r = 0)$,
4. psid.DAT: stream function at the disk $(z = 0)$,
5. streaml.DAT: 3D data for contour plots of streamline (by str-cont.py).

6.5.4.2   Description of GNUPLOT Scripts

1. SLS-vz-(Prop,WT).gpl and SLS-vr-(Prop,WT).gpl: plot shape and velocities of slipstream
2. vza.gpl and vzd.gpl: plot axial and radial velocity at disk
3. gamma.gpl: plot vortex strength
4. and more.

Some GNUPLOT scripts (waked.gpl and) have to be executed via:
   gnuplot -e "ct = 1.0; load 'waked.gpl'"
   to allow for input pf c_T.

6.5.4.3   One Python Script

1. str-cont.py: plot contour lines of stream function together with slipstream (red).

6.5.4.4   Run Time Estimation

As expected, CPU-time scales approximately $\sim NP^2$, see Fig. 6.8:

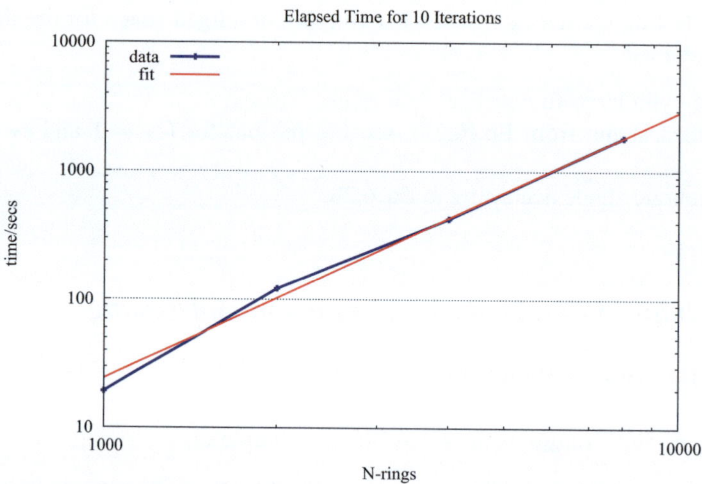

Fig. 6.8 Elapsed CPU time for 10 slipstream iterations. Hardware: Intel i3-4160 at 3.6 GHz. Operating system: openSUSE Leap 42.3 Linux, Kernel 4.12.14-lp151.28.44-default

6.5.5 *Results for Sample Cases*

6.5.5.1 Sample Case 1: Propeller with $c_T = 1$

As can be seen from the numbers above, this case is only moderately non-linear.
Slipstream contraction is down to approx. 0.93 and $c_T = 1.207$, compared to a linear
approximation [6], Eq. (9.107), which gives $c_{P,lin} = 1 + c_T/4 = 1.25$
  Some sample figures:

1. induced axial velocity at the slipstream, Fig. 6.9,
2. vortex strength at the slipstream, Fig. 6.10,
3. stream lines and slipstream, Fig. 6.11.

with van Kuik [21] data.

6.5.5.2 Sample Case 2: (Betz-)Turbine with $c_T = -8/9$

Same plots (with one addition) as shown in Subsect. 6.5.5.1:

1. induced axial velocity at the slipstream, Fig. 6.12,
2. induced radial velocity at the slipstream, Fig. 6.13,
3. vortex strength at the slipstream, Fig. 6.14,
4. stream lines and slipstream, Fig. 6.15.

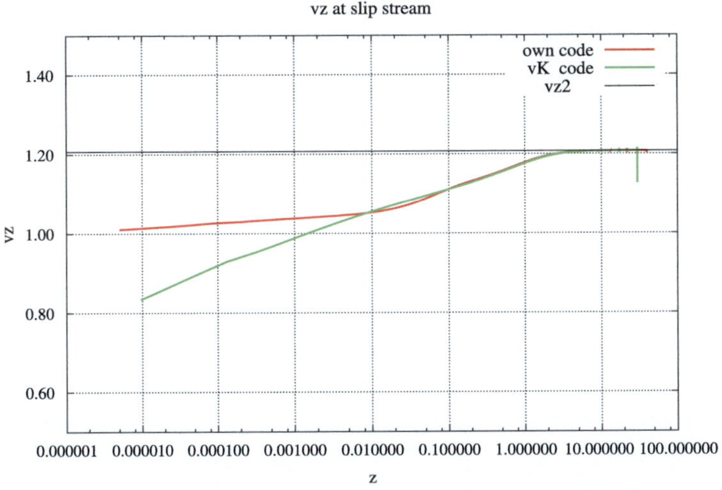

Fig. 6.9 Calculated axial distribution of induced axial velocity, $cT = 1$

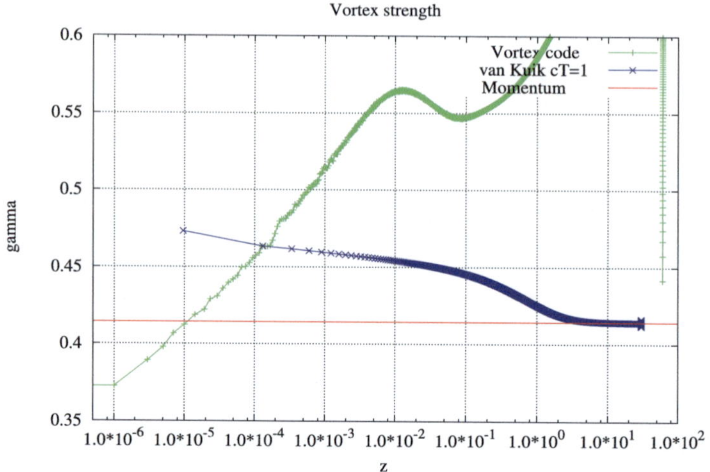

Fig. 6.10 Calculated axial distribution of vortex strength, $cT = 1$

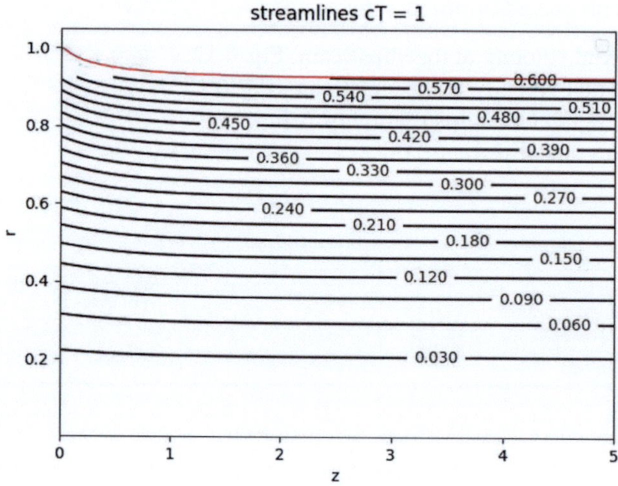

Fig. 6.11 Calculated stream lines (lines of $\Psi = $ const.) for $c_T = 1$. In red: wake boundary. As can be seen, the edge streamline coincides well with $\Psi_{wake} = 0.60355339$

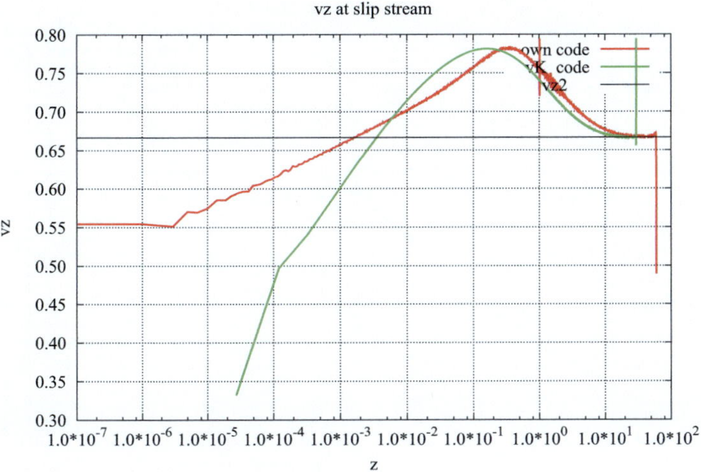

Fig. 6.12   Calculated axial distribution of induced axial velocity, Betz case

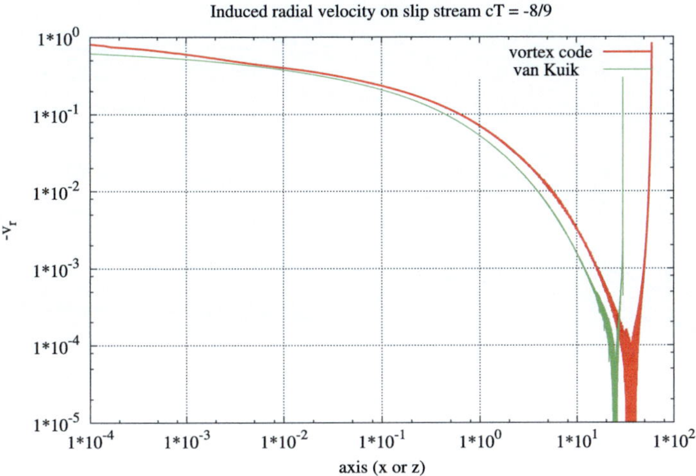

Fig. 6.13   Calculated axial distribution of induced radial velocity, Betz case

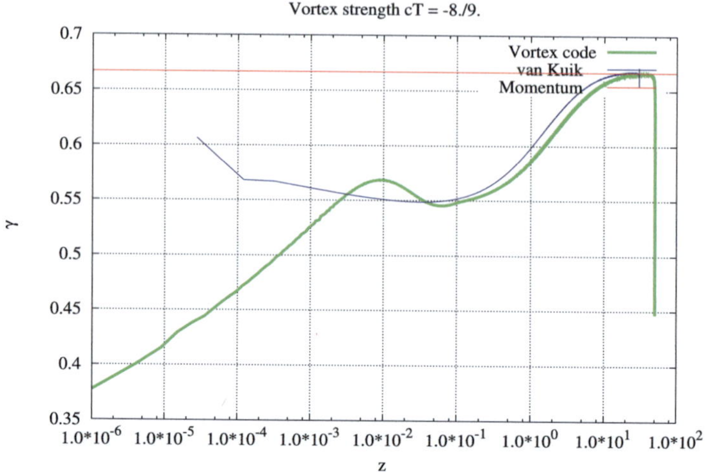

Fig. 6.14 Axial distribution of vortex strength, Betz case

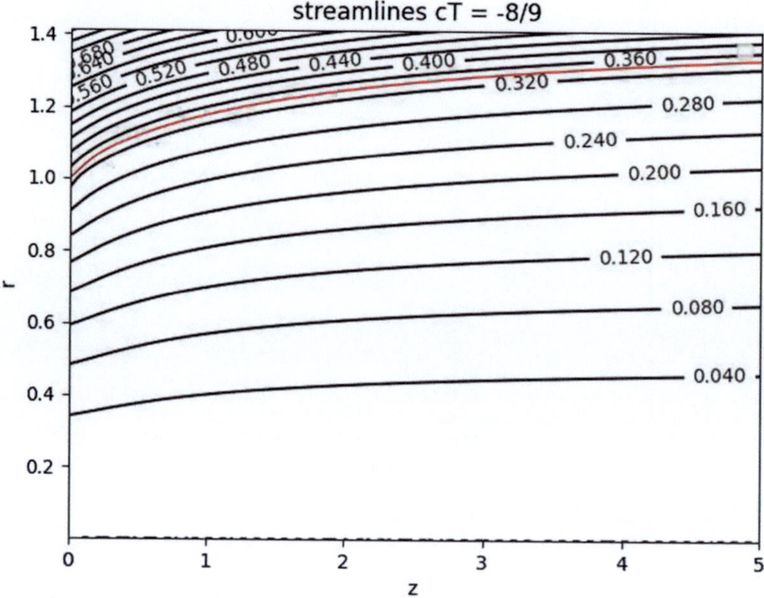

Fig. 6.15 Streamlines for the Betz case $c_T = -8/9$ red: slipstream, $\Psi_{wake} = 0.33333$

Table 6.3 Error in c_P as function of number of rings

Number of rings (NP)	Error in cP
4	$1.5 \cdot 10^{-4}$
8	$1.7 \cdot 10^{-5}$
16	$1.6 \cdot 10^{-6}$

6.5.6 Summary

This section is about an implementation of a vortex ring/cylinder model for investigating the constantly loaded axisymmetric Actuator Disk in the spirit of Øye, van Kuik, Bontempo and Manna. We confirm that the following model parameters influence its accuracy:

1. Number of rings,
2. Location of semi-infinite vortex cylinder,
3. Number of integration points at the disk,
4. Model for self-induction,
5. Shape of slipstream,
6. Number and distribution of vortex rings close to the disk edge.

We therefore confirm findings from Gijs van Kuik [58] that some inter-relationship between parameters (see Table 6.1) of the model exists. Nevertheless, it seems that an increase of number of rings always leads to an increase of accuracy.

To summarize:

1. Model accuracy: Slipstream iteration was stopped when a special combination of model parameters results in deviations for c_P from momentum theory $< 10^{-6}$. Table 6.3 gives an example of accuracy reached (with regard to c_P from momentum theory) as function of number of rings:
2. Edge singularity: Van Kuik [56, 58] and Chattot [9] speculated about a possible edge singualrity. Our findings, see, for example, Fig. 6.14, indicate a different behavior. As van Kuik already mentions in [58], Chattot [9] assumes some case of a constant vortex strength, which seems not to be the case. Nevertheless, Fig. 6.14) for $10^{-6} < z < 10^{-2}$ (four decades) shows a logarithmic **decrease** in voretx strength. Both v_z Fig. 6.12 and v_r Fig. 6.13 tend to finite values for $z \to 0^+$.

6.6 Helical Vortex Sheets

Vortex sheets were introduced in Sect. 3.6.4. As for the vortex line, Eq. 6.3, we may define formally:

$$\omega = \kappa \delta(n) \tag{6.29}$$

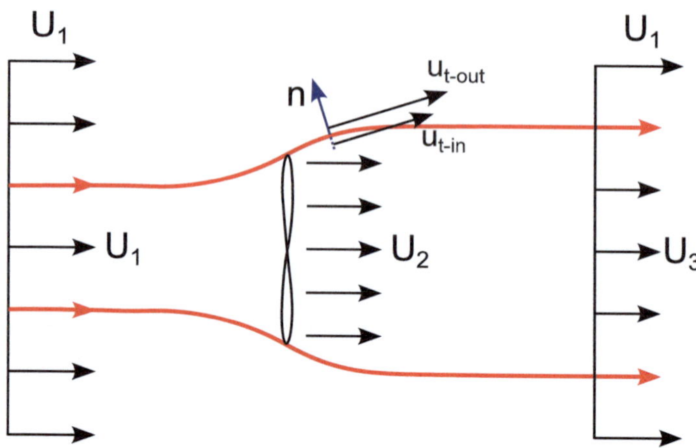

Fig. 6.16 Example of emergence of a simple vortex sheet (seen as jump in tangential velocity) for an actuator disk with constant loading

where **n** is the local normal and κ is the local strength of the sheet. As the flow normal to that sheet must be constant, only the tangential velocity may change. This is, in fact, the case at the border of the 1D-slipstream, where we have outside u_1 but inside all velocities from u_1 to u_3. Also, a vortex sheet may be thought of as an alignment of vortex lines. This is a very useful concept when we discuss finite wings either translating (as for a airplane) or rotating (as for propellers and turbines). In any case—because of the Kutta-Joukovski Theorem 3.55—circulation can be related to the chord c and local lift coefficient c_L of a wing:

$$\Gamma = \frac{c_L}{2} \, w \, c \tag{6.30}$$

Now if c and/or c_L are changing, Γ must also change. This means that, in case of $d\Gamma/dr \neq 0$, a vortex sheet must necessarily occur. This has significant consequences on the shape of the wake: In case of a constant Γ, the bound vortex must not end, and due to the rotation of the rotor and advection by the inflow, a helical vortex line like in Fig. 6.6 from [24] emerges.

This is in sharp contrast to the case of varying circulation. Inspired by Prandtl's *Tragflügeltheorie, (lifting-line theory)* [36, 37],[3] Betz investigated the problem of how to design a ship propeller with given thrust and minimal power. He found that the wake—a vortex sheet, see Fig. 6.17—must be of constant *pitch*.

---

[3] The most famous outcome was a quantitative description of the phenomenon of induced drag $c_{D,ind} = c_L^2/(\pi AR)$ which is minimized by an elliptical distribution of lift only. Corten [16] shows similarities and differences for the case of wind turbines. He found an AOA correction of $c_L/(8\pi) \cdot (Bc/r) \cdot \lambda_r$.

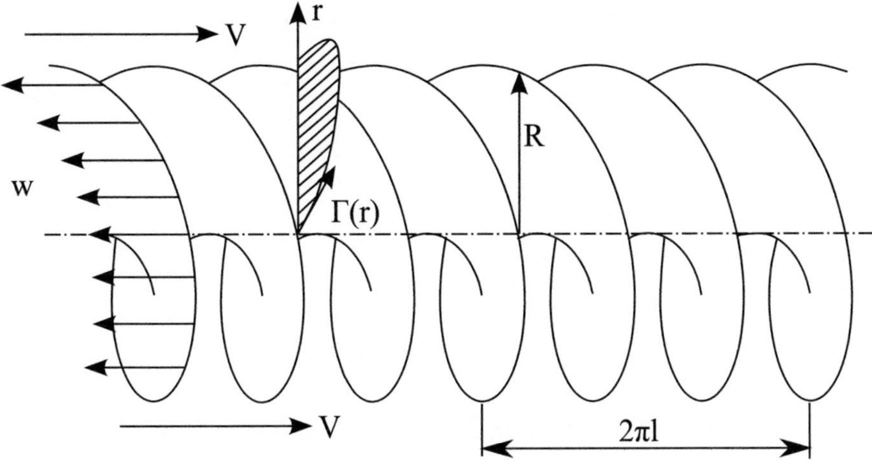

Fig. 6.17 Betz Wake

We follow (Chap. 1) [6] for a reliable derivation. (Compare to [5, 51], Chap. 12 for inclusion of drag—both from Sect. 5.8) From the Kutta-Joukovski Theorem, we have for Thrust T and Torque M (density of fluid $\rho = 1$):

$$T = \int_0^{R_{tip}} \Gamma \cdot \omega r(1 + a')dr \;, \tag{6.31}$$

$$M = u_1 \cdot \int_0^{R_{tip}} \Gamma(1 - a)r \, dr \tag{6.32}$$

As we want to be $P = \omega \cdot M \to min$ with $T = const$, we introduce a *Lagrange Multiplier* ℓ and use *calculus of variations* that for infinitely small amounts (variation) δT and δM

$$\delta T - \ell \cdot \delta M = 0 \tag{6.33}$$

must hold. Taking the variation of Eq. (6.32)

$$\delta T = \int_0^{R_{tip}} \delta \Gamma \cdot \omega r(1 + a') - \Gamma \frac{\partial a'}{\partial \Gamma} \delta \Gamma dr \;, \tag{6.34}$$

$$\delta M = u_1 \cdot \int_0^{R_{tip}} \delta \Gamma(1 - a) - \Gamma \frac{\partial a}{\partial \Gamma} \delta \Gamma r \, dr \tag{6.35}$$

To make Eq. (6.35) solvable, we have to connect the induced velocities to the circulation. For the tangential component, we have by Biot-Savart's Law [13]:

$$a' = \frac{\Gamma(r)}{4\pi \omega} \tag{6.36}$$

(note the extra factor 2 in the denominator !) and (compare again with [13] and (Chap. 1) [6]:

$$a = \frac{\Gamma(r)\omega}{4\pi u_1^2} \tag{6.37}$$

With both inserted into Eq. (6.33), it follows:

$$\int_0^{R_{Tip}} \delta\Gamma \underbrace{\left(\omega r(1 + 2a') - \ell u_1(1 - 2a)r\right)}_{(*)} dr \tag{6.38}$$

Now Eq. (6.38) may only hold for arbitrary $\delta\Gamma$ if

$$(*) = \omega r(1 + 2a') - \ell \cdot u_1(1 - 2a)r \tag{6.39}$$

vanishes, so that

$$\ell = \frac{\omega}{u_1}\frac{1 + 2a'}{1 - 2a} \tag{6.40}$$

must hold independent from r.

Putting Eqs. (6.36) and (6.37) into Eq. (6.40), we arrive at:

$$\Gamma_{Betz,opt} = 2\pi\,(\omega - \ell u_1)\,\frac{r^2}{1 + \ell \cdot \left(\frac{\omega}{u_1}\right)r^2} \tag{6.41}$$

Betz (Chap. 5) [2] conceived this in a slightly different manner. Figure (6.18) shows the original Betz circulation distribution together with the empirical tip correction in Eq. (5.59) which was given by Prandtl in the same paper as an appendix Fig. 6.18. It has to be added that—like in momentum theory—it is implicitly assumed:

$$a(x = x_{disk}, r) = \frac{1}{2}a(x \to \infty, r) \text{ and} \tag{6.42}$$

$$a'(x = x_{disk}, r) = \frac{1}{2}a'(x \to \infty, r)\,. \tag{6.43}$$

Betz' derivation is valid only for $B \to \infty$ and $c_T \to 0$ and shows no decrease in circulation close to the tip. When using Prandtl's simple correction, Eq. (5.59), further approximations are introduced, creating a need for a more precise derivation from first principles.

This was done first by Goldstein [22] (still valid for the case of lightly loaded propeller) and further improved by Theodorsen [54] for moderate loading.

His solution for the velocity potential was given as an infinite series of *Bessel functions*, see [1]. He was able to give figures for two- and four-bladed propellers only. With the advent of digital computers in the 1950s, more accurate and faster calculations were possible, see Tibery and Wrench [55].

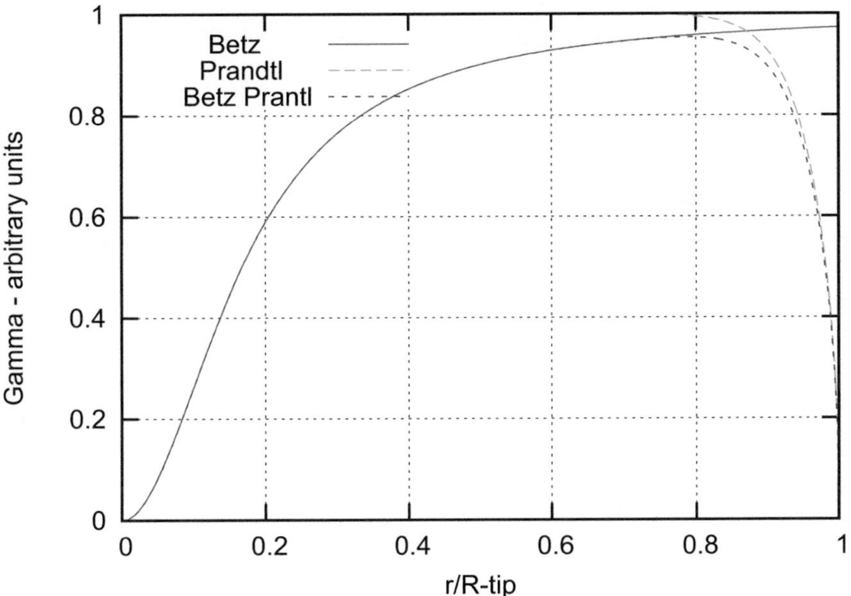

Fig. 6.18 Betz' optimum circulation distribution for a propeller, B $= 3$, TSR $= 6$

Okulov [33], based on his solution [19, 30] for the stability problem of flow of helical vortices, improved the numerical schema further. Figure 6.19 gives an example distribution for B $= 3$ and TSR $= 6$. The connection of $G(x, \ell)$ to power is given by [33]:

$$c_P = 2\bar{w}\left(1 - \frac{1}{2}\bar{w}\right)\left(I_1 - \frac{1}{2}I_3\right) \tag{6.44}$$

with

$$I_1 = 2\int_0^1 G(x, \ell)x\,dx, \tag{6.45}$$

$$I_3 = 2\int_0^1 G(x, \ell)\frac{x^3}{x^2 + \ell^2}\,dx \, . \tag{6.46}$$

The connection between ℓ, $\bar{w}$ and our common tip speed ratio $\lambda = \omega r_{tip}/u_1$ is given by

$$\ell = \frac{1 - \bar{w}}{\lambda} \, . \tag{6.47}$$

and $\bar{w}$ the circumferential averaged translation speed of the sheet (see Fig. 6.16).

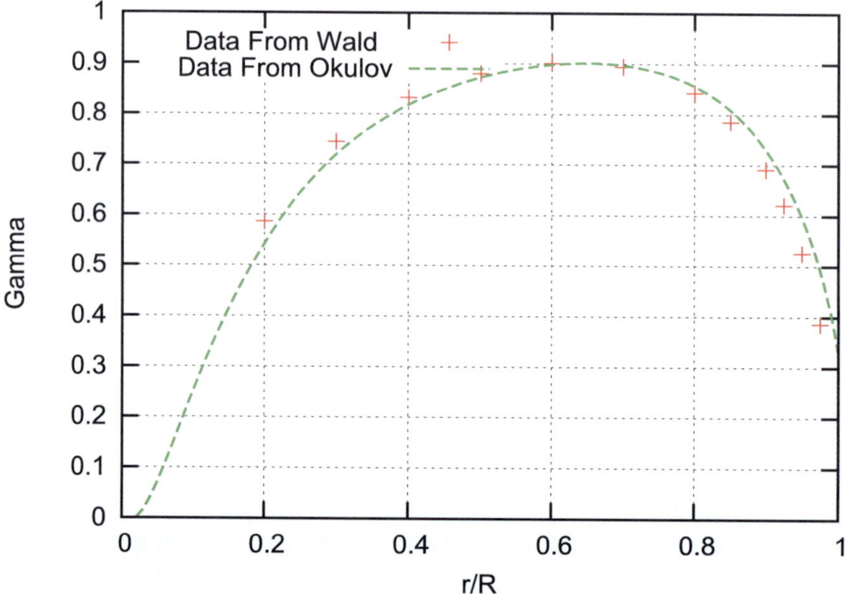

Fig. 6.19 Goldstein's optimum circulation, B = 3, $\lambda = 6$. Data from [31, 33, 64]

In the limit $B \to \infty$, Okulov is able to show [33]:

$$G_\infty = \frac{x^2}{x^2 + \ell^2}\,.$$
(6.48)

6.7   Stream Function Vorticity Formulation

6.7.1   Non-linear Actuator Disk Theory

In a series of papers, Conway [12–14], based on [66], formulated a semi-analytical
theory for actuator disks ($B \to \infty$) with vortex rings $\omega_{\varphi(r,z)}$ as the basic objects.
The slipstream (see Fig. 6.20) is thought to be consisting of **azimuthal vorticity**; this
being equivalent to a *volume distribution of ring vortices*.

In cylindrical coordinates, the axisymmetric velocity fields induced at a general
observation point (r, z) by a single ring vortex of radius r' and strength Γ placed at
axial position z' can be constructed by superposition from the elementary solutions
of Laplace's equation. The results are

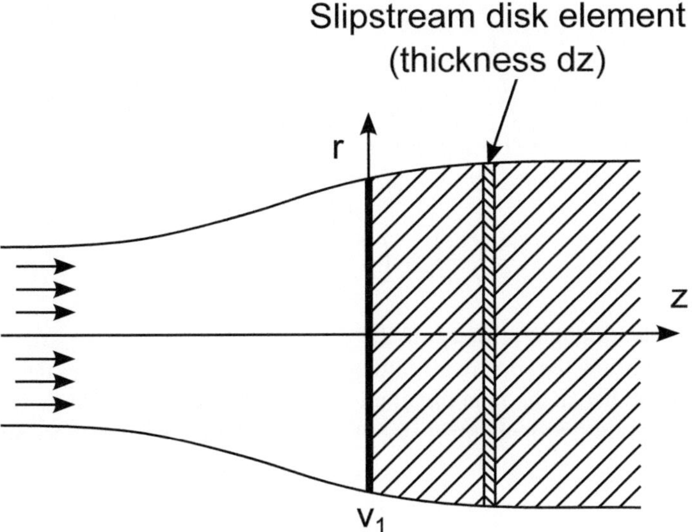

Fig. 6.20   Conways picture of an actuator disk

$$\Psi(r, z) = \frac{\Gamma r r'}{2} \int_0^\infty J_1(sr')J_1(sr)e^{-s|z-z'|}ds \tag{6.49}$$

$$V_r(r, z) = \frac{-\Gamma\, sign(z - z')r'}{2} \int_0^\infty s J_1(sr')J_1(sr)e^{-s|z-z'|}ds \tag{6.50}$$

$$V_z(r, z) = \frac{\Gamma r'}{2} \int_0^\infty s J_1(sr')J_0(sr)e^{-s|z-z'|}ds. \tag{6.51}$$

By superposition, the flow fields induced by the azimuthal vorticity distribution $\omega_\phi(r, z)$ in the slipstream are given by

$$\Psi(r, z) = \frac{r}{2} \int_0^\infty \int_0^{R(z)} \int_0^\infty \omega_\phi(r', z')r'J_1(sr')J_1(sr)e^{-s|z-z'|}dsdr'dz', \tag{6.52}$$

$$V_r(r, z) = \frac{1}{2} \int_0^\infty \int_0^{R(z)} \int_0^\infty \pm\omega_\phi(r', z')r' s J_1(sr')J_1(sr)e^{-s|z-z'|}dsdr'dz', \tag{6.53}$$

$$V_z(r, z) = \frac{1}{2} \int_0^\infty \int_0^{R(z)} \int_0^\infty \omega_\phi(r', z') r' s J_1(sr') J_0(sr) e^{-s|z-z'|} ds\, dr'\, dz'. \tag{6.54}$$

The vorticity within the slipstream is related to the specific enthalpy h, which is constant along streamlines, by the equation

$$\frac{\omega_\phi(r, z)}{r} = \left(\frac{h - h_0}{(\Omega r)^2} - 1 \right) \frac{dh}{d\Psi}, \tag{6.55}$$

where Ω is the angular velocity of the propeller and h_0 the free-stream specific enthalpy. Specifying the slipstream enthalpy as a function of Ψ defines the circulation distribution and hence the load distribution along the propeller blades. For the case of a counter-rotating propeller, Eq. (6.55) is replaced by the simpler equation:

$$\frac{\omega_\phi(r, z)}{r} = -\frac{dh}{d\Psi}. \tag{6.56}$$

This equation is also obtained from (6.55) in the limit as $\Omega \to \infty$ with $(h(\Psi) - h_0)$ held constant.

Provided the circulation has everywhere a finite slope, we can represent $h(\Psi)$ as a polynomial in (Ψ / Ψ_e) where Ψ_e is the stream function at the slipstream edge:

$$h - h_0 = \sum_{m=0}^{M} a_m \left(\frac{\Psi}{\Psi_e} \right)^m. \tag{6.57}$$

For most of the examples, we found that a reasonable representation of a generic rotor circulation distribution is given by the simple polynomial

$$h - h_0 = a \left[\left(\frac{\Psi}{\Psi_e} \right) - \left(\frac{\Psi}{\Psi_e} \right)^2 \right] \tag{6.58}$$

and this has been found to give good agreement between the actuator disk theory and the experiment. To evaluate (6.52), (6.53) and (6.54), the slipstream vorticity is represented as a function of the form:

$$\frac{\omega_\phi(r, z)}{r} = \sum_{n=0}^{N} A_n(z) \left[1 - \left(\frac{r}{R(z)} \right)^2 \right]^n. \tag{6.59}$$

This allows the radial integrals to be performed using the integral below which plays a role in the actuator disk theory, analogous to the Glauert integral in elementary aerodynamics.

$$\int_0^R r'^{m+1}(R^2 - r'^2)^n J_m(sr')dr' = 2^n n! R^{m+n+1} s^{-(m+1)} J_{m+n+1}(sR).$$
(6.60)

The remaining integrals with respect to s are all of the form

$$I_{(\lambda,\mu,\nu)}(R(z'), r, z - z') = \int_0^\infty s^\lambda J_\mu(sR(z')) J_\nu(sr) e^{-|z-z'|} ds,$$
(6.61)

and for λ, μ and ν integers, integrals of this form can always be evaluated in closed form in terms of complete elliptic integrals. This reduces equations (6.52), (6.53) and (6.54) to integrals representing the influence of an axial distribution of vortex disks, as illustrated in Fig. 1. After including the additional stream function due to the free-stream U_∞, equation (6.52) yields

$$\Psi(r, z) = \frac{U_\infty r^2}{2} + r \int_0^\infty \sum_{n=0}^N A_n(z') 2^{n-1} R^{2-n}(z') I_{(-(n+1),n+2,1)}(R(z'), r, z - z') dz'.$$
(6.62)

The coefficients $A_n(z)$ are weak functions of z and $R(z)$ and the $A_n(z)$ can be determined iteratively by employing equations (6.55) and (6.57) and enforcing the boundary condition $\Psi(R(z), z) = \Psi_e$. From Eq. (6.62) this gives

$$\Psi_e = \frac{U_\infty R(z)^2}{2} + R(z) \int_0^\infty \sum_{n=0}^N A_n(z') 2^{n-1} R^{2-n}(z') I_{(-(n+1),n+2,1)}(R(z'), R(z), z - z') dz'.$$
(6.63)

The corresponding formulae for the axial and radial velocities are readily obtained. The azimuthal velocity component for a single-rotating propeller is given directly by:

$$\frac{V_\phi(r, z)}{U_\infty} = \frac{h(\Psi(r, z)) - h_0}{\Omega r U_\infty}.$$
(6.64)

[44] and from a different viewpoint [39] the results were compared with CFD computations. The general expression was found to be very accurate except for the radial velocities close to the disk, which, in the numerical approach, has to be modeled with finite thickness (usually less then 10^{-2} times the disk radius, see Fig. 6.21.

6.7.2   *Application to Wind Turbines*

In [43, 44], the above mentioned ideas were applied to wind turbines, which in Conway's language corresponds to the negative sign case. Note that *advance ratio* J and tip speed ratio λ are related via

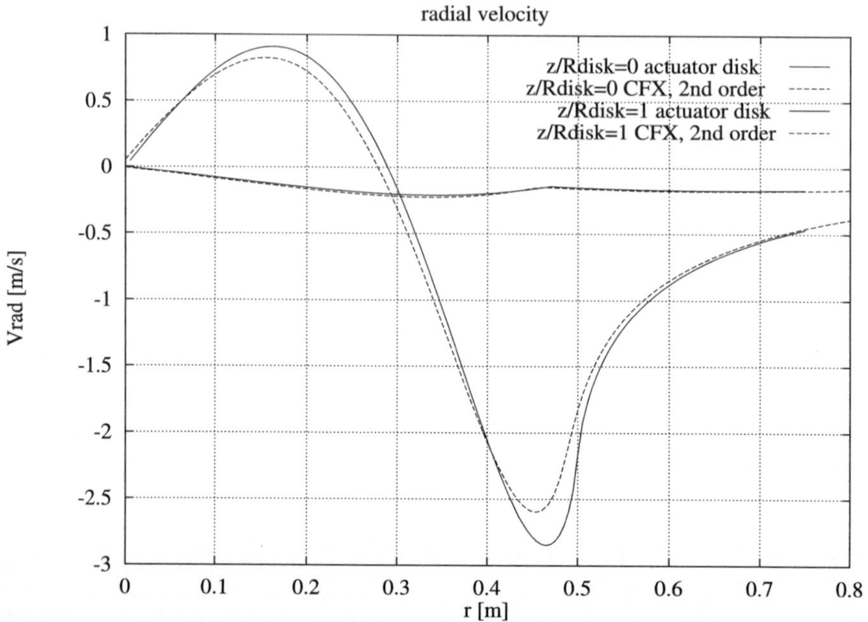

Fig. 6.21 Comparison of Conway's actuator disk theory to numerical solution of Euler's equation with load function of type (Eq. 6.58) and $c_T = 0.95$ (Propeller case)

$$J = \frac{\pi}{\lambda} . \tag{6.65}$$

Streamlines for the full slipstream iterated case with parabolic load $l(x) = -b \cdot (1 - r^2)$ with $b = -2.5$, corresponding to $c_T = 0.442$ and $c_P = 0.363$ are shown in Fig. 6.22. The slipstream expansion is from 0.97 (far upstream) to 1.11 (far downstream).

In addition, Conway [15] was able to compare his exact method (when $J = 0$ corresponding to $\lambda \to \infty$) to a simple approximation using momentum theory for given shapes of wake deficits. Figure (6.23) shows examples of such far-wake velocity deficit distributions which are possible to model (except in the case of constant load). It is clearly seen that a constant velocity deficit has much more influence outside the hub (Fig. 6.23).

Conway assumes

$$\left(\frac{R_{far-Slipstream}}{R_{tip}} \right)^2 = \frac{1 - a}{1 - 2a} \tag{6.66}$$

which becomes singular for $a \to 1/2$ For a parabolic shaped deficit, it follows

$$c_{P,parabolic} = 4a \cdot (1 - a) \cdot (1 - 2a) . \tag{6.67}$$

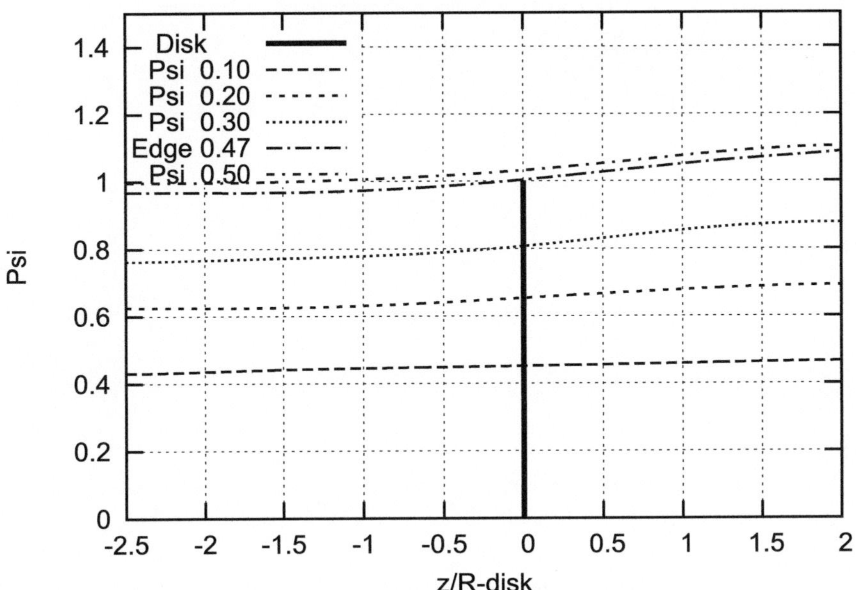

Fig. 6.22 Streamlines from Conway's actuator disk with parabolic load, $b = 2.5$

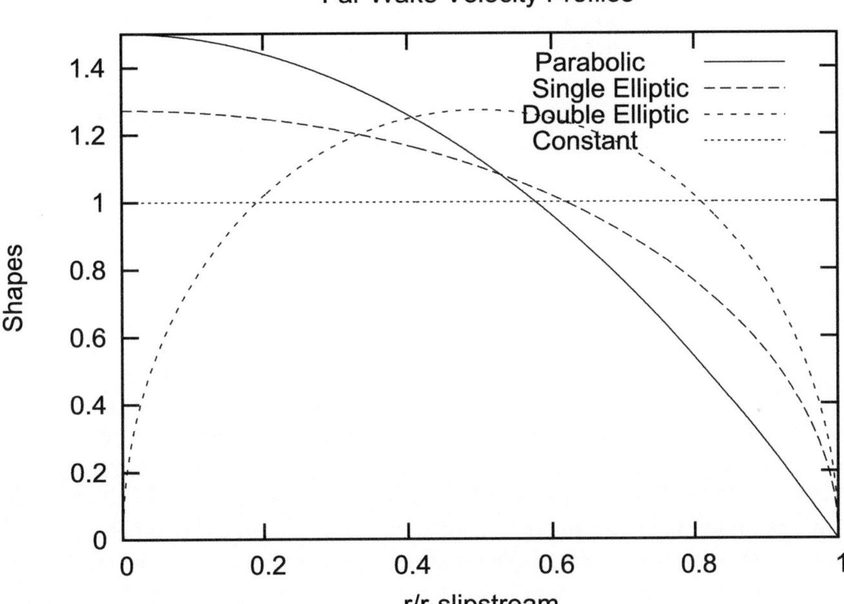

Fig. 6.23 Various shapes of far wake deficits

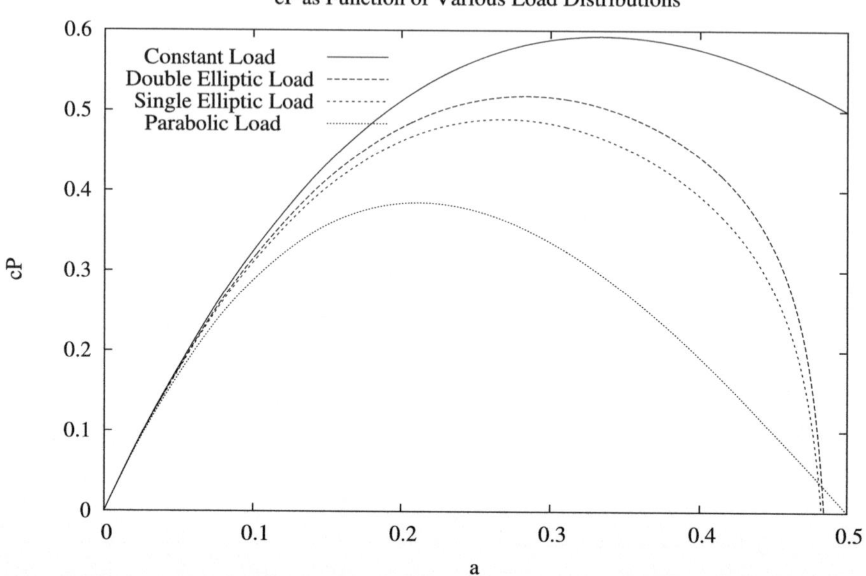

Fig. 6.24 Variation of c_P against induction factor from various wake deficit shape types from [15]

The elliptic shaped deficits are calculated in the same manner and all results are shown in Fig. (6.24). Close to $a = 0.5$, the variation of $c_P \rightarrow 0$ seem to indicate the breakdown of the simplified method. This is also confirmed when trying to perform a full slipstream iteration (Fig. 6.24).

To summarize, it may be said that full vortical slipstream seems to be in agreement with the numerical solution of the basic fluid equations, but so far, no additional insights have be gained.

6.8 Optimum Rotors II

As we have seen, constant and variable circulation distributions lead to very different vortical flow patterns in the wake. To be more precise, we have for the Joukovski rotor when $\Gamma = const$; a single approximately 1D structure, whereas in the Betz' case, $\Gamma(r) = const. \cdot r^2/(\lambda^2 + r^2)$ a 2D vortex sheet. In Table 6.4, results from [33, 34] are compared from both cases together with Glauert's approach. As it is clearly seen, the improvement is on the order of approximately 5%, but the practical feasibility of such a rotor remains undetermined. For any practical design, however, this is by far the simplest possible principle and was also deduced from momentum theory, presented in Sect. 5.3, by Sharpe (Chap. 5) [49].

Table 6.4 Numerical Values of c_P for optimal rotors, B = 3

TSR	Glauert	Joukovski	Betz/Goldstein	Improvement %
5	0.570	0.565	0.495	14
6	0.576	0.569	0.512	11
7	0.579	0.574	0.524	9
10	0.585	0.577	0.548	5
15	0.588	0.566	0.562	1

In the meantime, further investigations [35, 48] indicate that slipsteam expansion does not alter the results of Table 6.4 notably.

Case $\lambda = 7$ has been investigated in some detail by von Eitzen [63] with CFD using the Actuator Line (see Sect. 7.7) method implemented in the SOWFA distribution [11] and constant distribution of circulation along the blade was used. His results were $c_P = 0.578$ for the pure, and $c_P = 0.577$ for the implementation of Burton Sect. 5.8 [7] ($a = frac13, a' = \frac{2}{9}\lambda_r^{-1}$), which is remarkalbly close to Glauert's and Joukovsky's values.

Given $a = 1/3, a' = 0$ and restrict TSR> 3 a simple engineering equation for chord follows which sometimes is called *Betz' optimum blade shape*:

$$\frac{c}{R} \cdot \lambda^2 \cdot B \cdot c_L^{des} = \frac{16\pi}{9} \cdot x^{-1} \, , \quad x = r/R \, . \tag{6.68}$$

Interesting complimentary information can be found in Sect. 8.3 of [47] from Sect. 1.2.

6.9 Problems

Problem 6.1 Wilson's et al. [65] model, see (Chap. 5),

A vertical axis wind turbine of *H-Darrieus*-type may be analyzed with a model consisting of a simple rotating airfoil of length/height ($= H = 1$) and chord c. As the acting force, we assume only lift with $c_L = 2\pi \cdot sin(\alpha)$. With help of Fig. 6.25 and ϕ being the circumferential angle show:
(a) The circumferential averaged axial induction is

$$a = \frac{B \, c}{2R_{tip}} \frac{\Omega R_{tip}}{u_1} |cos(\theta)| \, . \tag{6.69}$$

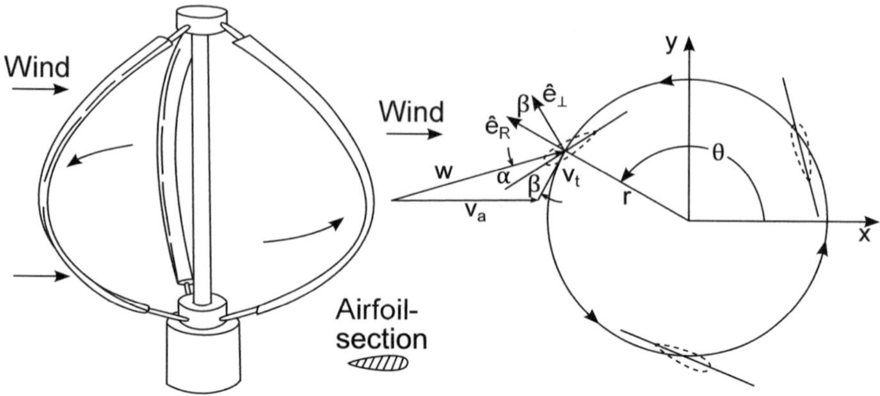

Fig. 6.25 H-Darrieus rotor

(b) The averaged power of a rotor of B blades:

$$c_P = \pi x \left(\tfrac{1}{2} - \tfrac{4}{3\pi} x + \tfrac{3}{32} x^2 \right) \text{ with} \tag{6.70}$$

$$x \qquad := \sigma \lambda \qquad\qquad \text{and} \tag{6.71}$$

$$\sigma \qquad = \frac{Bc}{R_{tip}} \qquad\qquad \text{and} \tag{6.72}$$

$$\lambda \qquad = \frac{\Omega R_{tip}}{u_1} \ . \tag{6.73}$$

(c) Show further that:

$$c_{P,max} = 0.554 \text{ at} \tag{6.74}$$

$$\frac{x_{max}}{2} = a_{max} = 0.401 \ . \tag{6.75}$$

(d) Find a corresponding expression for the thrust-coefficient c_T.
(e) Discuss your finding!

Problem 6.2 Holmes Darrieus Model From [23]: Use thin airfoil theory (Chap. 5 of (Chap. 3) [25]) to show that the model of Problem 6.1 in the limit $B \to \infty$ together with $c \to 0$ may be transformed in a stationary flow problem using a vortex sheet of strength:

$$\gamma(\phi) = \frac{1}{2} k s q_R(\phi) \ . \tag{6.76}$$

Expanding

$$q_R(\phi) = \sum_{n=1}^{\infty} (a_n cos(n\phi) + b_n sin(n\phi)) \ . \tag{6.77}$$

For example, c_P may be shown to be

$$c_P = \sum_{n=1}^{\infty} (a_n^2 + b_n^2) \, . \tag{6.78}$$

Problem 6.3 Calculate the vorticity of the Scully vortex, Eq. 6.4.

Problem 6.4 Show that the flow field of Hill's spherical vortex, Eq. 6.5, obeys Euler's equation, Eq. 3.31, provided that the pressure is given by

$$p = p_\infty - \frac{2 A^2 a^7}{225} \left(\frac{r^2 - 2z^2}{\left(r^2 + z^2\right)^{3/2}} + \frac{a^3 \left(r^2 + 4z^2\right)}{\left(r^2 + z^2\right)^4} \right) \, . \tag{6.79}$$

Derive H inside the sphere $r^2 + z^2 < a^2$ by integrating

$$\nabla H = \mathbf{u} \times \omega \tag{6.80}$$

to

$$H = \frac{1}{10} A^2 r^2 (z^2 + r^2) \, . \tag{6.81}$$

Problem 6.5 Run the vortex patch code attached in (A.6) to get comparable results as in Fig. 6.3.

References

1. Abramowitz M, Stegun I (1964) Handbook of mathematical functions. Dover Publications, Mineola, NY, USA
2. Betz A (1919) Schraubenpropeller mit geringstem energieverlust mit einem zusatz von l.prandtl. In: Nachr d Königl Gesell Wiss zu Göttingen, Math-phys Klasse, pp 198–217. Königl. Gesell. Wiss. zu Göttingen
3. Bontempo R, Manna M (2018) A ring-vortex free-wake model for uniformly loaded propellers. Part I—model description. In: Proceedings of the 73rd conference of the Italian thermal machines engineering association (ATI 201), Harwell-Lab, ETSU, UK, Sept 2018
4. Bontempo R, Manna M (2018) A ring-vortex free-wake model for uniformly loaded propellers. Part II—solution procedure and analysis of the results. In: Proceedings of the 73rd conference of the Italian thermal machines engineering association (ATI 201), Pisa, Italy, Sept 2018
5. Branlard E (2017) Wind turbine aerodynamics and vorticity-based methods. Springer, International Publulishingwin
6. Breslin A, Andersen P (1996) Hydrodynamics of ship propellers. Cambridge University Press
7. Burton T, Sharpe D, Jenkins N, Bossanyi E (2011) Wind energy handbook, 2nd edn. Wiley, Chichester
8. Carson B, Notis E (1981) Algorithm 577; algorithm for incomplete elliptic integrals. ACM Trans Math Softw 7:398
9. Chattot J-J (2020) On the edge singularity of the actuator disk model. J Sol Eng 143:1:014502
10. Chorin A (1994) Vorticity and turbulence. Springer, New York, USA
11. Churchfield M, Lee S, Moriarty P (2012) Overview of the simulator for wind farm application (sowfa). Technical report, NREL, Golden, CO, USA
12. Conway JT (1995) Analytical solutions for the actuator disk with variable radial distribution of load. J Fluid Mech 297:327–355

13. Conway JT (1998) Exact actuator disk solution for non-uniform heavy loading and slipstream contraction. J Fluid Mech 365:235–267
14. Conway JT (1998) Prediction of the performance of heavily loaded propellers with slipstream contraction. CASJ 44:169–174
15. Conway JT (2002) Application of an exact nonlinear actuator disk theory to wind turbines. In: Proceedings of the ICNPAA, Florida, USA. Melbourne
16. Corten G (2001) Aspect ratio correction for wind turbine blades. In: Proceedings of the IEA joint action, aerodynamics of wind turbines, Athens, Greece, Technical report
17. de Vries O (1979) Fluid dynamic aspects of wind energy conversion. Technical report AGAR-Dograph, 242, NATO, Neuilly sur Seine, France
18. Ferreira CS (2009) The near wake of the VAWT. PhD thesis, TU Delft, The Netherlands
19. Fukumoto Y, Okulov V (2005) The velocity field induced by a helical vortex tube. Phys Fluids 17:107101–107101–19
20. Gallavotti G (2002) Foundations of fluid mechanics. Springer, Berlin, Heidelberg, Germany
21. van Kuik G (2021) Private communication
22. Goldstein S (1929) On the vortex theory of screw propellers. Proc Roy Soc Ser A 123:440–465
23. Holme O (1976) A contribution to the aerodynamic theory of the vertical-axis wind turbine. In: Proceedings of the international symposium on wind energy systems, pp C-C4-72, UK, pp 4–55
24. Joukovski N (1929) Théorie tourbillonnaire de L'hélice. Gauthier-Villars, Paris, France
25. Katz J, Plotkin A (2001) Low-Speed aerodynamics, 2nd edn. Cambridge University Press, Cambridge, UK
26. Leishman J (2006) Principles of helicopter aerodynamics, 2nd edn. Cambridge University Press, Cambridge, UK
27. Leweke T et al (2013) Local and global pairing in helical vortex systems. In: Proceedings of ICOWES
28. NN. https://gcc.gnu.org/fortran/
29. Oeye S (1990) A simple vortex model. Technical report. In: Proceedings of the 3rd IEAwind Annex XI, Meeting, Harwell, UK
30. Okulov VL (2004) On the stability of multiple helical vortices. J Fluid 521:319–342
31. Okulov VL (2010) private communication
32. Okulov VL, Sørensen JN (2007) Stability of helical tip vortices in a rotor far wake. J Fluid Mech 576:1–25
33. Okulov VL, Sørensen JN (2008) Refined Betz limit for rotors with a finite number of blades. Wind Energy 11:415–426
34. Okulov VL, Sørensen JN (2010) Maximum efficiency of wind turbine rotors using Joukovsky and Betz approaches. J Fluid Mech 649:497–508
35. Okulov VL, Sørensen JN, van Kuik GAM (2013) Development of the optimum rotor theories. R & C Dyn, Moscow-Izhvsk
36. Prandtl L (1918) Tragflğeltheorie. I. mitteilung. In: Nachr DK, Gesell W, Zu G (eds), pp 451–477. Königl. Gesell Wiss zu Göttingen, Göttingen, Germany
37. Prandtl L (1919) Tragflğeltheorie. II. mitteilung. In: Nachr DK, Gesell W, Zu G (eds), pp 107–137. Königl. Gesell. Wiss. zu Göttingen, Göttingen, Germany
38. Press W, Teukolsky A, Vetterling W, Flannery B (1992) Numerical recipes in fortran, 2nd edn. The art of scientific computing. Cambridge University Press, Cambridge, UK
39. Réthoré P-E, Sørensen NN, Zahle F (2010) Validation of an actuator disc model. In: Proceedings of the EWEC 2010, Brussels, Belgium
40. Saffman PG (1992) Vortex dynamics. Cambridge University Press, Cambridge, UK
41. Saffman PG (1992) Vortex dynamics. Cambridge University Press, Cambridge, UK
42. Schaffarczyk AP (2011) Komponentenentwicklung und konstruktion wiko urban 5, aerodynamische auslegung des diffusors, des blattes und eines passiven pitch-systems (design and development of wiko urban 5: aerodynamic design of the diffusors, the blade and a passive pitch system). Internal confidential report No 81, Kiel, Germany

43. Schaffarczyk AP, Conway JT (2000) Application of a nonlinear actuator disk theory to wind turbines. In: Aerodynamics of wind turbines, Boulder, CO, USA, Dec 2000. Proceedings of IEA joint action
44. Schaffarczyk AP, Conway JT (2000) Comparison of nonlinear actuator disk theory with numerical integration including viscous effects. CASJ 46:209–215
45. Schepers JG, Boorsma K, Munduate X (2012) Final results from mexnext-i: analysis of detailed aerodynamic measurements on a 4.5 diameter rotor places in the large german dutch wind tunnel dnw. Technical report, Proceedings of the 4th conference of the science of making torque from wind, Oldenburg, Germany
46. Schepers JG et al (2012) Final report of iea task 29, mexnex (phase 1): analysis of Mexico wind tunnel measurements. Technical report, ECN-E-01–12-004, ECN, The Netherlands
47. Schmitz S (2020) Aerodynamics of wind turbines. Wiley, Hoboken, USA
48. Segalini A, Alfredson PH (2013) A simplified vortex model of propeller and wind-turbine wakes. J Fluid Mech 725:91–116
49. Sharpe DJ (2004) A general momentum theory applied to an energy-extracting actuator disk. Wind Energy 7:177–188
50. Sørensen JN (2016) General momentum theory for horizintal axis wind turbines. Springer International Publishing Switzerland
51. Sørensen JN (2016) General momentum theory for horizontal axis wind turbines. Springer, Cham, Switzerland
52. Strickland JH, Webster BT, Nguyen T (1979) A vortex model of the darrieus turbine: an analytical and experimental study. In: Paper 79-WA/FE-6, ASME, USA
53. Tao T (2016) Finite time blowup for langranian modifications of the three-dimensional euler equation. Ann PDE 2(9)
54. Theodorsen T (1948) Theory of propellers. McGraw-Hill, Ney York, USA
55. Tibery CL, Wrench JW (1964) Tables of the goldstein factor. Technical Report 1534, D Taylor Model Basin, Washington, DC, USA
56. van Kuik G (2003) The edge singularity of an actuator with a constant normal load. In: Proceedings of the 22nd AIAA/ASME wind energy symposium, volume AIAA-2003-0356, Reno, USA, Jan 2003
57. van Kuik G (2013) On the generation of vorticity in rotor & disc flows. In: Proceedings of the ICOWES2013, Copenhagen, Denmark
58. van Kuik G (2022) The fluid dynamic basis for actuator disc and rotor theories, revised, 2nd edn. IOS Press BV, Amsterdam, The Netherlands
59. van Kuik G (2022) On (non-)conservative body forces, vorticity generation and energy conversion in ideal flows. J Fluid Mech 941:A46
60. van Kuik G, Liganorolo L (2016) Potential flow solutions for energy extracting actuator disc flows. Wind Energy 19:1391–1406
61. van Kuik G, Sørensen J, Okulov V (2015) Rotor theories by professor Joukovsky: momentum theories. Prog Aero Sci 73:1–18
62. van Kuik GAM (2018) The fluid dynamic basis for actuator disc and rotor theories. IOS Press, TU Delft, The Netherlands
63. von Eitzen A (2018) Optimum circulation distribution on wind turbine rotor using sowfa code. Master's thesis, The Danish Technical University, Lyngby, Denmark
64. Wald QR (2006) The aerodynamics of propellers. Prog Aero Sci 42:85–128
65. Wilson RE, Lissaman PBS, Walker SN (1976) Aerodynamic performance of wind turbines. Technical report, Corvallis, Oregon, USA
66. Wu TY (1964) Flow through a heavily loaded actuator disc. Schiffstechnik 9(47):134–138
67. Young LA (2003) Vortex core size in the rotor near-wake. NASA/TM- 5:2003–21227

Chapter 7
Application of Computational Fluid Mechanics

You can't calculate what you haven't understood.[1]

7.1 Introduction

As we have seen in the previous chapters, due the non-linear behavior it is very difficult—if not impossible—to get simple analytical solutions of the basic fluid dynamics equations in a systematic way. Therefore it has become normal to use (massive) numerical methods for solving them. In an ideal situation this would mean that only Eqs. (3.37) from Chap. 3 are used (of course adapted to a suitable form for computers) together with a geometrical description of the problem (a wind turbine wing, for example) and a surrounding control volume for setting the boundary conditions for the unknown fields (pressure and velocity).

Unfortunately this is, even today, far too ambitious for most of the interesting engineering cases in which the Reynolds numbers are so high that an important part of the flow is turbulent. By now CFD = Computational Fluid Dynamics, meaning numerical solving of any fluid mechanical problem is an industry of its own, and we must refer the reader to the specific literature for most of the details [2, 12, 13, 21].

As is common practice, CFD is a three step process:

1. construct a 3D mesh from the geometry given (pre-processing)
2. solve the basic equations on (or in) this *finite volumes* (solving)
3. visualize and/or check the results (post-processing).

Everybody should know from undergraduate mathematics how to use at least two numerical schemes in calculus:

---

[1] Originally thought to be from P. W. Anderson.

© The Author(s), under exclusive license to Springer Nature Switzerland AG 2024
A. P. Schaffarczyk, *Introduction to Wind Turbine Aerodynamics*, Green Energy
and Technology, https://doi.org/10.1007/978-3-031-56924-1_7

(a) finding zeros from the *false position method* and
(b) numerical 1D integration by *Newton's Method* and/or *trapezoidal rule* and *Simpson's rule*.

Primary output from any fluid dynamics calculation are the forces on the structure, in most cases on the blade. Because pressure is related *non-locally* to the velocity field by Eq. (3.60), the whole 3D flow field must be calculated. It is immediately clear that a very high-performance computer must be available if CFD for any fluid-mechanical engine should be meaningful. Fortunately, since the first ideas of using digital computers for (turbulent) flow calculations by Johann von Neumann [69], computing possibilities have been grown almost constantly by *Moore's law*; see Fig. 7.1. The energy consumption is enormous, today reaching more than 18 MW for the computer and cooling power of the same order of magnitude.

First CFD (RANS = Navier-Stokes Equations + empirical differential equations for turbulent field quantities; see Sect. 7.5) calculation appeared in the academic world around the end of the 1990s [15, 43]. Table 7.1 gives a personal overview to

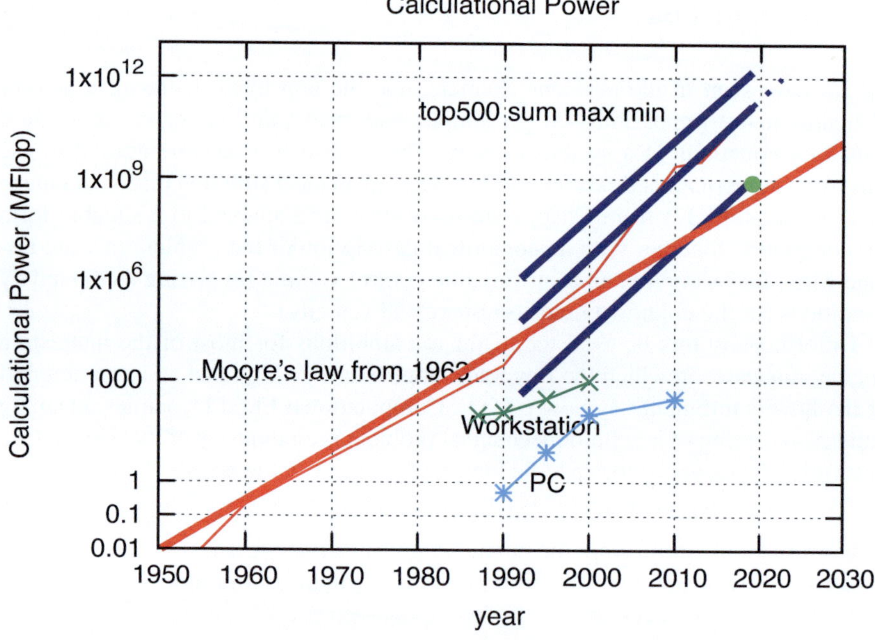

Fig. 7.1 Growth of Computing Power. Thin red line shows the development (absorbed by the middle blue line of TOP500's list) and the thick red one give Moore's extrapolation. Data from the TOP500 list is shown as three (sum, maximum and minimum) blue lines. The dotted extension gives an indication when the fastest computer will reach exa10^{18}-flop-scale. The green dot gives an indication of computational power accessible for an ordinary user at a local HPC, occasionally just on top the minimum-TOP500 list

Table 7.1 Some CFD codes

Code	Meshing	Special feature	Used by the author	Parallel computing	Language
Phoenics	1 Block structured	First code	1989–1993	No	FORTRAN77
CFX-4	Multiblock structured	Multiblocking	1995–2000	No	FORTRAN77
Fluent	Unstructured(U)	(U), easy to use	1997–	Yes	C
FLOWer	Multiblock	e^N method transition	2002–2012	Yes	FORTRAN77
Tau	Unstructured	(U)	2010–	Yes	C
OpenFOAM	Unstructured	Open source	2012–	Yes	C++

computer codes used by the author since the late 1980s. In the following discussion we will try to introduce the reader to this technique as it applies to wind turbine blades.

7.2 Pre-processing

Meshing a wing simply means putting a mesh around it as shown in Fig. 7.3. The grid points usually define the locations where the field variables are calculated. The distance of a grid point should be chosen in such a way that gradients are sufficiently resolved to have a defined accuracy. But here is the problem: As we know (see Fig. 7.2) sometimes pressure variation may be very large across small distances. To accurately resolve these large gradients without wasting computational efforts—most portions have very low gradients—one has to have an idea of the solution which the CFD model will produce. This is clearly a contradiction. Even worse each reasonable CFD computation has to be proven to be *grid-independent*—that means that if the grid is changed (finer or coarser) the solution must not change.

Mesh quality largely influences not only the accuracy but also the speed at which this accuracy is reached.

Two major measures are used:

1. aspect ratio, meaning that all cell edges have similar lengths; see Fig. 7.8
2. orthogonality. See Fig. 7.8.

Figure 7.4 shows a mesh with mostly orthogonal cells which was coded by Trede [68] along the lines of Sørensen [60]. Hyperbolic in this context means that the generating 2D partial differential equation system is of hyperbolic type (like the wave equation $u_{tt} - u_{xx} = 0$, the speed of the wave $c = 1$).

Three larger groups of mesh types are known, the last one being a mixture of the first two:

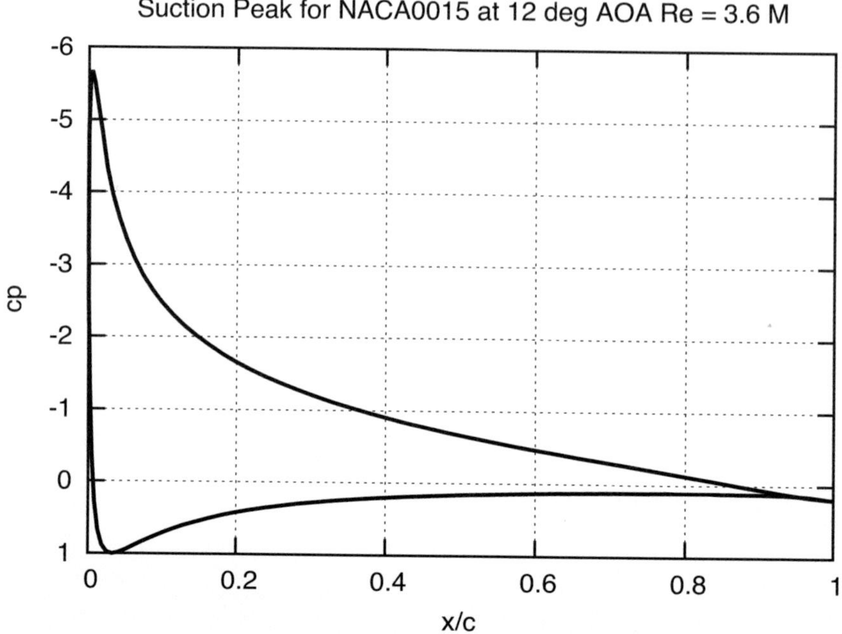

Fig. 7.2 Narrow Suction peak close to the nose around a thin airfoil at large angle of attack

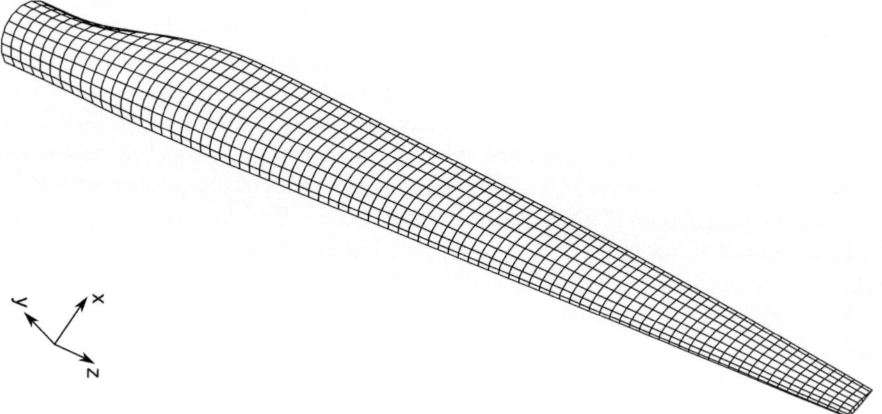

Fig. 7.3 Coarse Mesh on a wind turbine wing

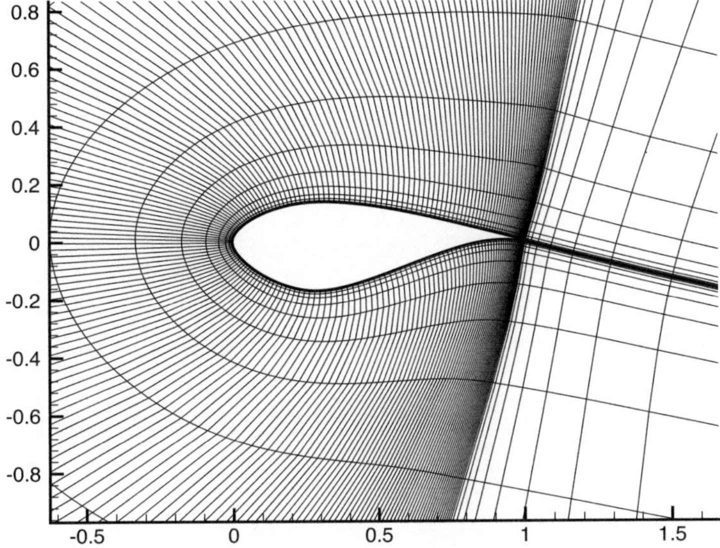

Fig. 7.4 Mesh perpendicular to a wind turbine wing, generated by a hyperbolic mesh generator [68]

Fig. 7.5 Mesh type 1

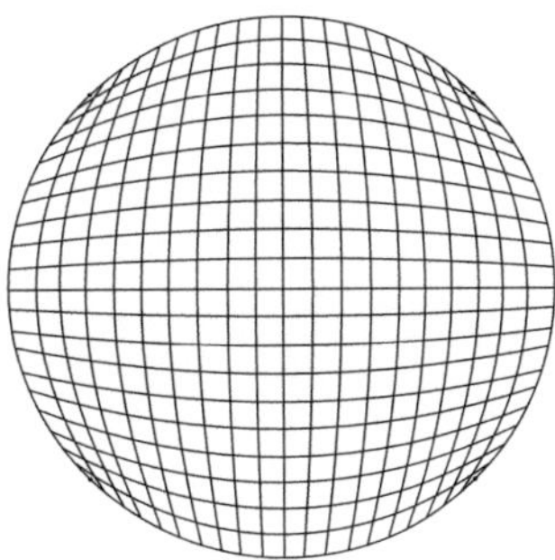

1. structured meshes meaning they have a fixed number of neighboring nodes; see Fig. 7.5
2. unstructured meshes with a free number of adjacent nodes, Fig. 7.7
3. mixtures of that, Fig. 7.6.

To summarize: mesh generation is still the most time-consuming part of a CFD job. Although unstructured grids are much easier and faster to generate than structured ones, often and in particular for boundary layer flow, structured grids still have to be used at least for specific parts.

A comparable new technique—the so-called *Chimera technique*—simply overlays two separate meshes on each other. Figure 7.29 shows an example of an wind turbine rotor consisting of three blades and a cylindrical hub.

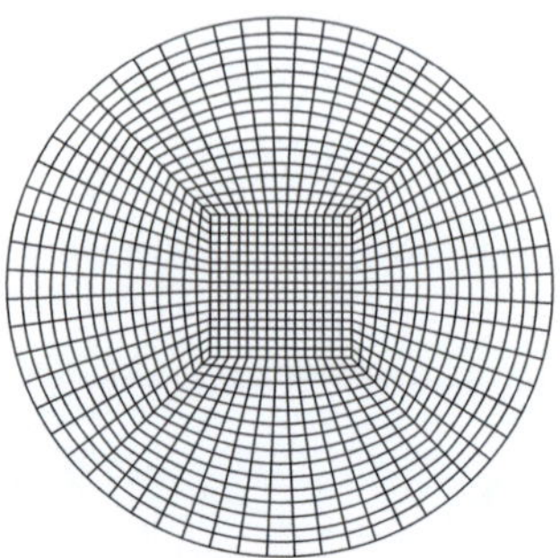

Fig. 7.6   Mesh type 2: structured H-type mesh

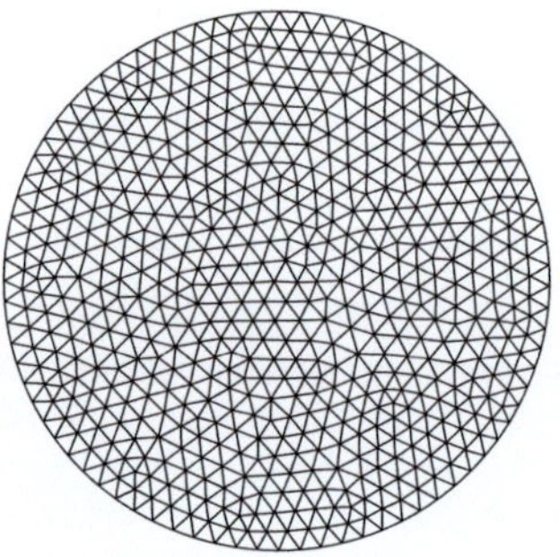

Fig. 7.7   Mesh type 3: unstructured mesh

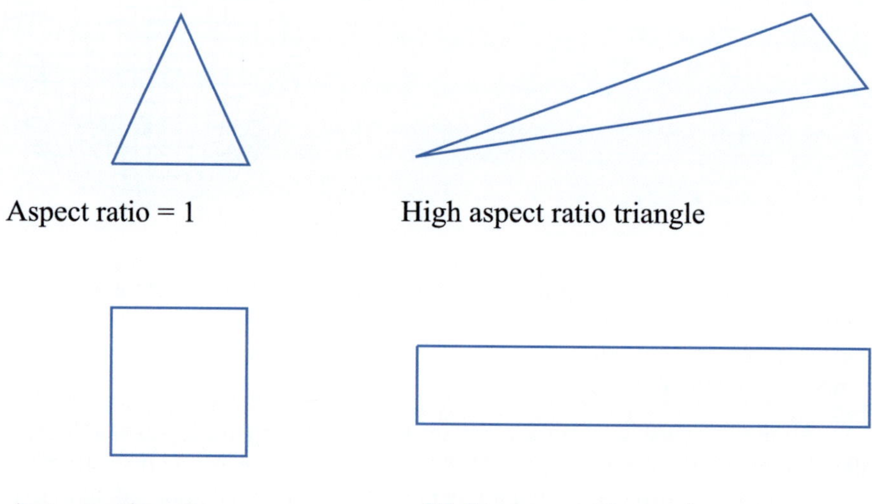

Fig. 7.8   Orthogonality and aspect ratio for triangles (upper row) and quads (lower row)

7.3 Solving the Numerical Equations

7.3.1 Discretization of Differential Equations

Discretization is to some extent the opposite procedure of the following *limits*:

$$\frac{dy(x)}{dx} = \lim_{\Delta x \to 0} \frac{\Delta y}{\Delta x} \, . \tag{7.1}$$

Methods based on generalizations for this trivial example therefore are called method of *Finite Differences*. What on a glance seems to be easy to 1D cases is it not at all in 2D and 3D.

Therefore more sophisticated concepts based on the general conservation laws discussed in Chap. 3 have been used to derive formulations suitable for digital computing.

7.3.2 Boundary Conditions

If the computational domain could be infinitely large, the only *boundary conditions* (BoCos) would be $\mathbf{u} = const$ and $p_0 = 0$. At solid walls it would be simply $\mathbf{u} = 0$. Other BoCos were introduced to decrease the extent of the computational domain, for example bi-symmetry considerations.

As the flow must be forced at least from one part of the domain in many cases *inflow BoCos* are used. They are very similar to the above-mentioned *far field* BoCos.

Complementary to inflow conditions the conditions defining an *outlet* simply specify that the flow does not change anymore, meaning that all gradients are set to zero.

The influence of BoCos both on the numerical implementation and on the accuracy of the result may not be underestimated. For example, a wind turbine blade (modeled as a 120 degree segment with *periodic boundary conditions*) uses a far field domain of 10 to 20 rotor diameters.

7.3.3 Numerical Treatment of the Algebraic Equations

After everything has been reduced to a huge number of discrete equations, efficient numerical algorithms are needed. The impressive improvement in speed and size of CFD calculations originated partly from increased computer power (as shown in Fig. 7.1) and partly in more efficient solving algorithms. As a well-known example the multi-grid algorithm is mentioned here.

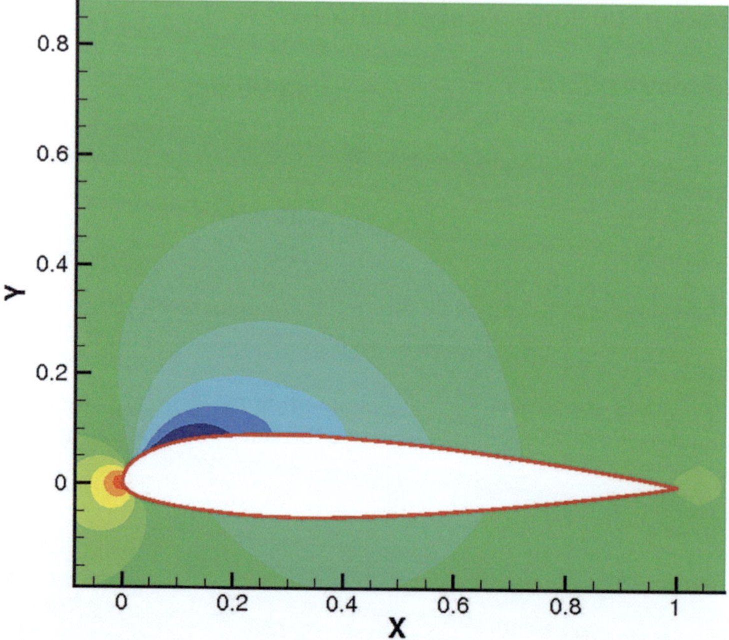

Fig. 7.9 Colored contour plot of pressure coefficient around a wind turbine airfoil; red: high pressure regions; blue: low pressure regions; AOA = 5°

7.4   Post Processing: Displaying and Checking the Results

Once a *converged solution* which is a solution with a prescribed tiny change (typically 10^{-6} or at least 10^{-3}) after one last iteration is reached, the problem of checking the results starts. It is used to be done by colored plots for velocity and pressure fields (see Fig. 7.9) for stationary flow, but with the emergence of unsteady methods like LES (Sect. 7.5.12) movies have become more fashionable. Any achievement of accuracy seems to be a delicate and highly non-trivial task. We will come back to this in more detail in Sect. 7.8.2.

7.5   Turbulence Models for CFD

Turbulence (and now more and more transition) modeling is at the core of CFD. If the flow is laminar, CFD is capable of generating as accurate numerical solutions as desired. This is (unfortunately) not at all the case for turbulent flow, which in almost all cases interesting for wind turbine applications is present. Therefore it has to modeled in an efficient way suitable for numerical solving of non-linear partial differential equations of the *transport equation* type.

Most models start from the statistical approach of Reynolds averaging and there-fore CFD is in most cases $RANS$ $=$ Solution of Reynolds Averaged Navier-Stokes Equations. Splitting into average and deviation using

$$u_\alpha := U_\alpha + u'_\alpha \,, \tag{7.2}$$
$$< u'_\alpha >= 0 \,, \tag{7.3}$$

we have further

$$\partial_t U_\alpha + U_\beta \partial_\beta U_\alpha = -\partial_\alpha p + \partial_\beta \left(\nu S_{\alpha\beta} - < u'_\alpha u'_\beta > \right) \,. \tag{7.4}$$

In Eq. 7.4 the so-called *Reynolds stresses*:

$$\tau_{\alpha\beta} :=< u'_\alpha u'_\beta > \tag{7.5}$$

are introduced which display the influence of velocity-correlations. These quanti-ties have to be precisely modeled. In addition, $S_{\alpha\beta}$ from Eq. 3.35 is defined as the *deformation tensor* , and is comprised of averaged velocities.

7.5.1 *Prandtl's Mixing Length Model*

One of the first empirical models of turbulence came from Prandtl [40, 71] and was inspired by kinetic gas-theory. It was postulated that *lumps of turbulence* (*Turbu-lenzballen* in German) are mixed up after a certain length ℓ:

$$< u'_\alpha u'_\beta >:= -\ell^2 \left(\partial_\beta U_\alpha \right). \tag{7.6}$$

Modeling now has to be applied to ℓ. Karman proposed close to a wall: $\ell = \kappa \cdot y^+$ with $\kappa \approx 0, 4$. Quite successful was the application to jets [71].

An ongoing discussion is about the so-called *universality* of κ meaning that it is a constant of turbulent flow independent of special boundary conditions.

Generalizations of the mixing length model are

- the already mentioned von Karmans approach inside boundary layers $\ell = \kappa y^+$,
- *van Driest's* boundary layer damping

$$\ell_{mix} = \kappa y (1 - \exp(-y^+/A_0^+)) \tag{7.7}$$
$$\text{with} \qquad A_0^+ \approx 26, \tag{7.8}$$

- Cebessi-Smith's two-layer model
- Smagorinsky's *sub-grid-scale model* for LES (see Sect. 7.5.12),
- the Baldwin-Lomax model with $v_T = \ell_m^2 \cdot \sqrt{\omega_\beta \omega_\beta}$ with $\boldsymbol{\omega} = \nabla \times \mathbf{u}$, the vorticity vector.

7.5.2   One Equation Models

Equation 7.6 is called an algebraic turbulence model, because only one turbulent quantity is connected to the averaged flow by an algebraic equation. Applied to jets it gave some accurate results, but it fails for more complex situations. Therefore it seems natural to proceed with the diagonal component of the stress tensor, the turbulent kinetic energy from Eq. 3.120.

The equation for the turbulent kinetic energy derived from the Navier-Stokes equation reads:

$$\partial_t k + U_\beta \partial_\beta k = \tag{7.9}$$

$$\tau_{\alpha\beta} \partial_\beta U_\alpha - \epsilon + \partial_\beta \left(\nu \partial_\beta k - \underbrace{\frac{1}{2} < {u'_\alpha}^2 u'_\beta >}_{*} - \underbrace{< p' u'_\beta >}_{**} \right), \tag{7.10}$$

where ϵ from Eq. 7.11 defines turbulent dissipation:

$$\epsilon := \frac{\nu}{2} \cdot \sum_\alpha \sum_\beta \left\langle \frac{\partial u'_\alpha}{\partial x_\beta} + \frac{\partial u'_\beta}{\partial x_\alpha} \right\rangle^2 . \tag{7.11}$$

Equation 7.11 is not closed, because of triple-correlations (*) $< {u'_\alpha}^2 u'_\beta >$ and velocity-pressure-correlations.

Further we need a link between ϵ and k.

If we set

$$(*) + (**) := -\frac{\mu_T}{\sigma_k} \partial_\beta k \tag{7.12}$$

$$\epsilon := C_D k^{3/2} / \ell. \tag{7.13}$$

we arrive a single differential equation model:

$$\partial_t k + U_\beta \partial_\beta k = \tag{7.14}$$

$$\tau_{\alpha\beta} \partial_\beta U_\alpha - \epsilon + \partial_\beta \left((\nu + \mu_T / \rho \sigma_k \partial_\beta k) \right) \tag{7.15}$$

$$\text{with } \mu_t = \rho k^{1/2} \ell \tag{7.16}$$

$$+ \text{ empirical model for } \ell. \tag{7.17}$$

Equation 7.17 is the weak link. Again we have to introduce empirical approximation which in many cases are from dimensional analysis, together with new constants which have to be determined experimentally.

7.5.3 *Spalart-Allmaras Model*

This model is *popular* in 2D airfoil aerodynamics. It starts from a turbulent viscosity which is given by

$$\nu_t = \tilde{\nu} f_{v1}, \quad f_{v1} = \frac{\chi^3}{\chi^3 + C_{v1}^3}, \quad \chi := \frac{\tilde{\nu}}{\nu}. \tag{7.18}$$

The transport equations for the modified viscosity $\tilde{\nu}$ are

$$\frac{\partial \tilde{\nu}}{\partial t} + u_j \frac{\partial \tilde{\nu}}{\partial x_j} = \tag{7.19}$$

$$C_{b1}[1 - f_{t2}]\tilde{S}\tilde{\nu} + \frac{1}{\sigma}\{\nabla \cdot [(\nu + \tilde{\nu})\nabla\tilde{\nu}] + C_{b2}|\nabla\nu|^2\} - \tag{7.20}$$

$$\left[C_{w1} f_w - \frac{C_{b1}}{\kappa^2} f_{t2} \right]\left(\frac{\tilde{\nu}}{d}\right)^2 + f_{t1}\Delta U^2 \tag{7.21}$$

$$\tilde{S} \equiv S + \frac{\tilde{\nu}}{\kappa^2 d^2} f_{v2}, \quad f_{v2} = 1 - \frac{\chi}{1 + \chi f_{v1}} \tag{7.22}$$

$$f_w = g\left[\frac{1 + C_{w3}^6}{g^6 + C_{w3}^6} \right]^{1/6}, \quad g = r + C_{w2}(r^6 - r), \quad r \equiv tu\frac{\tilde{\nu}}{\tilde{S}\kappa^2 d^2} \tag{7.23}$$

$$f_{t1} = C_{t1}g_t \exp\left(-C_{t2}\frac{\omega_t^2}{\Delta U^2}[d^2 + g_t^2 d_t^2] \right) \tag{7.24}$$

$$f_{t2} = C_{t3} \exp\left(-C_{t4}\chi^2 \right). \tag{7.25}$$

The rotation tensor $S = \sqrt{2\Omega_{ij}\Omega_{ij}}$ uses Ω_{ij} which is given by

$$\Omega_{ij} = \frac{1}{2}(\partial u_i/\partial x_j - \partial u_j/\partial x_i) \tag{7.26}$$

and d is the distance from the closest surface.

The model's constants are

$$
\begin{aligned}
\sigma &= && 2/3 \\
C_{b1} &= && 0.1355 \\
C_{b2} &= && 0.622 \\
\kappa &= && 0.41 \\
C_{w1} &= C_{b1}/\kappa^2 + (1 + C_{b2})/\sigma \\
C_{w2} &= && 0.3 \\
C_{w3} &= && 2 \\
C_{v1} &= && 7.1 \\
C_{t1} &= && 1 \\
C_{t2} &= && 2 \\
C_{t3} &= && 1.1 \\
C_{t4} &= && 2
\end{aligned}
\tag{7.27}
$$

The reader may decide by himself or herself if a model with such a large number of constants seems to be general relevance.

7.5.4 Two Equation Models

Kolmogorov [22] and Prandtl [41] (see also the paper of Spalding [67]) independently introduced a set of differential equations. Both emphasized the importance of k, the turbulent kinetic energy, but only one introduced a second quantity to use. Whereas Prandtl stayed at one equation, Kolmogorov used a kind of a frequency ω as an additional field.

As the k-equation (Eq. 3.120) includes the turbulent dissipation ϵ, one may hope to deduce a more realistic model if one uses an additional differential equation for that quantity. ϵ was defined in Eq. (7.11). From NSE we may deduce [71]

$$
(\partial_t + u_\beta \partial_\beta)\epsilon = \tag{7.28}
$$
$$
-2\left(< \partial_\gamma u_\alpha \partial_\gamma u_\beta > + < \partial_\alpha u_\gamma \partial_\beta u_\gamma >\right)\partial_\beta u_\alpha - 2\nu < u_\gamma \partial_\beta u_\alpha > \partial_{\beta\gamma} u_\alpha \tag{7.29}
$$
$$
-2\nu < \partial_\gamma u_\alpha \partial_\delta u_\alpha \partial_\gamma u_\beta > -2\nu^2 < \partial_{\gamma\delta} u_\alpha \partial_{\gamma\delta} u_\alpha > \tag{7.30}
$$
$$
+\partial_\beta \left[\nu \partial_\beta \epsilon - \nu < u_\beta \partial_\delta u_\alpha \partial_\delta u_\alpha > -2\nu/\rho < \partial_\delta p \partial_\delta u_\beta >\right]. \tag{7.31}
$$

Apparently this equation for ϵ is much more sophisticated than that for k. In more detail:

1. (7.29): production (of ϵ)
2. (7.30): dissipation (of ϵ)
3. (7.31): diffusion and turbulent transport (of $\epsilon)^2$

---

[2] Quadratic correlations may be replaced by a diffusion term because of the *Fluctuation-Dissipation Theorem* from linear and equilibrium statistical mechanics (Chap. 3) [39]. It states that the the variance of equilibrium fluctuations determines the strength of losses by small disturbances as well.

7.5.5 *Standard K-ϵ-model*

This model [36] sometimes is described as the *workhorse* of CFD and is one of the earliest models applied to CFD. Usually it is defined as

$$\tau_{\alpha\beta} = \frac{2}{3}k\delta_{\alpha\beta} - \nu_T S_{\alpha\beta} \qquad (7.32)$$

$$\nu_T = C_\mu \frac{k^2}{\epsilon} \qquad (7.33)$$

$$\partial_t k + U_\alpha \partial_\alpha k = -\tau_{\alpha\beta}\partial_\beta - \epsilon + \partial_\alpha\left(\frac{\nu_T}{\sigma_K}\partial_\alpha k\right) + \nu\partial_{\alpha\alpha}k \qquad (7.34)$$

$$\partial_t \epsilon + U_\alpha \partial_\alpha \epsilon = \underbrace{-C_{\epsilon 1}\frac{\epsilon}{k}\tau_{\alpha\beta}\partial_\beta U_\alpha}_{(1)} \underbrace{-C_{\epsilon 2}\frac{\epsilon^2}{k}}_{(2)} + \qquad (7.35)$$

$$\underbrace{\partial_\alpha\left(\frac{\nu_T}{\sigma_\epsilon}\partial_\alpha \epsilon\right) + \nu\partial_{\alpha\alpha}\epsilon}_{(3)}$$

and closure constants:

$$C_\mu = 0,09 \ , \ C_{\epsilon 1} = 1,44 \ , \ \text{and} \ C_{\epsilon 2} = 1,92; \ \sigma_K = 1,0; \ \sigma_\epsilon = 1,3. \qquad (7.36)$$

Terms (1), (2) and (3) in the ϵ-equation, Eq. (7.36), are the most unsavory part of the k-ϵ-model.

As a guiding line to truncate the higher order correlations of u_α, p, k, ϵ, only dimensional correctness is used and newly emerging *closure constants* 7.36 which have to be determined by experiment. Which one has to be used is discussed in [71]. As this model has been used in the early time of CFD in non-aerospace applications, a number of other different ones emerged, one of them being the $k - \omega$-model and derivative.

7.5.6 *K-ϵ-model from Renormalization Group (RNG) Theory*

A method well-known from quantum-field-theory and later applied to quantum and classical non-relativistic many-body systems was also applied to turbulence; see (Chap. 3) [31]. While there is still some discussion about the range of validity to turbulence, its outcome was a modified k-ϵ-model [56, 74]:

$$\partial_t k + u_\alpha \partial_\alpha k = \tau_{\alpha\beta} \partial_\beta U_\alpha - \epsilon$$
$$+ \partial_\alpha (\alpha(\nu_T) \nu \partial_\alpha k) \tag{7.37}$$
$$\partial_t \epsilon + u_\alpha \partial_\alpha \epsilon = \frac{\nu_T}{2} \sqrt{\epsilon} C_c Y_\epsilon \tau_{\alpha\beta} \partial_\beta U_\alpha$$
$$+ \epsilon^{3/2} Y_\epsilon + \partial_\alpha (\alpha(\nu_T) \nu \partial_\alpha \epsilon). \tag{7.38}$$

In addition the turbulent viscosity ν_T is being set into connection with quantities like $k/\sqrt{\epsilon}$, Y_ϵ and σ_K:

$$\frac{d}{d\nu_T} \frac{k}{\sqrt{\epsilon}}(\nu_T) = 1,72 \ (\nu_T^3 + C_c - 1)^{-1/2 \cdot \nu_T} \tag{7.39}$$

$$\frac{dY_\epsilon}{d\nu_T} = -0,5764 \ (\nu_T^3 + C_c - 1)^{-1/2} \tag{7.40}$$

$$\frac{1}{\nu_T} = \left(\frac{1,3929 - \alpha(\nu_T)}{0,3929}\right)^{0,63} \left(\frac{2,3929 - \alpha(\nu_T)}{3,3929}\right)^{0,37} \tag{7.41}$$

and constants:
$$C_\mu = 0,084; \ C_{\epsilon 1} = 1,063; \ C_{\epsilon 2} = 1,72;$$
$$\sigma_K = 0,7179; \ \sigma_\epsilon = 0,7179; \ C_c \approx 75. \tag{7.42}$$

Further discussion [56, 57] clarified some of the very controversially discussed items, especially on the status of the ϵ-equation 7.38 and the model (closure) constants.

7.5.7 *Boundary Conditions for Turbulent Quantities Close to Walls*

Any solution of a partial differential equation has to be supplemented by *boundary conditions*. Special attention must be spent on the new quantities like k and ϵ. Applicable boundary conditions are

$$\lim_{y^+ \to 0} k(y^+) = \tag{7.43}$$

$$= \lim_{y^+ \to 0} \epsilon(y^+) = 0 . \tag{7.44}$$

From DNS [64] one may compare the standard model and others. See Figs. 7.10 and 7.11.

From that we see that close to the wall k^+ has a maximum at approximately $y^+ \approx 20$, but ϵ^+ seems to reach a constant value only very close to the wall.

In addition one interpolates velocity field by the well-known logarithmic velocity profile, Eq. 3.149. It has to be noted that meshing depends on some of the turbulence models used.

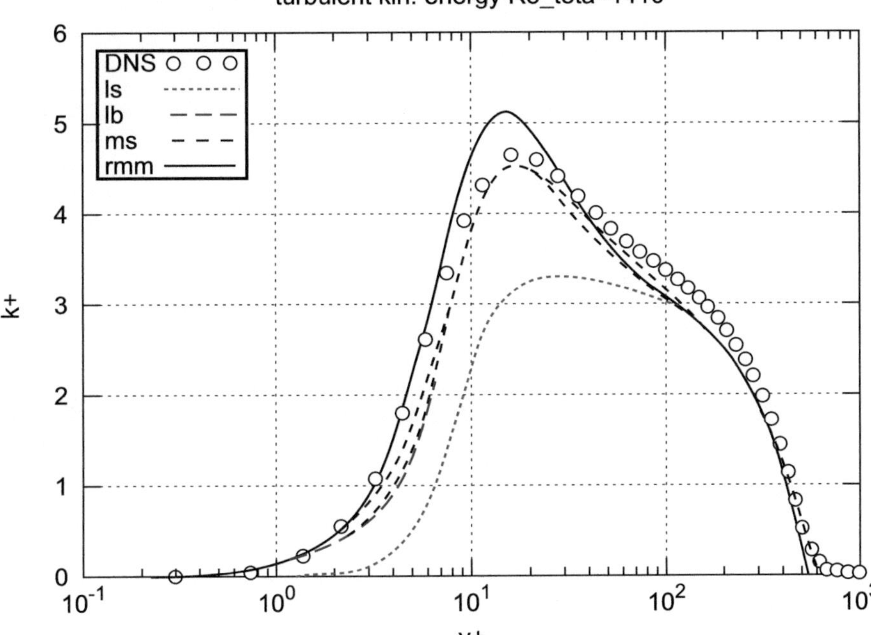

Fig. 7.10 Turbulent kinetic energy close to a wall for a flat plate boundary layer. DNS means Direct Numerical Simulation. ls refers to the standard k-ϵ model. Adapted from [35]

To summarize we may list some of the drawbacks of the original $k - \epsilon$-model:

1. only fully turbulent flow may be investigated,
2. no consistent extrapolation down to $y^+ \to 0$ possible,
3. so-called interpolating wall functions are valid only for non-separated flow.

Another drawback is that the k-ϵ-model may become inaccurate if strong inertia forces become important when used in rotating frames of reference.

7.5.8 *Low-Re Models*

Some of the above-mentioned turbulence expressions do not produce accurate results close to walls. Therefore correction for this so-called *Low-Reynolds number* has been developed [71]. A useful definition for a turbulent Re-number may be given in the following equation:

$$Re_t = \frac{\sqrt{k}\ell}{\nu} = \frac{k^2}{\nu_\epsilon} \cdot \frac{\nu_t}{\nu} \; . \tag{7.45}$$

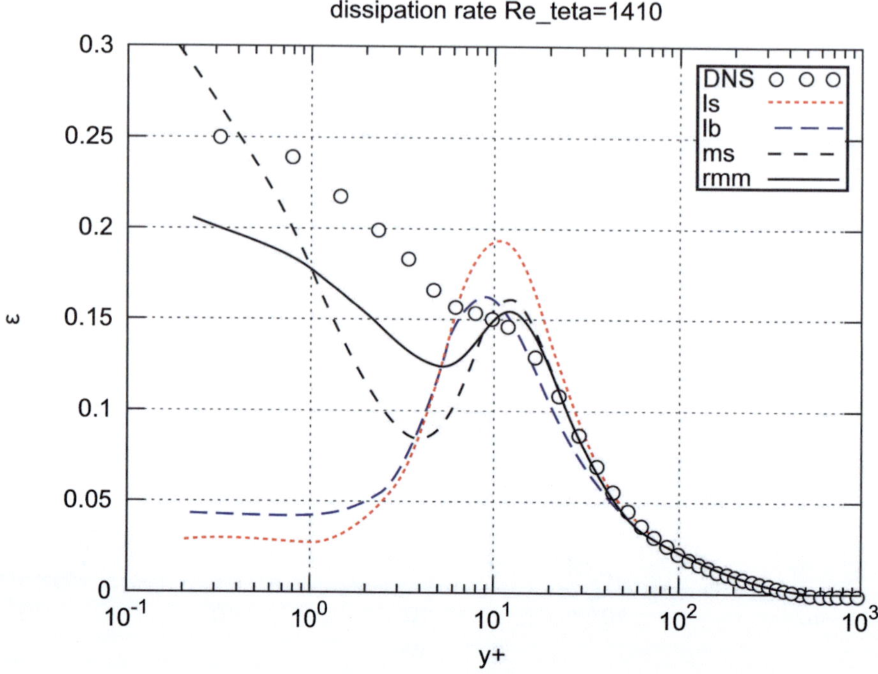

Fig. 7.11 Turbulent energy dissipation close to a wall. Same description as in Fig. 7.10

$Re_t < 100$ is **not** attained:

1. shortly after transition,
2. in heavily accelerated flow,
3. in the far field of wakes,
4. small wall distance.

In most cases damping- and blending-functions are used, and some progress has been seen as is shown in Figs. 7.10 and 7.11).

7.5.9   *Menter's Shear Stress Transport Model*

The aerospace industry uses ω (a local turbulent angular velocity) instead of ϵ as the second most important quantity; see Wilcox [71].

Menter [32, 33] developed a so-called S(hear)S(tress)T(ransport)-k-ω model which proved useful and is by now the most popular empirical engineering turbulence model for wind turbine flow. Its basic equations may be summarized as follows.

Kinematic Eddy Viscosity:

$$\nu_T = \frac{a_1 k}{\max(a_1\omega, SF_2)} \, . \tag{7.46}$$

Turbulence Kinetic Energy:

$$\frac{\partial k}{\partial t} + U_j \frac{\partial k}{\partial x_j} = P_k - \beta^* k\omega + \frac{\partial}{\partial x_j}\left[(\nu + \sigma_k \nu_T)\frac{\partial k}{\partial x_j}\right]. \tag{7.47}$$

Specific Dissipation Rate:

$$\frac{\partial \omega}{\partial t} + U_j \frac{\partial \omega}{\partial x_j} = \alpha S^2 - \beta\omega^2 + \frac{\partial}{\partial x_j}\left[(\nu + \sigma_\omega \nu_T)\frac{\partial \omega}{\partial x_j}\right] + 2(1 - F_1)\sigma_{\omega 2}\frac{1}{\omega}\frac{\partial k}{\partial x_i}\frac{\partial \omega}{\partial x_i}. \tag{7.48}$$

Closure Coefficients and Auxiliary Relations:

$$F_2 = \tanh\left[\left[\max\left(\frac{2\sqrt{k}}{\beta^*\omega y}, \frac{500\nu}{y^2\omega}\right)\right]^2\right] \tag{7.49}$$

$$P_k = \min\left(\tau_{ij}\frac{\partial U_i}{\partial x_j}, 10\beta^* k\omega\right) \tag{7.50}$$

$$F_1 = \tanh\left\{\left\{\min\left[\max\left(\frac{\sqrt{k}}{\beta^*\omega y}, \frac{500\nu}{y^2\omega}\right), \frac{4\sigma_{\omega 2}k}{CD_{k\omega}y^2}\right]\right\}^4\right\} \tag{7.51}$$

$$CD_{k\omega} = \max\left(2\rho\sigma_{\omega 2}\frac{1}{\omega}\frac{\partial k}{\partial x_i}\frac{\partial \omega}{\partial x_i}, 10^{-10}\right) \tag{7.52}$$

$$\phi = \phi_1 F_1 + \phi_2(1 - F_1) \tag{7.53}$$

$$\alpha_1 = \frac{5}{9}, \alpha_2 = 0.44 \tag{7.54}$$

$$\beta_1 = \frac{3}{40}, \beta_2 = 0.0828 \tag{7.55}$$

$$\beta^* = \frac{9}{100} \tag{7.56}$$

$$\sigma_{k1} = 0.85, \sigma_{k2} = 1 \tag{7.57}$$

$$\sigma_{\omega 1} = 0.5, \sigma_{\omega 2} = 0.856. \tag{7.58}$$

7.5.10   Reynolds Stress Models

From NSE we may even derive a complete set of equations for all of the Reynolds stresses, Eq. 7.5:

$$\partial_t \tau_{\alpha\beta} + U_\gamma \partial_\gamma \tau_{\alpha\beta} =$$

$$-\tau_{\alpha\gamma} \partial_\gamma U_\beta - \tau_{\beta\gamma} \partial_\gamma U_\alpha \tag{7.59}$$

$$+ \Pi_{\alpha\beta} - \epsilon_{\alpha\beta} - \partial_\gamma C_{\alpha\beta\gamma} + \nu \partial_{\gamma\gamma} \tau_{\alpha\beta} \tag{7.60}$$

with :

$$-\tau_{\alpha\gamma} \partial_\gamma U_\beta - \tau_{\beta\gamma} \partial_\gamma U_\alpha$$

Stress creation

$$\Pi_{\alpha\beta} = \langle p \partial_\beta u_\alpha + \partial_\alpha u_\beta \rangle \tag{7.61}$$

pressure-dilation correlation

$$\epsilon_{\alpha\beta} = 2\nu \langle \partial_\gamma u_\alpha \partial_\gamma u_\beta \rangle \tag{7.62}$$

dissipation rate correlation and

$$C_{\alpha\beta\gamma} = <u_\alpha u_\beta u_\gamma> + <pu_\alpha> \delta_{\beta\gamma} + <pu_\beta> \delta_{\alpha\gamma} \tag{7.63}$$

diffusions correlation of third order.

Compared to Eq. 3.120, these equations are far more complicated and use much more than 20—sometimes difficult to determine—new constants.

7.5.11   Direct Numerical Simulation

Direct numerical simulation (DNS) with no turbulence modeling at all is possible only for very few cases. [19] gives an account of the work on isotropic turbulence ($Re_\lambda = 1131$) on a 4096^3 grid.

Data for a 1024^3 ($Re_\lambda = 433$) grid is available even publicly [23].

For wall-bounded flow, only the simulation of Spalart [64] for a turbulent boundary layer on a flat plate $Re_\theta = 1410$ corresponding to $Re_x \approx 10^6$ is available. Some of his findings are shown in Figs. 7.10 and 7.11, and are compared to simpler RANS models there.

The Exa FLOW project (Enabling Exascale Fluid Dynamics) is coordinated by the Royal Institute of Technology—KTH, Sweden. These systems will have up to 10^9 processors, and it is expected that Reynolds number for homogeneous isotropic turbulence (HIT) will increase by a factor of 3 to 5 to about 7000. Spalart [65] conjectured 20 years ago that a full turbulence model-free CFD simulation of an airplane would need 1016 grid points and $10^{7.7}$ time steps. As seen from today, it seems we're likely to witness this event much earlier on an Exa-scale computer.

7.5.12 *Large and Detached Eddy Simulation*

An intermediate concept between DNS and RANS is *Large Eddy Simulation* where
a different concept of separation of velocities is used:

$$u = \bar{u} + u' \, , \tag{7.64}$$

$$\bar{u}, rt = \int G(r - \xi; \Delta) \cdot u(\xi, t) \, d\xi \, . \tag{7.65}$$

An important difference between averaging and filtering is

$$\bar{\bar{u}} \neq \bar{u} \, . \tag{7.66}$$

Δ is the usual filter size. As in the case of RANS, the smallest scales (typically
several ten times larger than the Kolmogorov scale) have to be modeled. Starting
point is the so-called Smagorinsky *sub-grid scale stress, SGS* :

$$\tau_{ij} = \mu_t \cdot \frac{1}{2} \left(\frac{\partial \bar{u}_i}{\partial x_j} + \frac{\partial \bar{u}_j}{\partial x_i} \right) \, , \tag{7.67}$$

$$\mu_t = \rho C_s^2 \Delta^2 \sqrt{S_{ij} S_{ij}} \quad \text{and} \quad C_S = 0.1 \ldots 0.24 \, . \tag{7.68}$$

Some properties of LES are

- LES is always unsteady
- LES is 3D
- Near walls the eddy viscosity has to be damped out.

To compare the computational effort (usually against DNS) we may refer to Pope,
Problem 13.29 (Chap. 3), [39].

The number of cells N^3 for DNS scales as $\approx 160 \cdot Re^3$ and for LES (depending of
the *near wall resolution* $\sim Re^{1.8}$. Both calculations are transient. For DNS the number
of time steps needed is roughly $15 \cdot Re_\lambda$ and for LES one order of magnitude smaller.

The total number of cells is smaller only by a factor of 10 to 20, and the CPU time
is one order of magnitude smaller.

If solid boundaries for any part of the application are avoided (by means of actuator
disk and actuator line methods; see. Sect. 7.7) a major application is wake aerodynamics of wind turbines and even wind farms [42, 73].

An interesting mixture between LES and RANS, *Detached-Eddy Simulation*, [66]
may be used especially for highly separated flow.

7.6　Transitional Flow

CFD codes in general only model fully turbulent flow. Laminar parts have to be separated *by hand.* As we have seen in Chap. 5 (Fig. 5.13), reasonable c_P-values around 0.5 need lift-to-drag ratios of more that 80. Keeping c_L fixed to moderate values around 1, the only chance is to decrease drag.

In the pure laminar case Blasius (see Eq. 3.102) has shown that

$$c_D = \frac{1.3}{\sqrt{Re_x}} \; , \tag{7.69}$$

whereas for the fully turbulent case

$$c_D = 2 \left[\frac{\kappa}{\ln Re_x} \right]^2 \; . \tag{7.70}$$

Therefore, simulation of high lift-to-drag ratios of airfoils necessarily relies on modeling all laminar, transitional and turbulent parts. In [24, 34, 61, 72] some of the details of currently used methods and approaches are described (Fig. 7.12).

It has to be noted that at least two important so-called *scenarios* are used to describe the mechanisms from laminar to turbulent flow on a wall:

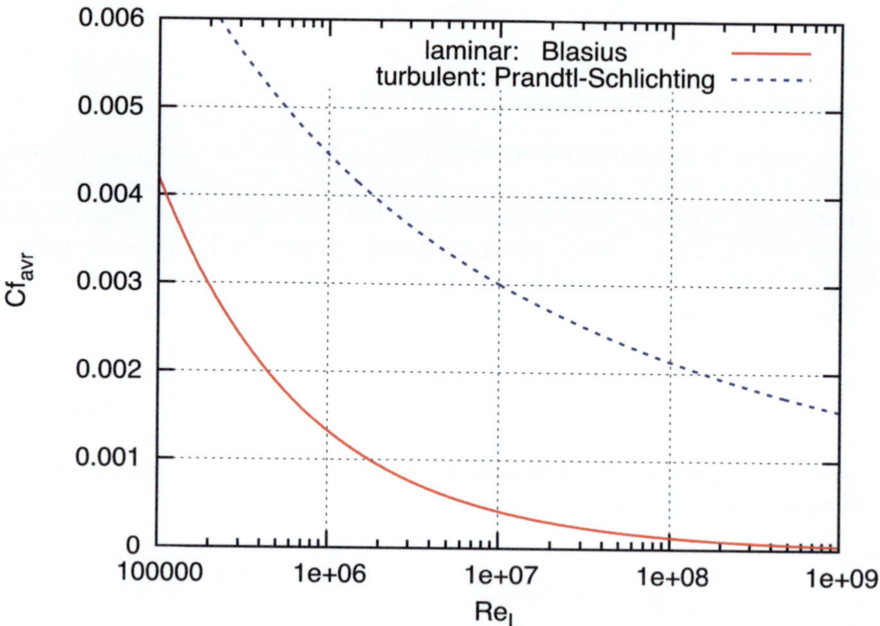

Fig. 7.12 Comparison of analytical approximation; see also Fig. 3.16

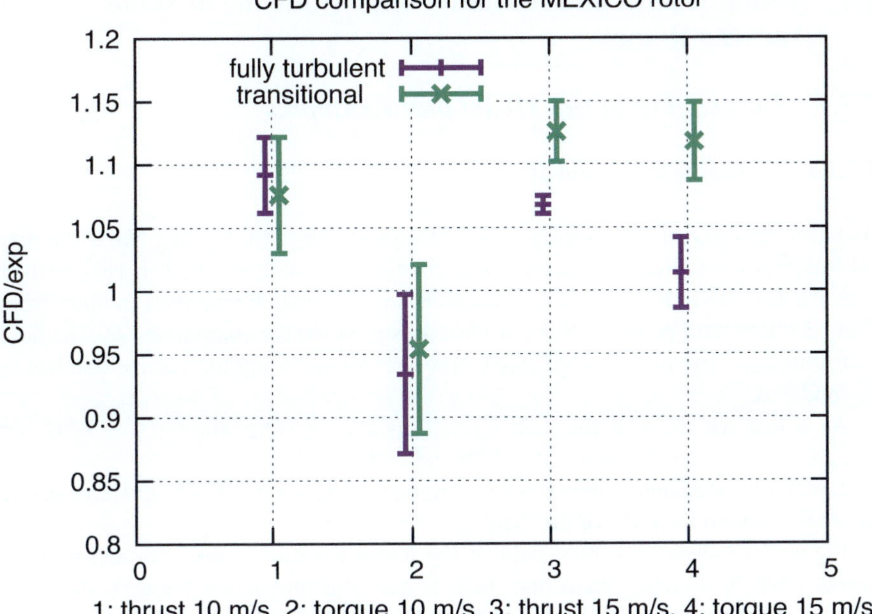

Fig. 7.13 Comparison of fully turbulent and transitional CFD simulation with experiment. From [44]

- Tollmien-Schlichting and
- Bypass [48].

The first one already described in Chap. 3 may be used for inflow conditions with less than 1% turbulence intensity. If more turbulence is present, this transition schema is *bypassed* .

Much effort has been made to describe and explain the findings during the NASA Ames Phase VI experiments (Chap. 8) [49]; see Sect. 8.5. Figures 7.21 and 7.20 show comparisons for the S809-airfoils used and Figs. 7.22 and 7.23 show the influence on the power curve.

During the IAEwind Task 29 (MexNext) Phase 3 (2015–2017), extensive investigations [26, 44] to detect laminar-turbulent frm MEXICO (2006) and NEW MEXICO (2014) experiments were undertaken. Result are shown in Fig. 7.13. It can be seen that using a transition model at least for the 15 m/s case gives results closer to the measured values. Nevertheless, deviations as well as error-bars are still in the 5% range.

7.7 Actuator Disk and Actuator Line Modeling of Wind Turbine Rotors

7.7.1 Description of the Model and Examples

7.7.1.1 Actuator Disk Models

As a first example of the usefulness of CFD, we present results using Navier-Stokes (or Euler) solvers when combined with Actuator Disk method [29, 53]. This method uses the full 3D (not 2D as in the simple Blade Element Momentum method from Chap. 5) flow field as input. It is particularly suited for the comparison of free-field configurations and more complicated structures surrounding the turbine that has to be modeled.

As a test we show in the next figures results for Betz' rotor ($c_T = 8/9$) and Glauert's rotor ($\lambda = 2$) (Figs. 7.14, 7.15 and 7.16).

It is clearly seen that radial velocity components are not negligible and have strong influence toward the edge of the disk.

Figure 7.17 shows the flow field of the Rotor for $v_{in=10}$ m/s with and without tunnel [29]. It is clearly seen that there is a strong interaction between the wind turbine and the wind tunnel.

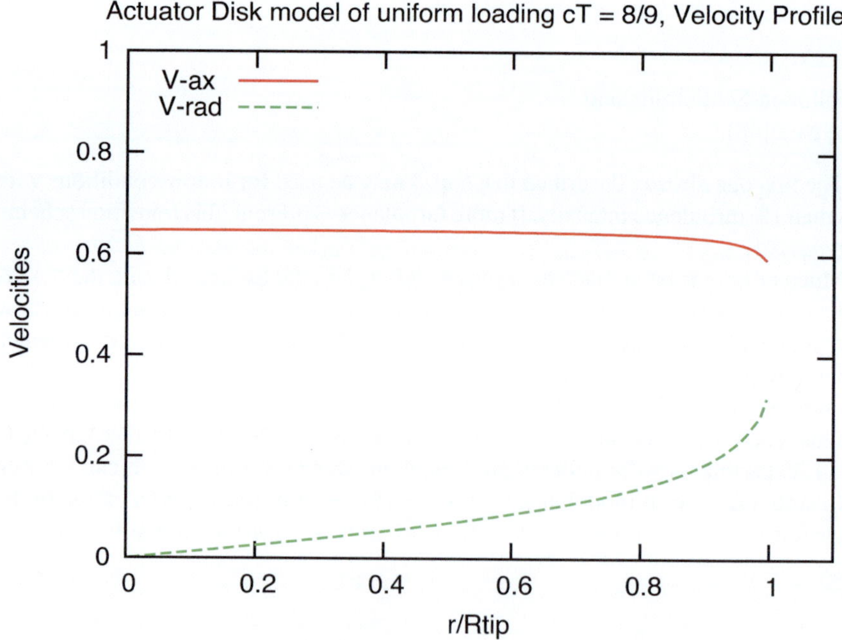

Fig. 7.14 Example of CFD-based actuator disk model: ax = axial; rad = radial

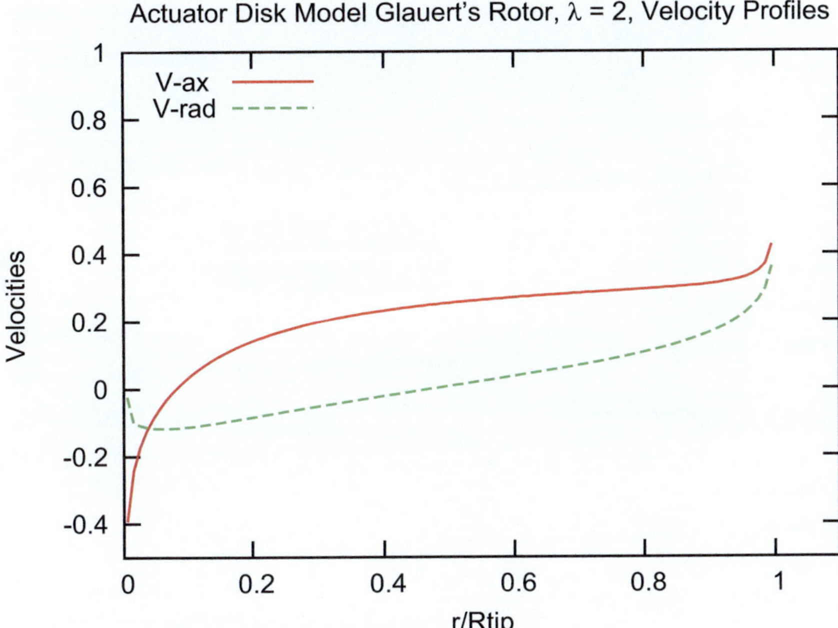

Fig. 7.15 Example of application of the actuator disk method to the Betz rotor

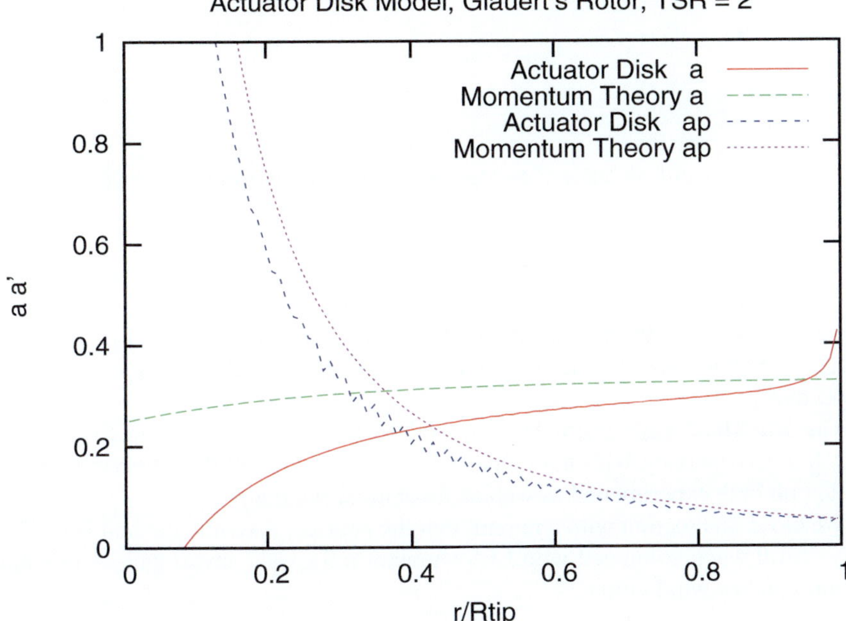

Fig. 7.16 Example of application of the actuator disk method to the Glauert rotor

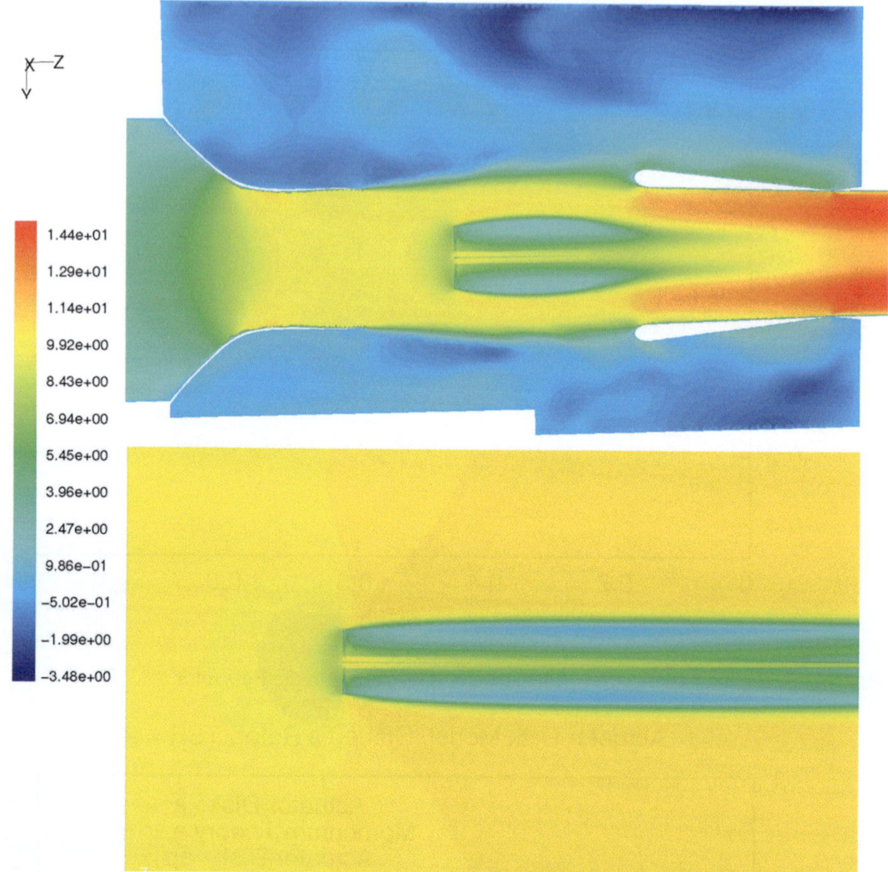

Fig. 7.17 Actuator disk modeling of the Mexico rotor with and without wind tunnel

Shen et al. [53] discuss the same measured data with their actuator line method, also with and without a wind tunnel. They found an influence of the order of 3 % to the axial velocity in the rotor plane. Knowing that thrust $\sim v^2$ and power $\sim v^3$, we may see that variations in these quantities magnify to about 6 % or even 9 %, respectively.

The MEXICO experiment is described in more detail in (Chap. 5) [46] and Sect. 8.6. Comparison with numerical results provided important additional information on both experimental as well as theoretical accuracy.

We close this section with a remark that the actuator disk/line method is particularly useful when combined with LES simulation for wake investigations of single turbines and/or wind farms.

7.7.2 *Improved Models for Finite Number of Blades in BEM Codes*

As we have seen in Chap. 5, there are three main sources of losses for an ideal wind turbine:

- Swirl: > 0
- Finite number of blades: $B < \infty$
- Drag of lift-generating airfoils: $d/L > 0$.

The all important ad hoc Prandtl correction for finite number of blades, Fig. 5.10 and Eq. (5.59), may be re-investigated with the aid of vortex models presented above [8] or with full 3D CFD models [18, 51, 52]. As a result a new model for BEM applications could be formulated but still today there are few agreements on which tip correction gives the most accurate results.

7.7.3 *Actuator Line Method*

As a further example of the usefulness of vortex models, we mention the application of an extension, the *actuator line* [59] method to the MEXICO Sect. 8.6 experiment. Here the forces are distributed on rotating lines modeling the blades. This usually is done using Gaussian smearing, and as a consequence a new parameter ϵ is introduced:

$$\eta_\epsilon(x, y, z) = \frac{1}{\epsilon^3 \pi^{3/2}} \cdot exp(-(x^2 + y^2 + z^2)/\epsilon^2) . \tag{7.71}$$

Consistent determination of this parameter—which unfortunately influences the accuracy of the method considerably—is not as easy as originally thought. See [10, 30].

It must be noted that both Actuator Disks and Lines rely on aerodynamic polars. With some justification AD/AL might be regarded as an extension of BEM. Therefore its shortcomings such as how to formulate tip-correction approaches, for example, remain.

7.8 Full-Scale CFD-Modeling of Wind Turbine Rotors

In this section we discuss some of the results reached so far on applying CFD[3] to full 3D wind turbine rotor flow.

---

[3] From now on in this book the term CFD is assumed to be a RANS simulation where only the full 3D geometry of the wind turbine (or its blades) is used.

7.8.1   NREL Phase VI Turbine

One of the first successful applications of CFD was for the so-called NASA-AMES blind comparison (see Sect. 8.5) around the year 2000. Here a very special rotor was chosen with two blades and a difficult S809 Profile. The main objective was to collect detailed load data from defined operating conditions in a wind tunnel, different and complementary to those gained in the free-field (Chap. 8) [55]. Figure 7.18 shows the shape and Fig. 7.19 measured polar data for Re$= 2 \cdot 10^6$ Fig. 7.19. In Figs. 7.20 and 7.21, two simulations including transition modeling are shown. However the first simulation (using a tabulated e^N method for transition prediction) is not able to capture c_D^{max}. Only one profile was used for the whole rotor, so it was of utmost importance to capture even the 2D deep stall characteristics of this profile for power (or low shaft torque) prediction which is shown in Figs. 7.22 and 7.23. The results from Fig. 7.22 show that including modeling transition produces a correct trend even in the highly separated case ($v_{wind} > 10$ m/s). The same was done by N. Sørensen [61] including typical turbulence intensities (0.5 ... 1.5 %) reported in (Chap. 8) [75]. A remarkably strong dependence on turbulence intensity, especially in the velocity range around 11 m/s, is clearly seen. This emphasizes again that accurate transition modeling is absolutely necessary to accurately predict and understand wind turbine flow on wind turbine blades.

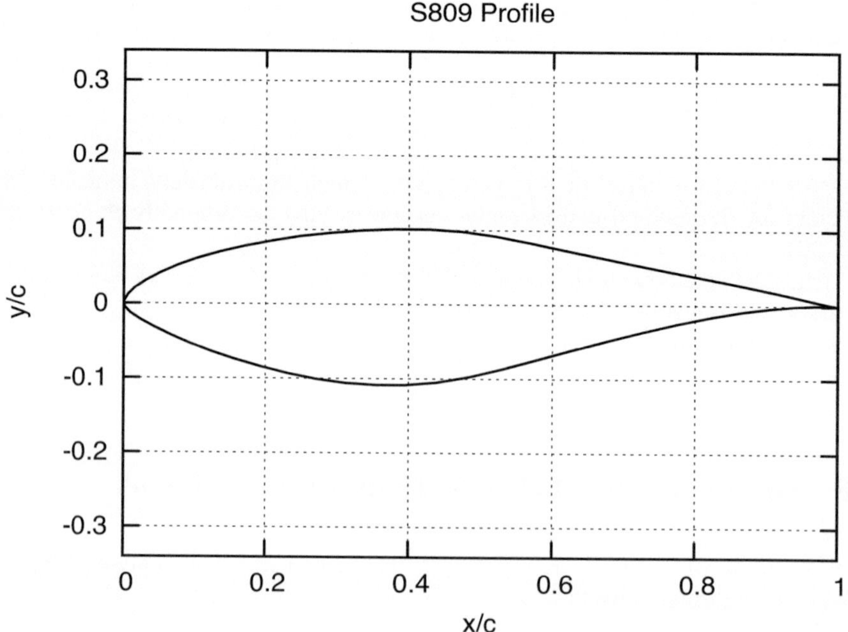

Fig. 7.18 Cross section of S809 airfoil used during the NASA Ames blind comparison

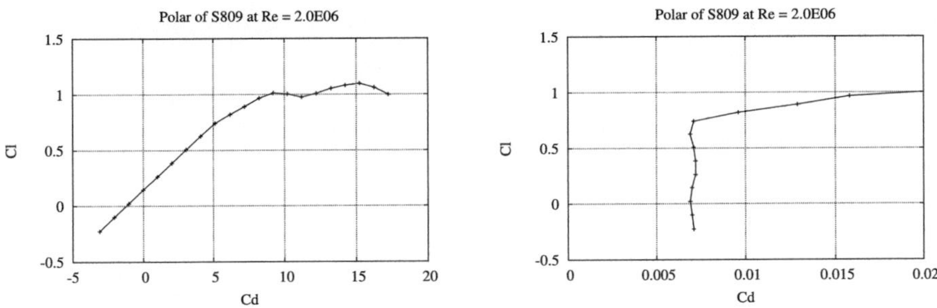

Fig. 7.19 S809: measured polar (Chap. 8) [49, 54]

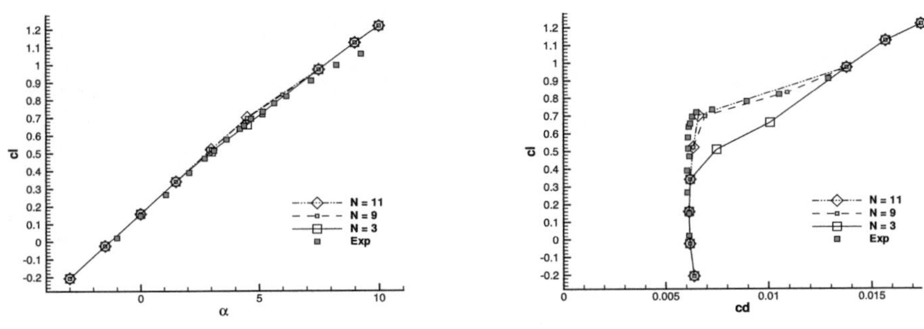

Fig. 7.20 S809: polar for $Re = 3 \cdot 10^6$, $Ma = 0.1$, [14, 72]

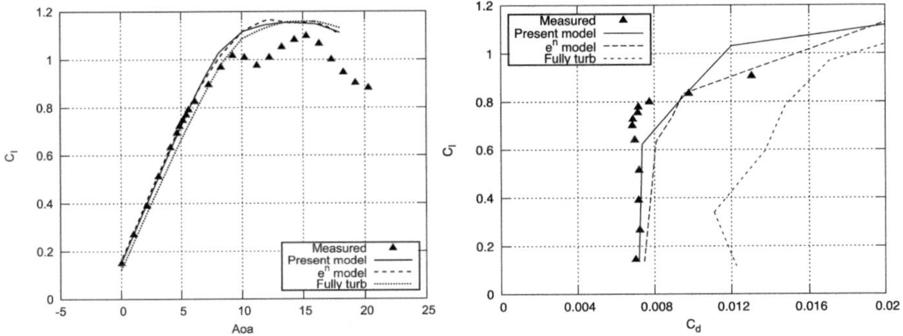

Fig. 7.21 S809: CFD simulation of [61]

7.8.2 The Mexico Rotor

During the IEAwind Annex 20 (2003–2008), the NASA-Ames unsteady experiment and blind comparison data were extensively investigated. In 2006 (see Sect. 8.6), [6] and (Chap. 5) [47] a European-Israel consortium constructed and measured a 3-bladed, 5 m diameter wind turbine in the European LLF wind tunnel (see Fig. 8.24). Shortly after, another IEAwind Annex (29, *MexNext*) was established. It was divided

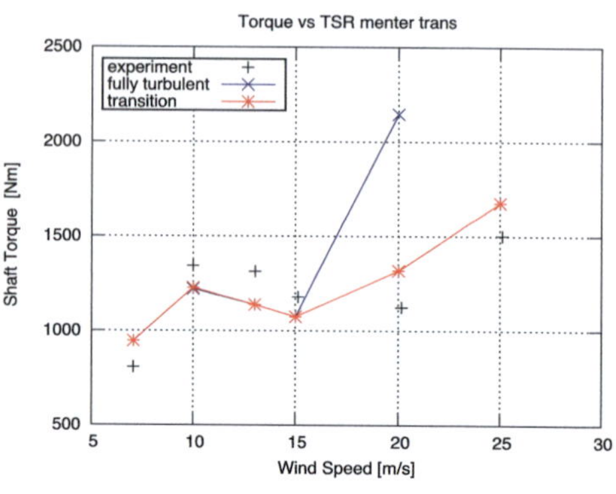

Fig. 7.22 NASA ames experiment. Torque versus Wind speed, measurements, fully turbulent and transitional flow modeling, [24, 34]

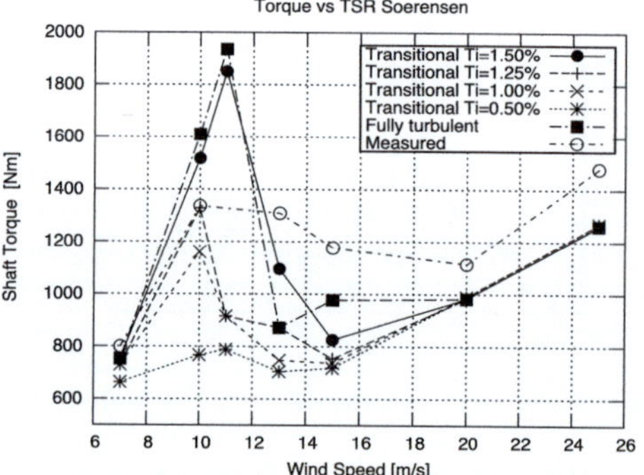

Fig. 7.23 Same as Fig. 7.22 but with different turbulence intensities [61]

into two phases (01.06.2008–01.06.2011 and 1.11.2012–31.12.2014). Extensive CFD (not only one or two as in the NREL unsteady experiment, Phase VI) simulation from at least 7 teams was used to extract useful information from the experiment. The blade was not equipped with only one profile, but 3 (DU91, Risoe-121 and NACA-64-4-18). See Figs. 7.24 and 7.25. A typical outcome is shown in Figs. 7.26 and 7.28 (from (Chap. 5) [46]). Again the results from Risoe-DTU are closest to the measurements. Having had typical mesh sizes of several millions [63] in connection with the IEAwind Annex 20 CFD-modeling, the new mesh sizes grew considerably during the ensuing 10 years from IEAwind Annex 20 to 29: Figs. 7.29 and 7.30 [20] show the Chimera mesh-topology and a typical vortical structure for TSR = 6.7 m/s and $v_{in} = 15$ m/s. The tau-code from DLR , the German Aerospace Association was used. It consists of 150 M cells and used the e^N transition modeling [37]. A typical job takes about 50 h on a 64-node parallel machine. Recently [9] reported a transient wake calculation for the MEXICO-rotor with almost 1 G cells (850 M) with emphasis on investigations of the tip-vortex breakdown phenomenon.

As a last example we show in Fig. 7.31 our full set of applied computational methods to the MEXICO experiment. It shows the same amount of scatter as in Figs. 7.26 and 7.28. Again in the attached flow case (v <15 m/s inflow velocity), agreement for thrust (torque) is on the order of 10 (30)% only. It is further increased

Fig. 7.24 Lift data of
profiles: Cl versus AOA

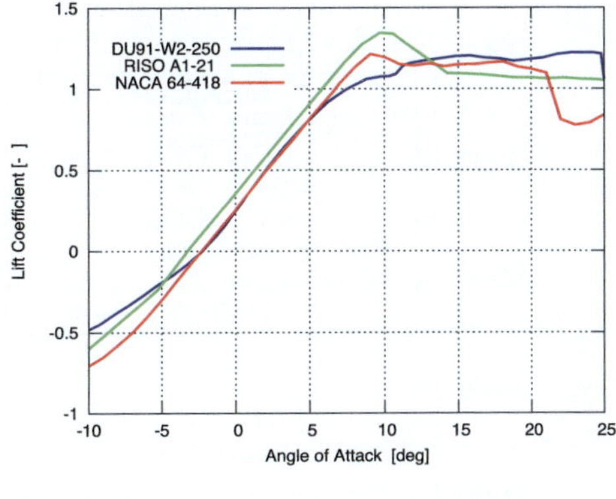

Fig. 7.25 Lift data of
profiles: Cd versus Cl

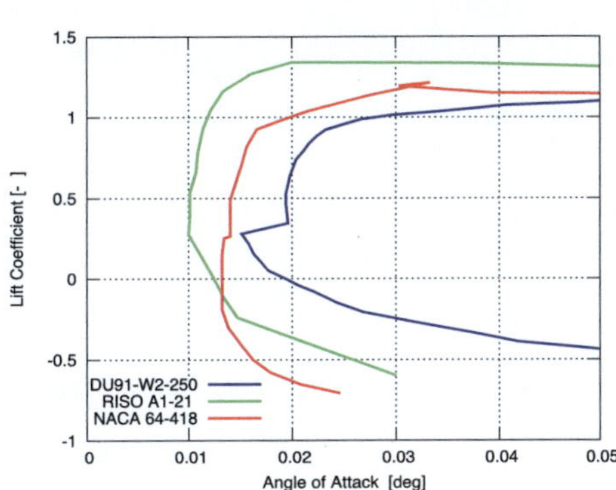

when stall occurs (v >15 m/s). An argumentation on how to explain these findings
could be conducted as follows:

The experiments should be accurate in a range of 3 to 5% [53]. BEM (light blue)
use only the airfoil data given from 2D wind tunnel measurements. It should be
accurate in the attached flow regime. Actuator Disk models with Fluent (red) and
OpenFOAM (green) as flow-solver use 2D airfoil data as well but the flow field is
more complete. In principle this should reduce the error or at least the difference
between measurement and BEM. Full 3D CFD (tau-code, blue) uses only the geometry of the blade and no airfoil data at all. However, as could be seen, life is not that
easy.

The experiment was repeated in 2014 [5], and the results wer evaluated in IEAwind
Annex 29 (MexNext) Phase 2 (2012–2014) [45] and Phase 3 (2015–2015) [7]. Some
of the newer results are incorporated in Fig. 7.13, and some are shown in Fig. 7.27.

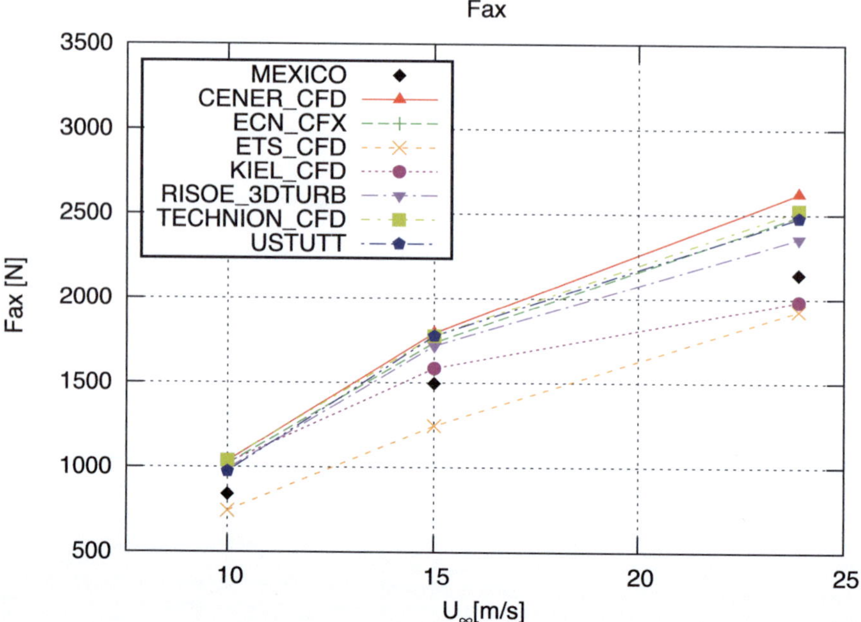

Fig. 7.26 Summary CFD (2014): thrust (axial force), from [46]

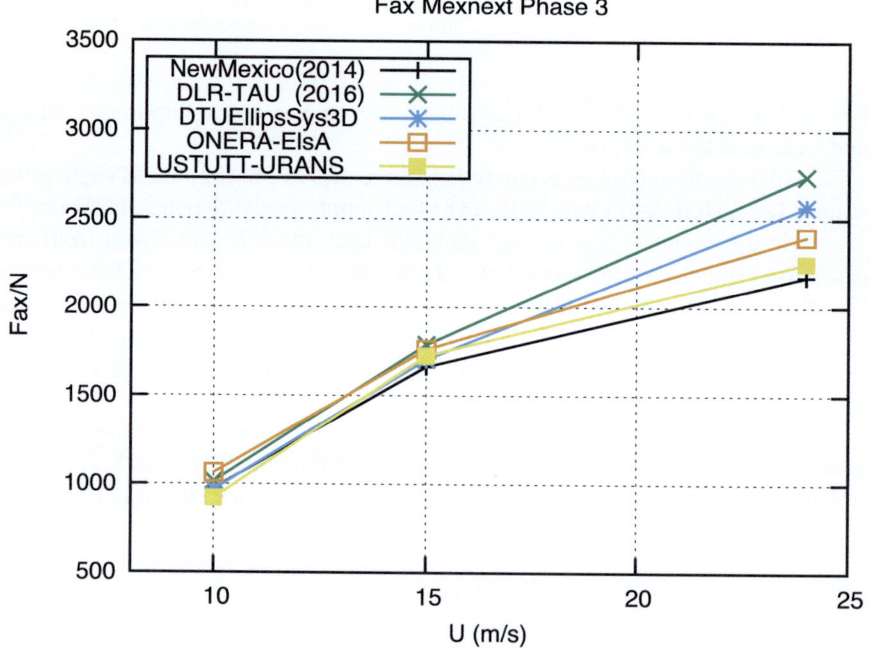

Fig. 7.27 Summary CFD (2019): thrust (axial force), from [46]

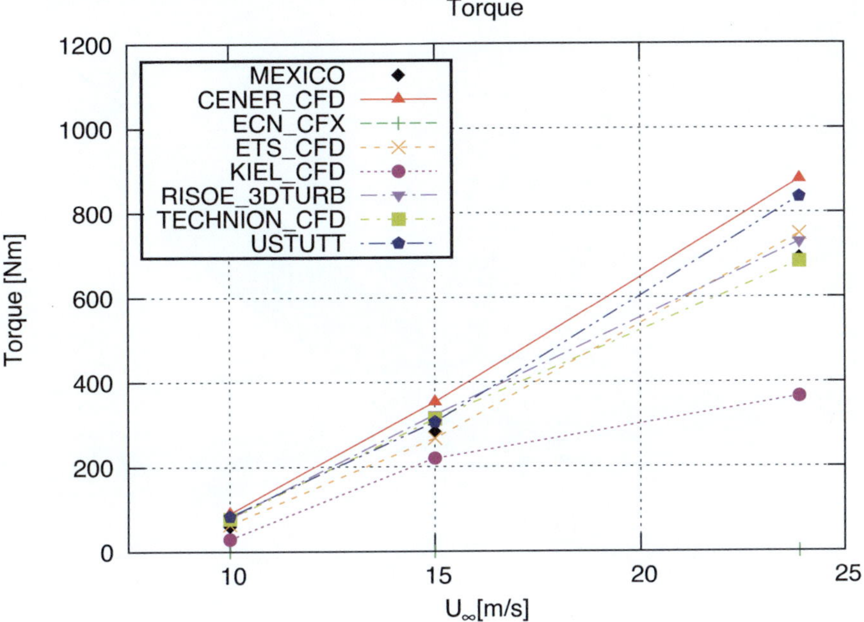

Fig. 7.28 Summary CFD: torque, from [46]

7.8.3 *IEAwind Task 47 TURBulent INflow Innovative Aerodynamics (turbinia)*

In this project (duration form 2021 to 2024) results from the DanAero experiment were further investigated. Recall that the MN80 turbine is equipped with LM38.8 blades. Figure 7.32 [3] shows the variation of the normal force at four different radial stations in sheared inflow (powerlaw with $N = 0.35$) and pitch $\vartheta = 3°$. Some more information can be found in Sect. 7.8.4.4 (Fig. 7.32).

7.8.4 *Commercial Wind Turbines*

We now enter a discussion about the use of CFD for commercial blades. As was stated several times, in the beginning of the first boom of wind turbine use (around 2000) the only reliable tool was BEM, sometimes with adapted *3D-corrected* airfoil data. Therefore it was tempting to apply CFD to these blades.

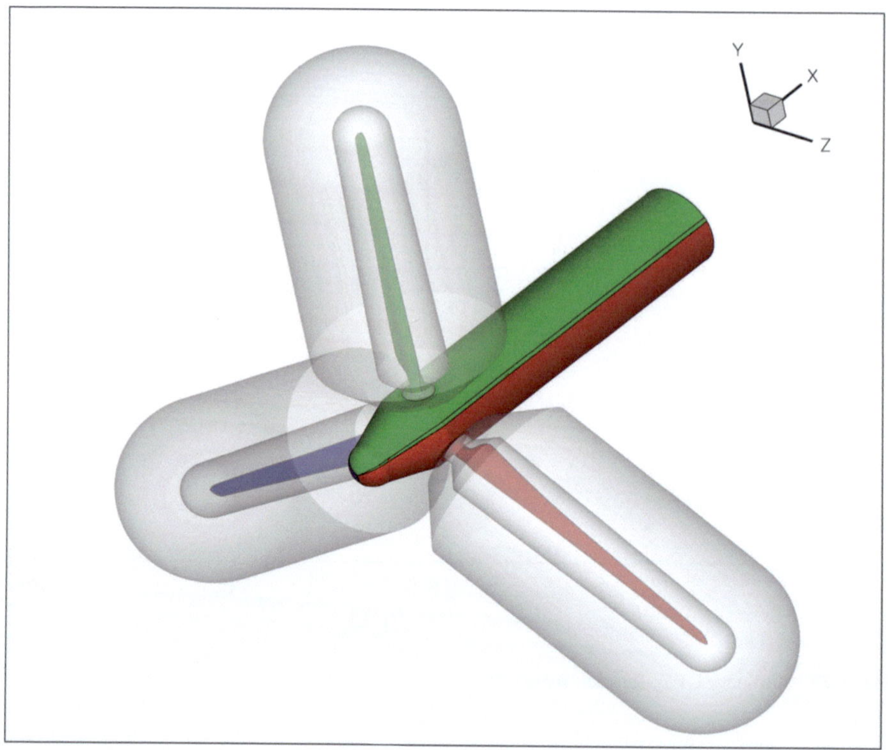

Fig. 7.29 Block structure of a 150 M cell mesh using Chimera technique

7.8.4.1 ARA48 Blades

This blade was designed and developed by aerodyn Energysysteme, Germany, together with Abeking & Rassmussen Rotoc. It was used in a 600/750 German wind turbine from *Husumer Schiffswerft*. Later on other companies, Jacobs Windturbinen, Heide (now Senvion S.A.) and Goldwind, Beijing, China, manufactured several thousands of this sub-Megawatt turbine [43, 50]. Later the ARA blade family was complemented and substituted by LM blades. Figure 7.33 shows two different BEM designs as well as a measured curve. More about standards and how to measure a power curve will be seen in Sect. 8.2. An important conclusion was that CFD could only predict more accurate aerodynamic properties of stall blades if the turbulence model was able to do so. Clearly the model used by [43] (k-ϵ model) was not appropriate [16].

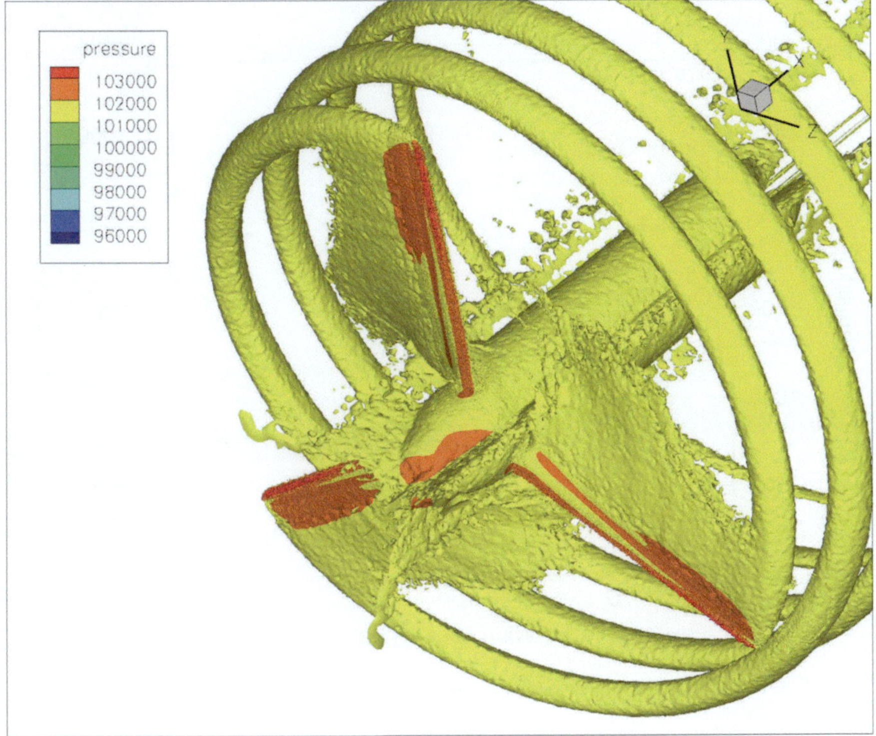

Fig. 7.30 Iso vorticity of the Mexico rotor for TSR of 6.7

7.8.4.2 NTK500

A comparable task for the Nortank NTK500 (stall) turbine [38] was undertaken somewhat earlier with the aid of the actuator disk model [15] and without 2D airfoil data [62] by the Danish group; see Fig. 7.34.

As the machine (and blade) was designed using flow-stall to limit output power, prediction of P_{max} was most important. Shortly after, the NASA Ames experimental data Sect. 8.5 took over becoming the model turbine for CFD-testing.

7.8.4.3 Siemens Wind Turbine

CFD found its way to designers and manufacturers of wind turbines only very slowly. The main cause was the increasing number of so-called load-cases (special conditions for inflow according to design rules) which demand very fast codes. Only BEM in its pure form is able to do so. In addition due to strong competitions, results were published very rarely. One remarkable exception was information disseminated at the second *torque* conference [25]. A summary of results from simulation of a

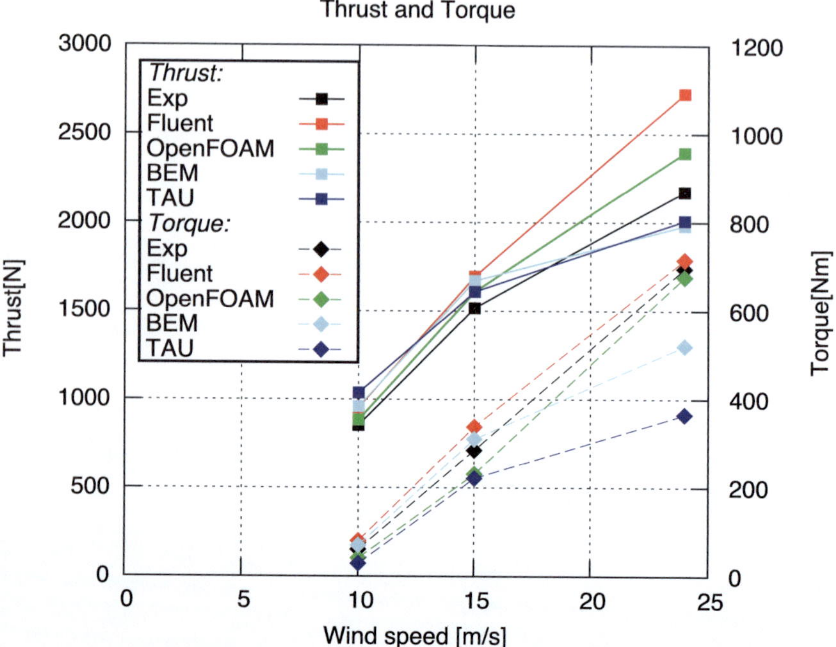

Fig. 7.31 Overview of some CFD methods applied to the MEXICO experiment

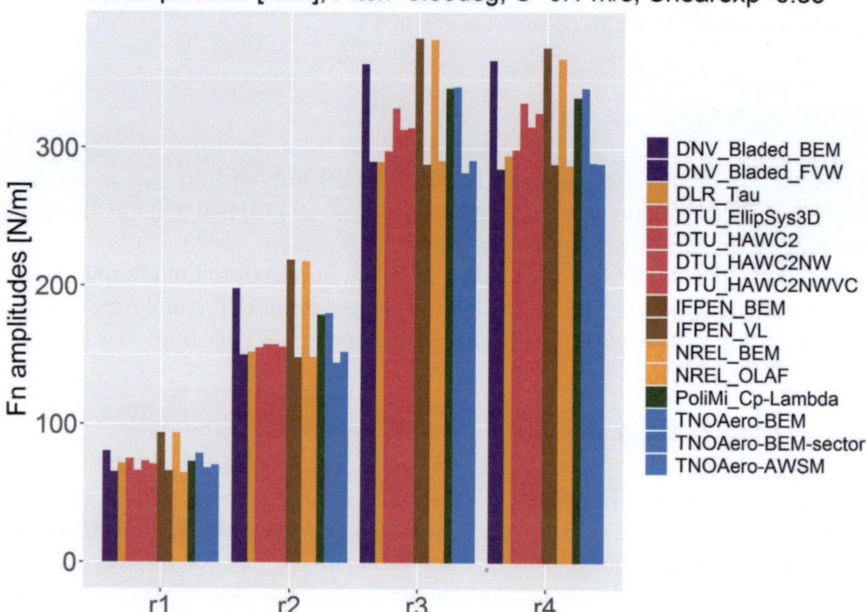

Fig. 7.32 Normal force from a variety of codes

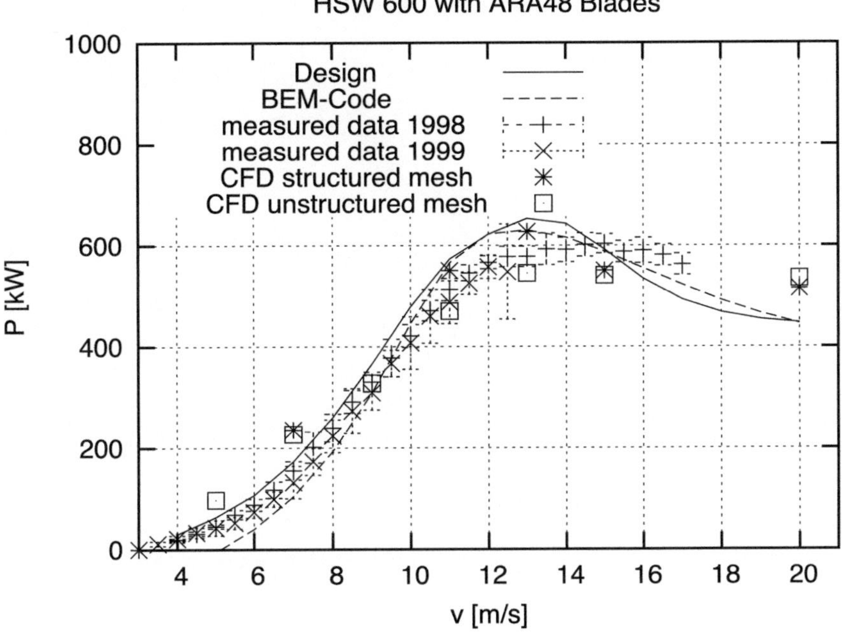

Fig. 7.33 Computed and measured power curve for a 400 kW stall machine, HSW 600 [43]

Table 7.2 Mechanical Power (kW) prediction for a commercial (SWT-2.3-93) Wind Turbine, from [25]

Wind (m/s)	RPM	Measured	ANSYS, fully turb	ANSYS, transitional	Ellipsis	BEM-code
6	10,5	400	396	428	392	408
8	13,5	986	950	–	945	967
10	16,0	1894	1853	1977	1850	1850
11	16,0	2323	2388	–	–	–

commercial 2.3 MW (93 m rotor diameter) is given in Table 7.2. ANSYS (CFX) is a commercial CFD code, and Ellipsis is the research code developed by Risø-DTU. The deviations seem to be on the order of few percentages only. Remarkably, as also seen in the cases discussed above, transition modeling does not improve accuracy in all cases. Clearly much more comparisons have to be undertaken. A somewhat similar comparison (using a different turbine) and using ANSYS-CFX and OpenFOAM was performed by Dose [11]. His findings agree with those from [25], with the exception that an OpenFOAM run needs much more CPU time (more than a factor of 10) so far.

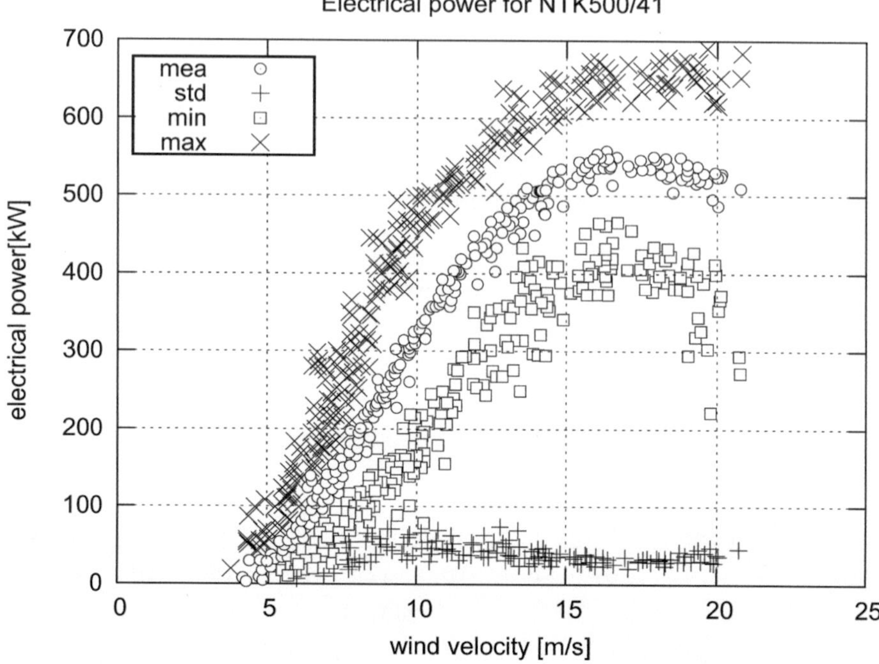

Fig. 7.34 Computed and measured power curve for a 400 kW stall machine, NTK 500 (Chap. 3) [17]

7.8.4.4   NM80 Wind Turbine—IEAwind Task 29, Phase 4

After Phase 3 (2012–2017) IEAwind Task 29 was renamed to *Analysis of Aerodynamic Measurements* and extended to a duration from 2018 to 2020. Based on the DAN-AERO experiment [1, 27, 28] (Chap. 8), measured data was made available to the participants and BEM as well as CFD calculations were carried out. The turbine has a rotor diameter of 80 m and blades from LM (LM38.8) were used. Figure 7.35 shows a preliminary comparison from CFD codes (first seven columns), BEM (next) and an estimation of measured power (last red box). Agreement between the first six codes is within few percent. Results from Column 7 are from a 3D panel code FUNAERO [4].

7.9   Concluding Remarks About Use (and Abuse) of CFD

This chapter could not be finished without some remarks on the state of the art and future progress on CFD methods applied to wind turbines. Clearly CFD—defined as numerical solution of the NAST including some turbulence modeling—is the only

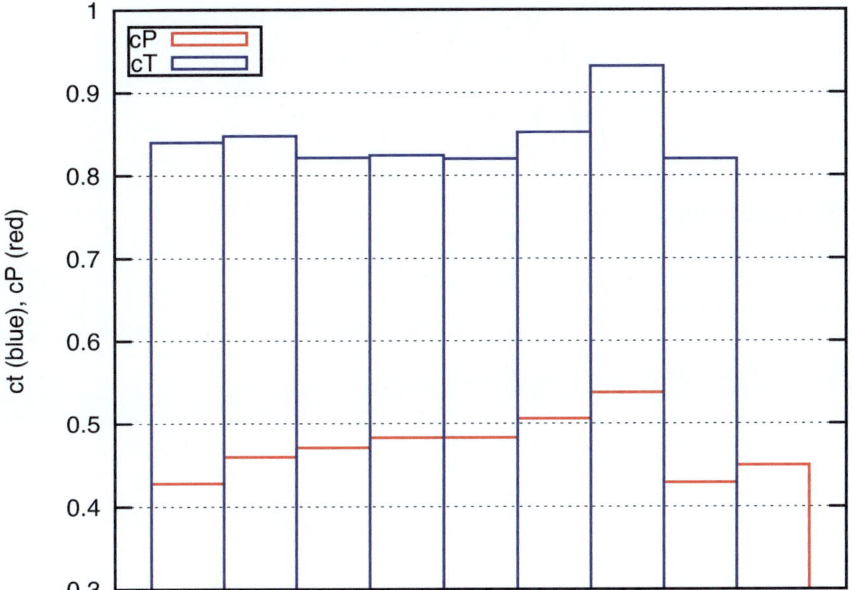

Fig. 7.35 Results from a comparison of CFD calculation for the NM80 rotor. Inflow is 6.1 m/s and RPM = 12.3

way to improve ordinary BEM. As reviewed in [67], two-equation models have a long history, only slightly surpassed in recent times by LES and hybrid RANS/LES models [66]. So by now the accuracy of any CFD calculation is limited by the (sometimes accidental) accuracy of the turbulence model. As explained in Chap. 5 many researchers agree (Chap. 5) [58] that only a deeper understanding of the nature of turbulence and the mathematical properties of the Navier-Stokes equations will lead to controlled improved turbulence models for use in wind turbine aerodynamic applications.

The only independent way for assessment is the use of experiments which will be described and discussed in the following Chap. 8.

7.10 Problems

Problem 7.1 Turbulent boundary layers for $20 < y^+ < 200$ show a logarithmic velocity profile:

$$u^+ = \frac{1}{\chi} \cdot ln(y^+) + 5.5, \quad \chi = 0.41. \tag{7.72}$$

Then k-ϵ simplifies to

$$-\partial_y \left(c_\mu \frac{k^2}{\epsilon} \frac{\partial k}{\partial y} \right) - c_\mu \frac{k^2}{\epsilon} E + \epsilon = 0 \,, \tag{7.73}$$

$$-\partial_y \left(c_\epsilon \frac{k^2}{\epsilon} \frac{\partial k}{\partial y} \right) - c_1 k E + c_2 \frac{\epsilon^2}{k} = 0 \,. \tag{7.74}$$

Show that with

$$k(y) \;=\; \frac{u^{*2}}{\sqrt{c_\mu}} \,, \tag{7.75}$$

$$\epsilon(y) = \frac{u^{*3}}{\sqrt{\chi y}} \ \text{and} \tag{7.76}$$

$$E(y) \;=\; \frac{u^{*2}}{\chi^2 y^2} \tag{7.77}$$

the above equations may be solved only if

$$c_\epsilon = \frac{\sqrt{c_\mu}}{\chi^2} \left(c_2 c_\mu - c_1 \right) (= 0.08) \,. \tag{7.78}$$

Remark: $u^+ := U(y)/u^*$, $y^+ := y/y^*$, $y^* = v/u^*$, $u^* = v = const.$

Problem 7.2 Decaying homogeneous turbulence.

In case $\mathbf{u}$ and $\nabla \mathbf{u}$ both are zero, the model k-ϵ-equations reduce to

$$\partial_t k + \epsilon = 0 \,, \tag{7.79}$$

$$\partial_t \epsilon + c_2 \frac{\epsilon^2}{k} = 0 \,. \tag{7.80}$$

Show that a polynomial decay in time:

$$k(t) = k_0 \cdot (1 + \lambda t)^{-n} \,, \tag{7.81}$$

$$\epsilon(t) = \epsilon_0 \cdot (1 + \lambda t)^{-m}, \tag{7.82}$$

obeys the equation. Use n = 1.3 from experiments to find a value for c_2.

Problem 7.3 Falkner-Skan Flow with simple Fortran code. At the edge of the boundary layer $u_e = u_0 \cdot (x/L)^m$. Define $\beta := 2m/(1+m)$. With

$$\eta = y\sqrt{\frac{u_0(m+1)}{2vL}} \left(\frac{x}{L}\right)^{\frac{m-1}{2}} \tag{7.83}$$

the BL DEQ reads

$$f''' + f \cdot f'' + \beta \left(1 - f'^2\right) = 0 \,. \tag{7.84}$$

Use the simple code (Chap. 3) [70] from Fig. 7.36 for solving.

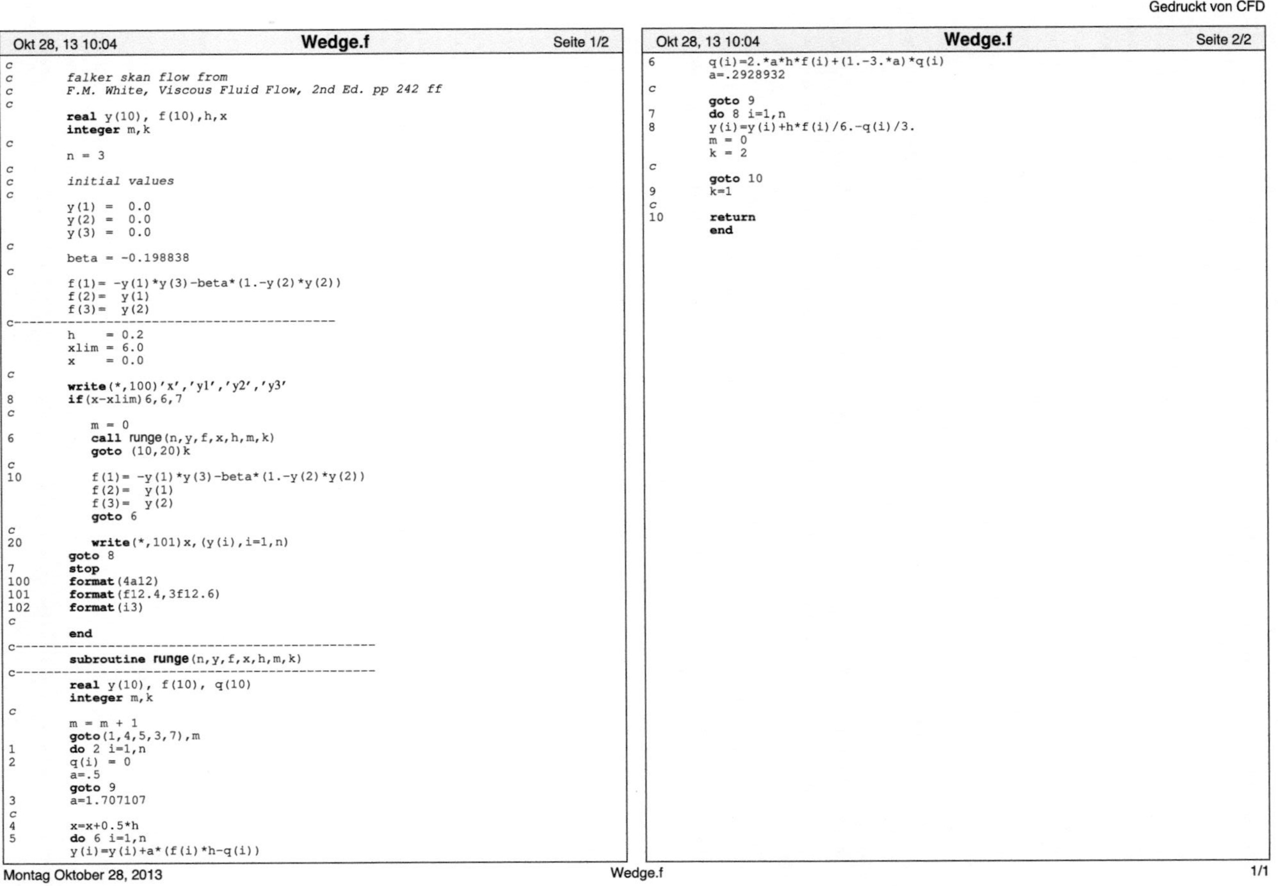

```
Okt 28, 13 10:04                    Wedge.f                    Seite 1/2
c
c     falker skan flow from
c     F.M. White, Viscous Fluid Flow, 2nd Ed. pp 242 ff
c
      real y(10), f(10),h,x
      integer m,k
c
      n = 3
c
c     initial values
c
      y(1) =  0.0
      y(2) =  0.0
      y(3) =  0.0
c
      beta = -0.198838
c
      f(1)= -y(1)*y(3)-beta*(1.-y(2)*y(2))
      f(2)=  y(1)
      f(3)=  y(2)
c-------------------------------------------
      h    = 0.2
      xlim = 6.0
      x    = 0.0
c
      write(*,100)'x','y1','y2','y3'
8     if(x-xlim)6,6,7
c
      m = 0
6     call runge(n,y,f,x,h,m,k)
      goto (10,20)k
c
10    f(1)= -y(1)*y(3)-beta*(1.-y(2)*y(2))
      f(2)=  y(1)
      f(3)=  y(2)
      goto 6
c
20    write(*,101)x,(y(i),i=1,n)
      goto 8
7     stop
100   format(4a12)
101   format(f12.4,3f12.6)
102   format(i3)
c
      end
c-------------------------------------------
      subroutine runge(n,y,f,x,h,m,k)
c-------------------------------------------
      real y(10), f(10), q(10)
      integer m,k
c
      m = m + 1
      goto(1,4,5,3,7),m
1     do 2 i=1,n
2     q(i) = 0
      a=.5
      goto 9
3     a=1.707107
c
4     x=x+0.5*h
5     do 6 i=1,n
      y(i)=y(i)+a*(f(i)*h-q(i))
Montag Oktober 28, 2013                              Wedge.f


Okt 28, 13 10:04                    Wedge.f                    Seite 2/2
6     q(i)=2.*a*h*f(i)+(1.-3.*a)*q(i)
      a=.2928932
c
      goto 9
7     do 8 i=1,n
8     y(i)=y(i)+h*f(i)/6.-q(i)/3.
      m = 0
      k = 2
c
      goto 10
9     k=1
c
10    return
      end
                                                               1/1
```

Gedruckt von CFD

Fig. 7.36 FORTRAN Code for Falkner-Skan Flow

Table 7.3 Coordinates for DU-W-300-mod

x/c	y/c	x/c	y/c	x/c	y/c	x/c	y/c
1.000000	0.000000	0.159531	0.125199	0.001989	−0.016639	0.279441	−0.165530
0.991346	0.002458	0.150115	0.123207	0.002724	−0.019462	0.290216	−0.164664
0.977819	0.006263	0.141097	0.121057	0.003584	−0.022332	0.301248	−0.163440
0.963559	0.010090	0.132472	0.118773	0.004565	−0.025252	0.312578	−0.161850
0.948539	0.013944	0.124231	0.116351	0.005673	−0.028222	0.324247	−0.159870
0.932787	0.017867	0.116366	0.113822	0.006920	−0.031240	0.336318	−0.157508
0.916224	0.021898	0.108870	0.111183	0.008300	−0.034310	0.348825	−0.154758
0.898905	0.026052	0.101728	0.108456	0.009817	−0.037435	0.361820	−0.151588
0.880975	0.030297	0.094937	0.105634	0.011477	−0.040615	0.375355	−0.148006
0.862597	0.034583	0.088475	0.102750	0.013287	−0.043851	0.389491	−0.143981
0.843925	0.038881	0.082343	0.099782	0.015248	−0.047149	0.404275	−0.139526
0.824935	0.043199	0.076513	0.096773	0.017361	−0.050511	0.419747	−0.134619
0.805770	0.047494	0.070988	0.093707	0.019636	−0.053939	0.435927	−0.129310
0.786591	0.051727	0.065750	0.090603	0.022078	−0.057432	0.452772	−0.123594
0.767284	0.055925	0.060785	0.087473	0.024701	−0.060990	0.470261	−0.117492
0.747636	0.060133	0.056091	0.084311	0.027514	−0.064609	0.488327	−0.111078
0.727658	0.064355	0.051649	0.081134	0.030525	−0.068291	0.506850	−0.104353
0.707496	0.068545	0.047455	0.077944	0.033739	−0.072040	0.525745	−0.097433
0.687207	0.072701	0.043491	0.074750	0.037163	−0.075853	0.544833	−0.090427
0.666950	0.076790	0.039759	0.071550	0.040815	−0.079720	0.564031	−0.083357
0.646893	0.080764	0.036237	0.068361	0.044700	−0.083639	0.583174	−0.076325
0.627015	0.084639	0.032925	0.065176	0.048834	−0.087601	0.602089	−0.069419
0.607140	0.088433	0.029805	0.062010	0.053224	−0.091602	0.620884	−0.062634
0.587170	0.092169	0.026875	0.058862	0.057878	−0.095632	0.639535	−0.055977
0.567157	0.095824	0.024128	0.055734	0.062809	−0.099682	0.658011	−0.049507
0.547197	0.099394	0.021557	0.052627	0.068024	−0.103741	0.676178	−0.043263
0.527321	0.102858	0.019155	0.049543	0.073527	−0.107800	0.693952	−0.037306
0.507499	0.106222	0.016908	0.046491	0.079329	−0.111843	0.711285	−0.031676
0.487780	0.109452	0.014820	0.043464	0.085432	−0.115858	0.728146	−0.026381
0.468108	0.112505	0.012876	0.040472	0.091832	−0.119839	0.744508	−0.021448
0.448540	0.115431	0.011080	0.037507	0.098528	−0.123764	0.760365	−0.016889
0.429194	0.118206	0.009427	0.034572	0.105522	−0.127617	0.775738	−0.012722
0.410155	0.120805	0.007911	0.031668	0.112804	−0.131382	0.790647	−0.008936
0.391510	0.123195	0.006526	0.028797	0.120371	−0.135039	0.805139	−0.005536
0.373327	0.125381	0.005275	0.025956	0.128212	−0.138565	0.819249	−0.002527
0.355667	0.127331	0.004160	0.023144	0.136323	−0.141944	0.833003	0.000131
0.338577	0.129045	0.003180	0.020361	0.144689	−0.145162	0.846429	0.002425
0.322114	0.130510	0.002330	0.017608	0.153290	−0.148215	0.859546	0.004367
0.306299	0.131712	0.001625	0.014883	0.162111	−0.151081	0.872385	0.005956
0.291143	0.132654	0.001066	0.012187	0.171136	−0.153743	0.884978	0.007192
0.276635	0.133321	0.000640	0.009524	0.180335	−0.156192	0.897349	0.008098
0.262759	0.133722	0.000331	0.006899	0.189687	−0.158416	0.909509	0.008675
0.249473	0.133830	0.000134	0.004315	0.199181	−0.160382	0.921472	0.008933
0.236726	0.133641	0.000026	0.001774	0.208793	−0.162099	0.933261	0.008858
0.224457	0.133149	0.000005	−0.000720	0.218516	−0.163537	0.944897	0.008433
0.212625	0.132392	0.000081	−0.003244	0.228346	−0.164679	0.956443	0.007607
0.201202	0.131376	0.000249	−0.005824	0.238282	−0.165522	0.967976	0.006337
0.190187	0.130133	0.000513	−0.008454	0.248333	−0.166030	0.979640	0.004523
0.179567	0.128679	0.000887	−0.011135	0.258532	−0.166201	0.991639	0.002067
0.169349	0.127025	0.001376	−0.013863	0.268893	−0.166045	1.000000	0.000000

Problem 7.4 DU-W-300-mod

(a) Install XFoil or JavaFoil on your computer. For the profile of Table 7.19:
(b) Extract a polar: c_L versus c_D for Re $= 1.15$, 3.6 and $10 \cdot 10^6$.
(c) Compare to measured data of Figs. A.31 and A.32 (Table 7.3).

Problem 7.5 (a) Install a CFD (RANS) code on your computer. For the same profile
(b) extract a polar (c_L versus c_D for Re $= 1.2$, 5.8 and $10.2 \cdot 10^6$.
(c) Compare it to measured data of Figs. A.31 and A.32.

References

1. Aagaard Madsen H et al (2009) The ADN-AERO MW experiment final report. Technical report Risø-R-1726(EN), Risø National Lab, Roskilde, Denmark
2. Anderson D (1995) Computational fluid dynamics. McGraw-Hill, New York, USA
3. Boorsma K (2023) Comparison rounds. In: IEAwind Task 47, annual meeting, Glasgow, UK. TNO, The Netherlands
4. Boorsma K, Greco L, Bedon G Rotor (2018) wake engineering models for aeroelastic applications. In: S JPC (ed) 062013, Proc TORQUE2018. Insitute of Physics
5. Boorsma K, Schepers G (2014) New Mexico experiment. Technical report ECN-E–14-048, Petten,The Netherlands
6. Boorsma K, Schepers JG (2011) Description of experimental setup—mexico measurements. Technical report ECN-X–11-120, ECN, Petten, The Netherlands
7. Boorsma K, Schepers JG, Gomez-Iradi S, Herraez I, Lutz T, Weihing P, Oggiano L, Pirrung G, Madsen HA, Shen WZ, Rahimi H, Schaffarczyk AP (2018) Final report of iea wind taks 29 mexnext (phase 3). Technical report ECN-E–18-003, Petten, The Netherlands
8. Branlard E (2013) Wind turbine tip-loss correction. Master's thesis, Master's thesis (public version), Risø DTU, Copenhagen, Denmark
9. Carrion M (2013) Understanding wind turbine wake breakdown using CFD. In: 3rd IEAwind MexNext. Pamplona, Spain
10. Davg KO, Sørensen J (2020) A new tip correction for actuator line computations. Wind Energy 23(2):148–160
11. Dose B (2013) CFD simulations of a 2.5 mw wind turbine using ansys CFX and openfoam. Master's thesis, Kiel University of Applied Sciences, Kiel , Germany
12. Ferziger JH, Perić M (2002) Computational methods for fluid dynamics, 3rd ed. Springer, Berlin
13. Fletcher CAJ (2005) Computational techniques for fluid dynamics. Springe, Berlin
14. Freudenreich K, Kaiser K, Schaffarczyk AP, Winkler H, Stahl B (2004) Reynolds number and roughness effects on thick airfoils for wind turbines. Wind Eng. 28(5):529–546
15. Hansen MEA (1997) A global navier-stokes rotor prediction model. AIAA 97–0970
16. Hansen MOL (1998) Private communication. Priv comm
17. Hansen MOL (2008) Aerodynamics of wind turbines, 2nd ed. Earthscan, London, UK
18. Hansen MOL, Johansen J (2004) Tip studies using CFD and comparison with tip loss models. In: Proceedings of 1st conferences on the science of making torque from wind, Delft, The Netherlands (2004)
19. Ishihara T, Gotoh T, Kaneda Y (2009) Study of high-reynolds number isotropic turbulence by direct numerical simulation. Annu Rev Fl Mech 41:165–180
20. Jeromin A, Schaffarcyzk AP (2012) First steps in simulating laminar-turbulent transition on the mexico blades. In: 2nd IEAwind MexNext Meeting, NREL, Golden, CO, USA (2012)
21. Jiyuan TU, Yeoh GH, Chaoqun L (2008) Computational fluid dynamics. Butterworth-Heinemann, Elsevier, Amsterdam, The Netherlands

22. Kolmogorov AN (1942) Equations of turbulent motion of an incompressible fluid. Izv Akad Nauk SSSR Ser Fiz VI(1,2):56–58
23. La Y et al (2008) A public turbulence database cluster and applications to study lagrangian evolution of velocity increments in turbulence. J Turb 9(31):1–20
24. Langtry RB (2006) A correlation-based transition model using local variables for unstructured parallelized CFD codes. PhD thesis, Dissertation, Universität Stuttgart
25. Laursen J, Enevoldsen P, Hjort S (2007) 3d CFD quantification of the performance of a multi-megawatt wind turbine. In: The science of making Torque from Wind, Denmark. Danish Technical University
26. Lobo BA, Boorsma K, Schaffarczyk AP (2018) Investigation into boundary layer transition on the mexico blade. In: CSJPCS I (eds). 052020, Proceedings of TORQUE2018
27. Madsen HA, Bak C (2012) The dan-aero mw experiment. Technical Report IEAwind Annex 29 (MeNext) annual meeting, NREL, Golden, CO, USA (2012)
28. Madsen HÅ et al (2010) The dan-aero mw experiment. Technical Report AIAA-2010-645, AIAA, Orlando, FL, USA
29. Mahmoodi E, Schaffarczyk AP (2012) Actuator disc modeling of the mexico rotor experiment. In: Oldenburg U (ed), Proceedings of the Euromech Coll.: wind Energy and the impact of turbulence on the conversion process, Oldenburg, Germany. University of Oldenburg
30. Martinez-Tossa LA, Meneveau C (2019) Filtered lifting line theory and application to the actuator line method. J Fluid Mech 863:269–292
31. McComb WD (1992) The physics of fluid turbulence. Clarendon Press, Oxford, UK
32. Menter F (1992) Improved two-equation k-ω turbulence models for aerodynamical flows. Technical report 103975, NASA Technical Memorandum, Moffett Field, CA, USA
33. Menter F (1994) Two-equation eddy-viscosity turbulence models for engineering applications. AIAA—J 32(8):1598–1605
34. Menter F, Langtry R (2006) Transitionsmodellierungen technischer strömungen (modeling of transitions in engineering flow). Technical report, ANSYS Germany, Otterfing, Germany
35. Michelassi V, Rodi W, Zhu J (1993) Testing a low reynolds number k-ϵ model based on direct simulation data. AIAA J 31(9):1720–1723
36. Mohammandi B, Pirronneau O (1994) Analysis of the K-Epsilon turbulence model. Wiley, Chichester, Uk
37. NN (2009) Transition module (v8.76) user guide (v1.0 beta)
38. Paulsen US (1995) Konceptundersøgelse nordtanks 500/41 strukturelle laster. Technical Report Risø-I-936(DA), Risø National Lab, Roskilde, Denmark
39. Pope SB (2000) Turbulent FLows. Cambridge University Press, Cambridge, UK
40. Prandtl L (1925) Bericht über untersuchungen zur ausgebildeten turbulenz. Z Angew Math und Mech 5
41. Prandtl L (1945) über ein neues formelsystem für die ausgebildete turbulenz. Nachr d Akad d Wiss in Göttingen, Math-nat Klasse, S, 6–20
42. Sanders B, van der Pijl SP, Koren B (2011) Review of computational fluid dynamics for wind turbine wake aerodynamics. Wind Energy 14:799–819
43. Schaffarczyk AP (1997) Numerical and theoretical investigation for wind turbines. In: Petten TNECN (ed), IEAwind, Annex X, Dec 1997
44. Schaffarczyk AP, Boisard R, Boorsma K, Dose B, Lienard C, Lutz T, Madsen HA, Rahimi H, Reichstein T, Schepers G, Sørensen N, Stoevesand B, Weihig P (2018) Comparison of 3d transitional cfd simulations for rotating wind turbine wings with measurements. In: I C S J P C S (eds). 022012, Proceedings of the TORQUE2018
45. Schepers JG, Boorsma K, Gomez-Iradi S, Schaffarczyk AP, Madsen HA, Sørensen NN, Shen WZ, Lutz T, Schulz C, Herraez I, Schreck S (2014) Final report of IEA wind task 29 mexnext (phase 2). Technical Report ECN-E–14-060, ECN, Petten, The Netherlands
46. Schepers JG et al (2012) Final report of iea task 29, mexnex (phase 1): analysis of mexico wind tunnel measurements. Technical report, ECN-E-01–12-004, ECN, The Netherlands
47. Schepers JG, Snel H (2007) Model experiments in controlled conditions, final report. Technical report, ECN-E-07-042, ECN, The Netherlands

48. Schlatter P (2005) Large-Eddy simulation of transitional turbulence in wall-bounded shear flow. PhD thesis, ETH, Zürich, Switzerland
49. Schreck S (2008) IEA wind annex xx: hawt aerodynamics and models from wind tunnel measurements. Technical report NREL/TP-500-43508, NREL, Golden, CO, USA
50. Schubert M, Schumacher K (1996) Entwurf einer neuen aktiv-stall rotorblattfamilie (design of a new family of active stall blades). In: Molly J (ed), Proc DEWEK, Wilhelmshaven, Germany. Deutsches Windenergie Institut
51. Shen WZ, Mikkelsen R, Sørensen JN, Bak C (2003) Validation of tip corrections for wind turbine computations. In: Madrid S (ed), Proceedings of the EWEC
52. Shen WZ, Mikkelsen R, Sørensen JN, Bak C (2005) Tip loss corrections for wind turbine computations. Wind Energy 8:457–475
53. Shen WZ, Zhu WJ, Sørensen JN (2012) Actuator line/navier-stokes computations for the mexico rotor: comparison with detailed measurements. Wind Energy 15(5):811–825
54. Simms D, Schreck S, Hand M, Fingersh LJ (2001) Nrel unsteady aerodynamics experiment in the nasa-ames wind tunnel: a comparison of predictions to measurements. Technical Report NREL/TP-500-29494, NREL, Golden, CO, USA
55. Simms DA, Hand MM, Fingersh LJ, Jager DW (1999) Unsteady aerodynamics experiment phases II–IV, test configurations and available data campaigns. Technical report NREL/TP-500-25950, NREL, Golden, CO, USA
56. Smith LM, Reynolds WC (1992) On the yakhot-orzag renormalization group method for deriving turbulence statistics and models. Phys Fluids 4:364 ff
57. Smith LM, Reynolds WC (1998) Renormalization group analysis of turbulence. Ann Rev Fluid Mech 30:275–310
58. Sørensen JN (2011) Aerodynamic aspects of wind energy conversion. Annual Rev Fl Mech 43:427–448
59. Sørensen JN, Shen WZ (2002) Numerical modelling of wind turbine wakes. J Fl Eng 124(2):393–399
60. Sørensen NN (1998) Hypgrid2d a 2-d mesh generator. Technical report Risø-R-1035(EN), Risø National Lab, Roskilde, Denmark
61. Sørensen NN (2009) CFD modeling of laminar-turbulent transition for airfoils and rotors using the $\gamma - \tilde{Re}_\theta$ mode. Wind energy 12:8:715–733
62. Sørensen NN, Hansen MOL (1998) Rotor performance prediction using a Navier-Stokes method. AIAA-98-0025, Jan 1998
63. Sørensen NN, Michelsen JA, Schreck S (2002) Navier-stokes prediction of the NREL phase vi rotor in the NASA Ames 80 ft × 120 ft wind tunnel. Wind energy 5:151–168
64. Spalart P (1988) Direct simulation of a turbulent boundary layer up to $r_\theta = 1410$. J Fl Mech 187:61–98
65. Spalart P (2000) Strategies for turbulence modelling and simulations. Int J Heat Fl Flow 21:225–263
66. Spalart P (2009) Detached-eddy simulation. Annu Rev Fl Mech 41:181–202
67. Spalding B (1991) Kolmogorov's two-equation model of turbulence. Proc R Soc 434:211–216
68. Trede R (2003) Entwicklung eines netzgenerators. Master's thesis, FH Westküste, Heide, Germany
69. von Neumann J (1949) Recent theories of turbulence. Technical report VI, The Naval Office, 1949. Collected Work, Ulam S (ed), pp 437–472
70. White FM (2005) Viscous fluid flow, 3rd edn. Mc Graw Hill, New York, USA
71. Wilcox DC (1993/94) Turbulence modeling for CFD. DCW Industries Inc
72. Winkler H, Schaffarczyk AP (2003) Numerische simulation des reynoldszahlverhaltens von dicken aerodynamischen profilen für off-shore anwendungen. Technical report 33, Kiel University of Applied Sciences, Kiel, Germany
73. Wu YT, Port-Agel F (2013) Modeling turbine wakes and power losses within a wind farm using les: an application to the horns-rev offshore wind farm. In: Proceedings of the ICOWES, Copenhagen, Denmark, 2013. Technical University of Denmark

74. Yakhot V, Orszag SA (1986) Renormalization group analysis of turbulence. I. Basic theory. J Sci Comp 1:1
75. Zell PT (1993) Performance and test section flow characteristics oft he national full-scale aerodynamics complex 80- by 120-foot wind tunnel. Technical report NASA Technical Memorandum, 103920, NASA, Moffett Field, CA, USA

Chapter 8
Experiments

Everybody believes in measurements - except the experimentalist.
Nobody believes in theory - except the theorist.[1]

This chapter describes methods and results that have been applied to gain insight into the physics of wind turbine flow by performing experiments.

8.1 Measurements of 2D Airfoil Data

As we have explained extensively in the previous chapters the blade-element-momentum method describes blades—aerodynamically—as a set of independent 2D airfoils. Therefore, wind tunnel measurements of 2D sections are the basis of all aerodynamical experiments for wind turbines. Fortunately, a lot of experience has been gained for airfoils of airplanes which could be used when special airfoils started to be designed [65]. The requirements for a wind turbine blade airfoil are different in several respects:

- a high lift-to-drag ratio is most important,
- c_L^{max} should be limited to values not far from $c_L^{L2D-max}$,
- thick ($>30\%$) airfoils have to be used at least in the inner part of the blade.

In Fig. 8.1, we show a profile and its set-up for measurement campaigns using a 30% thick airfoil (slightly modified DU97-W-300 profile from Technical University Delft [66]).

Some results [18, 66] are summarized in Figs. 8.2, 8.3 and 8.4.

---

[1] Unknown source.

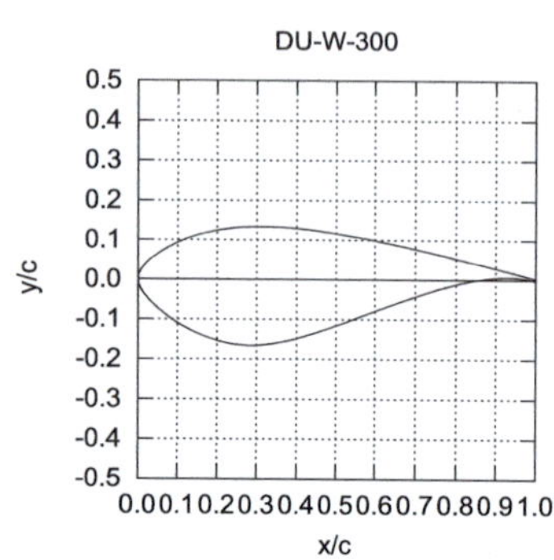

Fig. 8.1 Outline of measured profile DU97-W-300-mod and model set-up in the wind tunnel

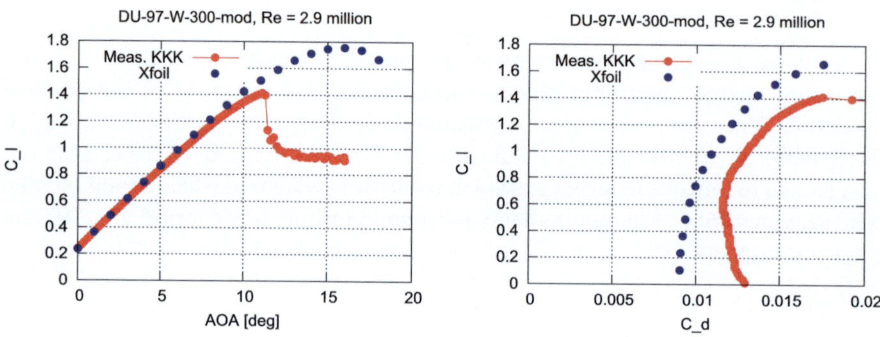

Fig. 8.2 DU97-W-300-mod measured at Kölner KK and XFoil, Re = 2.85 M

Comparing Figs. 8.2 and 8.3, the last performed at RN = 5 M, we see clearly the failure of XFoil (a popular open source panel-boundary-layer-code, see Chap. 3) to predict c_L^{max}.

c_L^{max} may only be predicted with appropriate turbulence models, e.g. Menter's (7) [29].

Usually, only global forces are measured by gauges and/or *wake rakes* only. A common but elaborate technique uses small holes (diameter $\varnothing < 0.5$ mm) on the surface which are connected to pressure transducers. Figure 8.4 shows one example

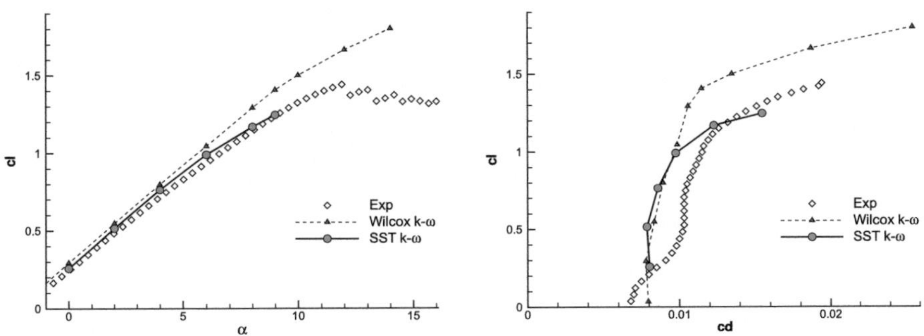

Fig. 8.3 DU97-300-mod simulated with FLOWer and SST k-ω turbulence model, Re = 5 M

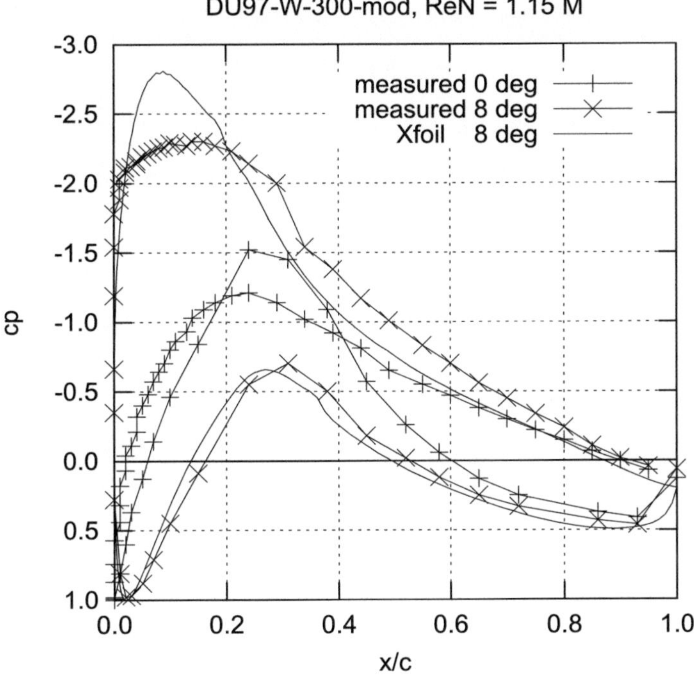

Fig. 8.4 Sample pressure distribution for a 30% thick wind turbine aerodynamic profile (DU97-W-300-mod) [67, 68]

of such a circumferentially resolved measurement. Small laminar separation bubbles or sometimes even the transition from laminar to turbulent flow (by the displacement effect) are visible (small bump at x/c = 0.2 for the XFoil data) in the c_p-distribution. Notice that the pressure does not indicate a stagnation point ($c_p = 1$) at the tail. Even thicker (from 35% up about 50%) [2, 41, 73] profiles are used in the very inner part of a blade for improving structural integrity (see Chap. 9). Figures 8.5 and 8.6 show

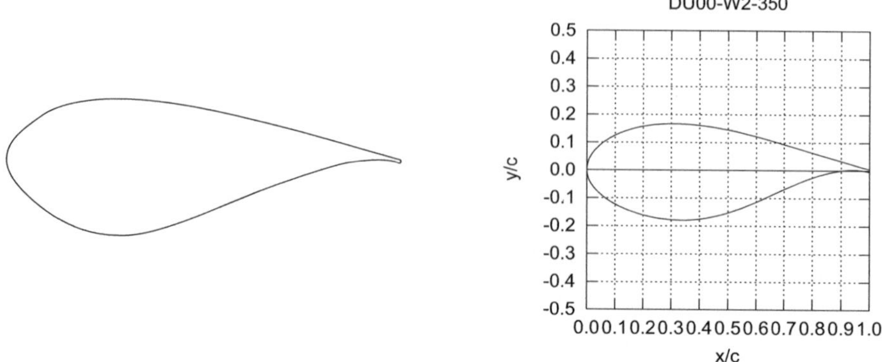

Fig. 8.5 Shapes of two 35% thick profiles: DU96-W-351 and DU00-W2-350

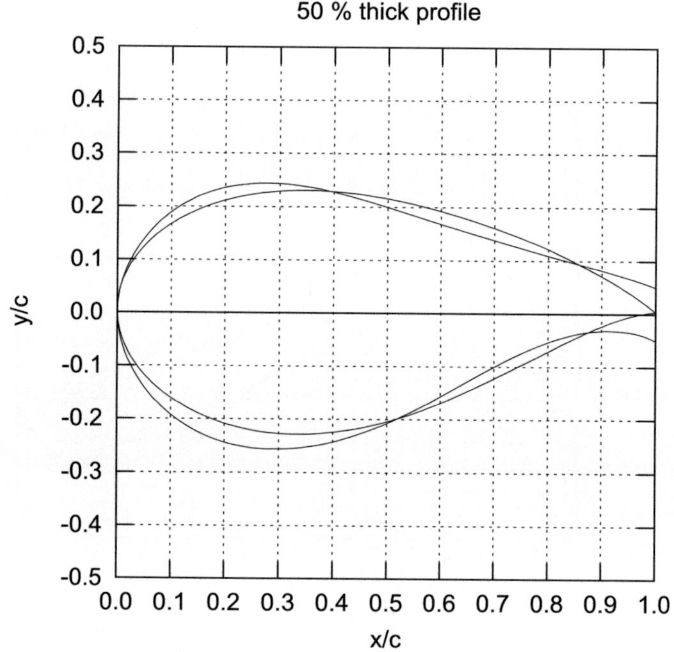

Fig. 8.6 Two examples for approximately 50% thick airfoils for root sections of wind turbine blades

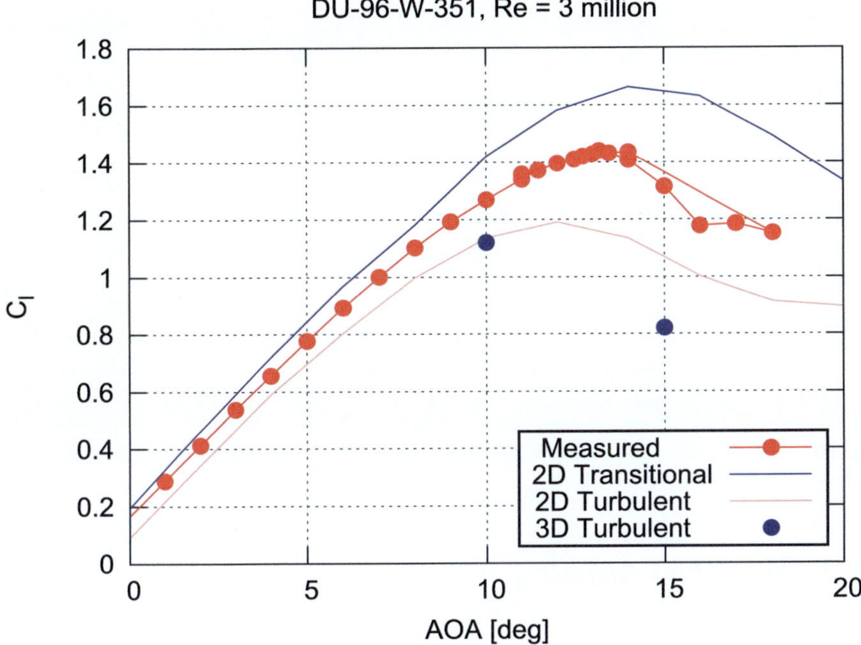

Fig. 8.7 Lift data for a 35% thick airfoil (DU96-W-351) including 3D CFD simulation, from [27]

shapes of two 35 (50)% profiles. Similar comparisons have been performed at the LM wind tunnel by [27], see Fig. 8.7.

Summarizing the findings from [2, 27, 41, 73], we may conclude: The capture of c_L^{max} is somewhat easier by measurements but very difficult by CFD even when using DES. It has to be noted that a high (>100) lift-to-drag ratio is possible only at Reynolds number $>5 \cdot 10^5$ [21]. This has to be taken into account for an airfoil to be used for blades in small wind turbines and wind tunnel models.

8.1.1 DU00-W-210

During the AVATAR project, a blind test was carried out for DU00-W-210 which has been measured at DNW-HDG. Several CFD codes as well as panel codes were asked to investigate Re- as well as TI-dependence of lift-and-drag data. Figure 8.8 shows one of the results. First of all, L2D data can be predicted with less than 3% (Ellipsys from DTU) accuracy. It seems that the e^N-method is more accurate in predicting transition than the correlation method. Finally, it is clearly seen that fully turbulent calculations underestimate max-L2D by more than 30%. TI-dependence (by Mack's correlation Eq. 8.3) can be modeled accurately as well.

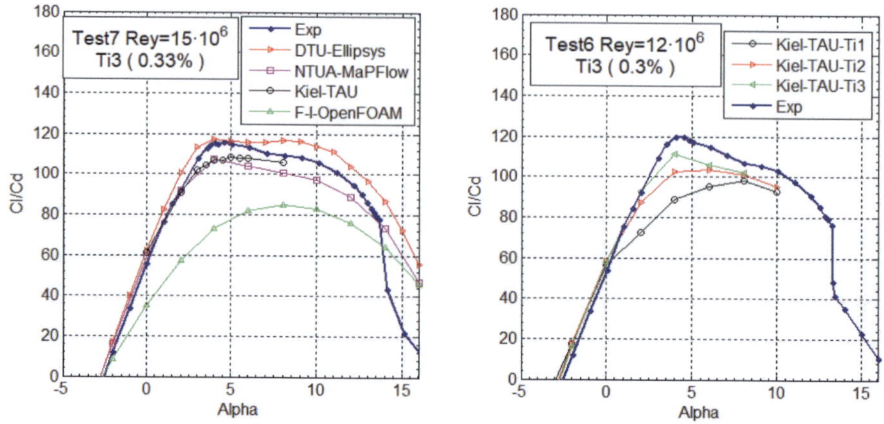

Fig. 8.8 CFD simulation for DU00-W-210 with transitional CFD codes. Left: various CFD codes. Right: TI-dependence From [33]

8.1.2   Very Thick Airfoils

Figure 8.9 (from [43]) shows c_L versus AOA data of a special airfoil-type profile comparable to those of Fig. 8.6. As an unusual feature, it exhibits regions of negative lift slope. It cannot be excluded that this phenomenon, probably due to a pair of counter-acting vortices, may be present on other very thick profiles.

In the meantime, this question has been addressed in more detail, see [40] (Figs. 8.10 and 8.11).

8.2   Measurement of Wind Turbine Power Curves

A reliably measured power curve of a wind turbine is of utmost importance for the economic success of a wind turbine as a product. We already have seen in Chap. 7 two examples, see Figs. 7.33 and 7.34. Standards [12] exist how the terrain (where the turbine or wind farm it to be places in) should be classified and which measurement equipment has to be used when a *certified* power curve $P(v)$ is desired. This is especially important when no (expensive) separate met mast is available. Then the usual (specially calibrated) nacelle anemometer is the only device for measuring the wind speed. Figure 8.14 shows a sample measured power curve for a small wind turbine (Figs. 8.13 and 8.15).

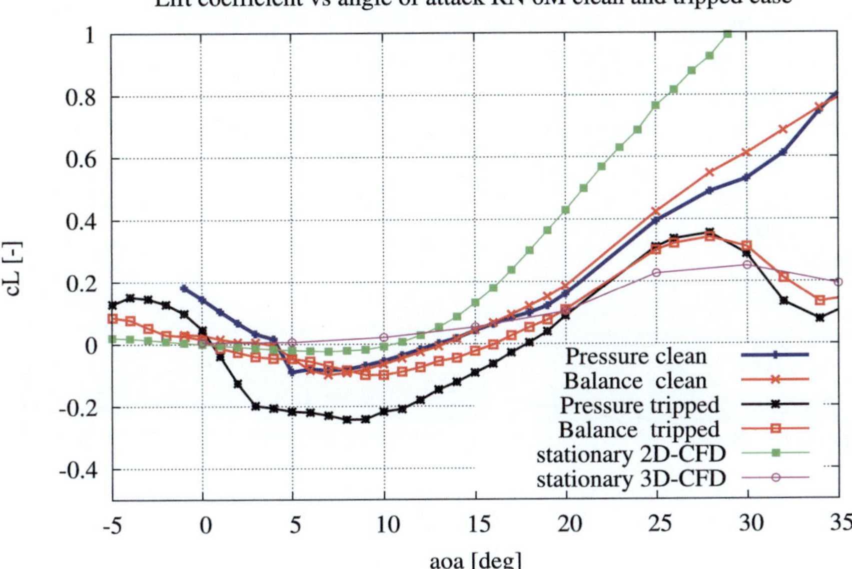

Fig. 8.9 Lift versus AOA of a special, 46% thick profile

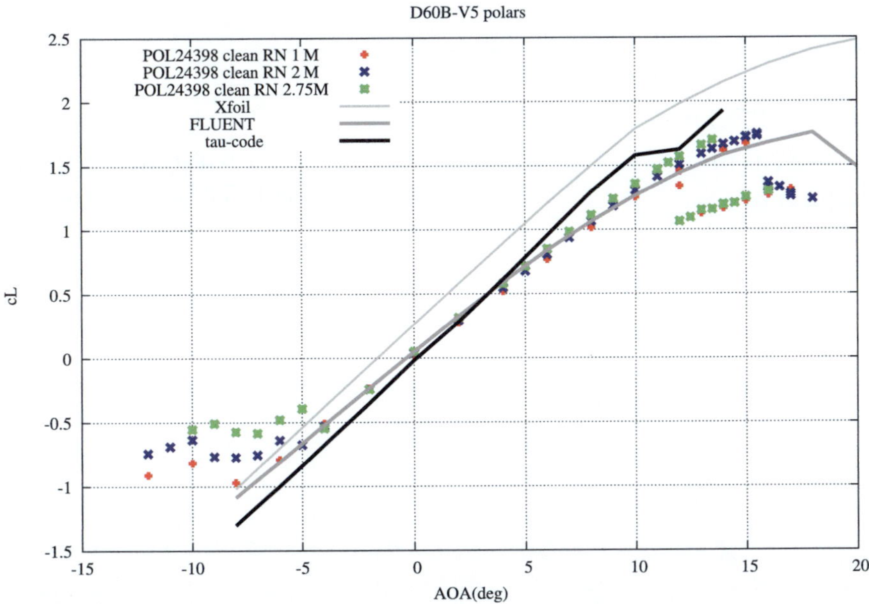

Fig. 8.10 Variation of lift-coefficient via angle of attack at Reynnolds number from 1 to 2.75 million. Measurements and CFD predictions from various codes

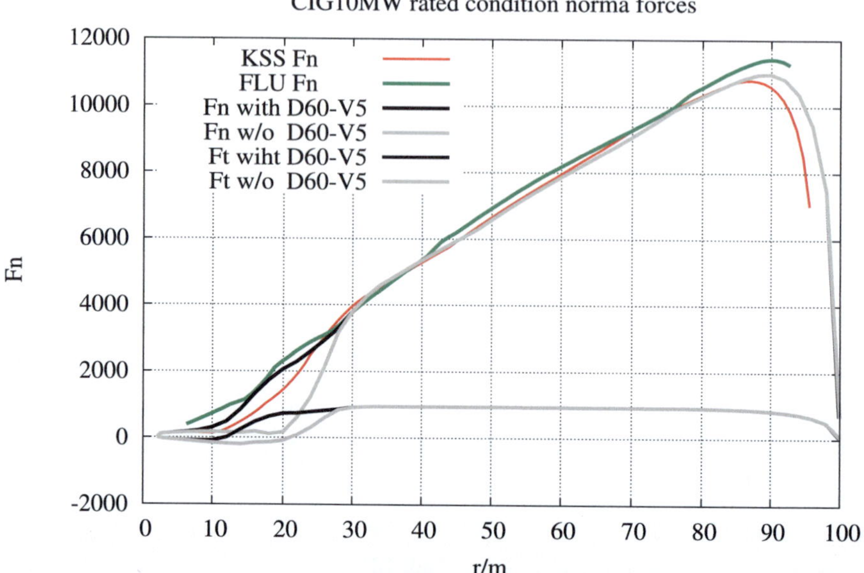

Fig. 8.11 Radial resolved forces using BEM (KSS) and CFD (FLUENT) for a generic 10MW blade CIG10MW with and without a new 60%-thick airfoil at r = 15 m

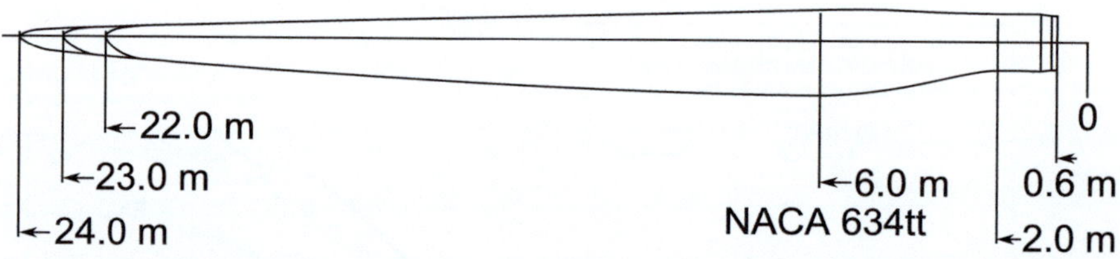

Fig. 8.12 Shape of the ARA-48 Blade

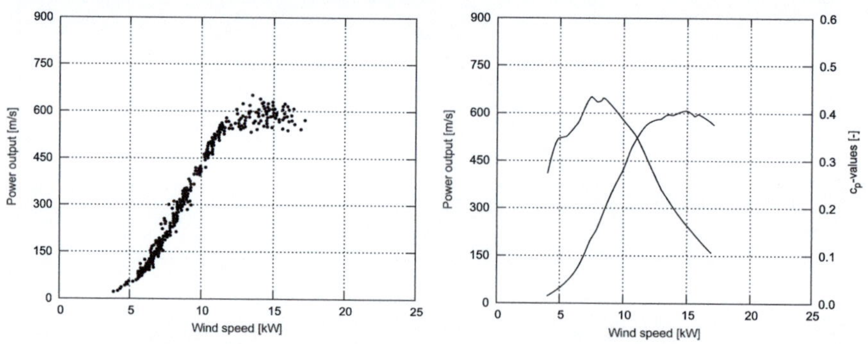

Fig. 8.13 Measured power of blade from Fig. 8.12. Left: raw data. Right: averaged power curve (right scale) and c_P (left scale) as function of wind speed

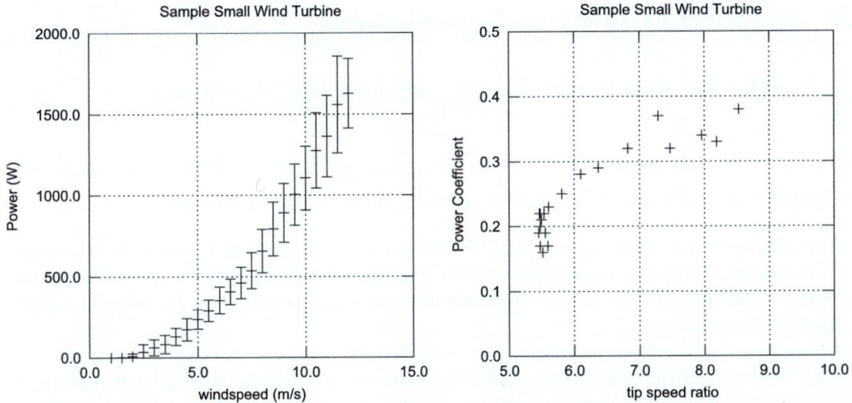

Fig. 8.14 Power versus wind speed (left) and c_P versus λ (right) for a small wind turbine

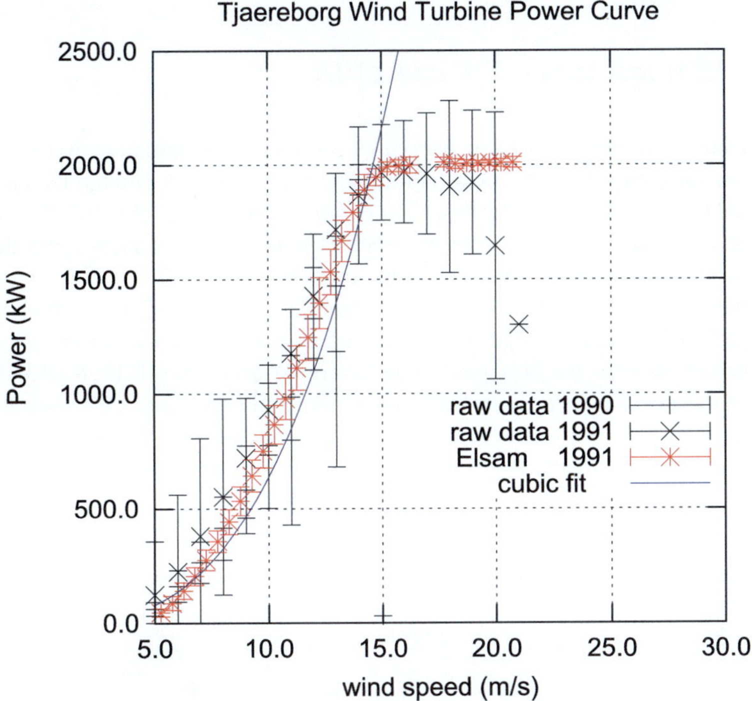

Fig. 8.15 Power versus wind speed for a 2 MW test wind turbine, based on 10-min averages, data from [5]

8.3 IEAwind Field Rotor Experiments

8.3.1 The International Energy Agency and IEAwind

The IEAwind is a branch of the International Energy Agency (IEA) and was founded
comparatively early (in 1974) and defines its work as quoted here:

> The International Energy Agency (IEA) Wind agreement is a vehicle for member countries
> to exchange information on the planning and execution of national large-scale wind system
> projects and to undertake co-operative research and development (R & D) projects called
> Tasks or Annexes.

It is therefore a unique place for scientists from all around the world to communicate
about wind turbine aerodynamics. First, results were published in the early 1980 and
one (Task XI *Base Technology Information Exchange*) is continuously working.

8.3.2 IEAwind Annex XIV and XVIII

Task 14 and 18 (for more complete overviews see [48, 72]) summarize first field measurements for comparatively small (around 25 m rotor diameter) wind turbines and
supported the re-invention of the so-called *Himmelskamp effect* (5) [20]. One important task was to compare the very different experimental approaches and thereby
identify the accuracy of the measurements (Figs. 8.16 and 8.17).

The Himmelskamp effect—in short—is an effect of delay of stall at high angles
of attack. Much higher c_L values than expected may be reached when the rotor is in
operation compared to the 2D case. Examples are shown in Figs. 8.18, 8.19 and 8.22.

These measurements are very time-consuming and expensive. One special point
is the long averaging period due to the heavily transient inflow conditions. Therefore,
wind tunnel measurements were planned. We will introduce this in more detail in
Sects. 8.4 and 8.5.

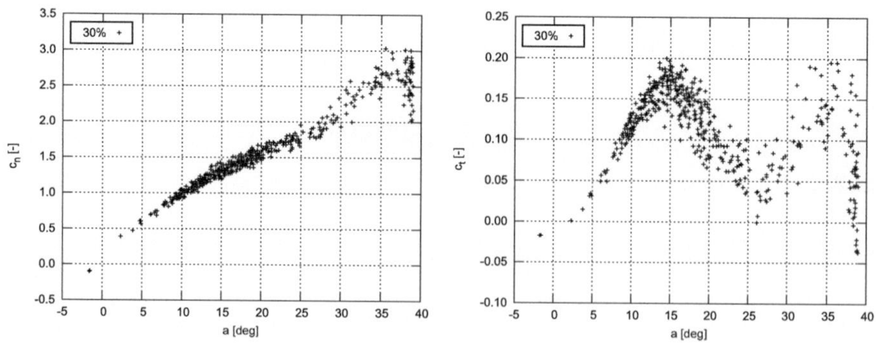

Fig. 8.16 Raw data for normal and tangential for r/R = 30%

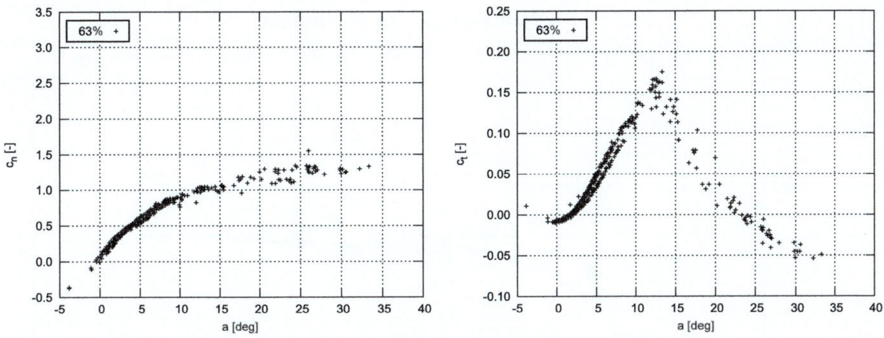

Fig. 8.17 Raw data for normal and tangential force r/R = 63%

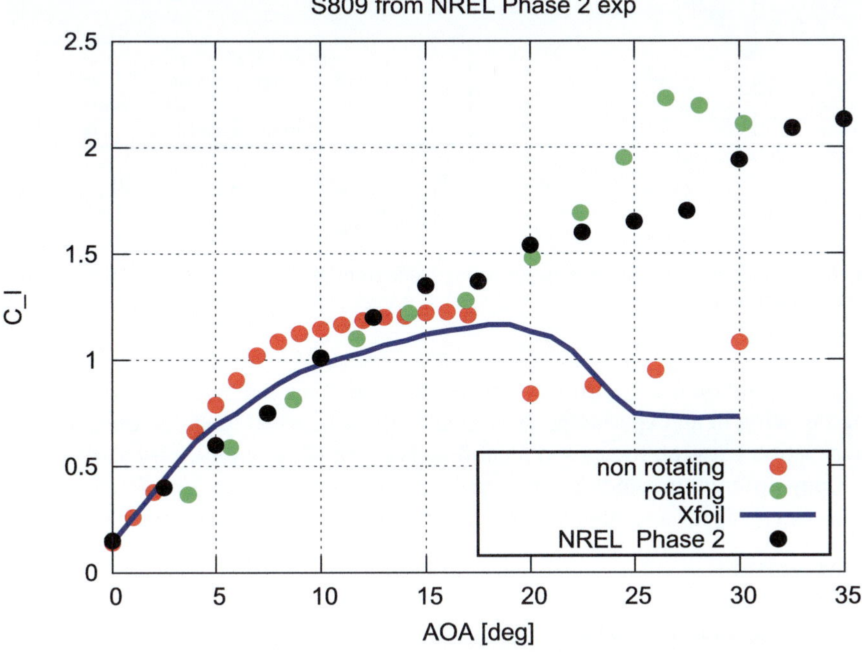

Fig. 8.18 Example of stall-delay on a rotating blade measured in the field, data from [4, 46]

8.3.3 *Angle of Attack in 3D Configuration*

Questions arising from instrumentation were discussed in Sect. 8.3 as well as the
important question of how to relate these data to the 2D polars used in BEM. One
special point is to convert measured normal (usually in wind direction) and tangential
(in the rotor plane) forces into lift and drag components which can only be done if a
meaningful (total) inflow velocity direction is known.

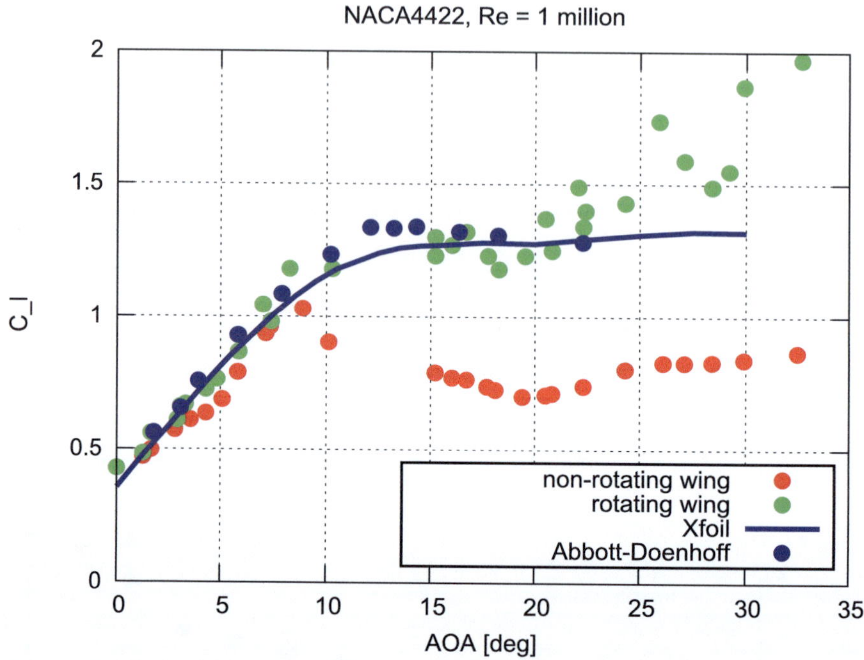

Fig. 8.19 Exampled of stall-delay on a rotating blade from the FFA/CARDC wind tunnel experiment. Data from [6]

In 2D wind tunnel experiment, this is guaranteed by geometric conditions. For a rotating wing in an experiment, as well as in the numerical model, there is a strong interrelation between forces and local flow because of Newton's third axiom.

Many different methods are used to extract this information from force measurements (see [55] for a summary of recent results). To mention only the most used:

- inverse BEM,
- Biot-Savart,
- stagnation point method and
- direct measurement of the velocity vector.

So far it can be said that no simple solution is possible and one should use more than one method if one wants to extract quasi 2D data from 3D configurations.

8.4 FFA/CARDC Wind Tunnel Experiments

One of the first wind tunnel measurements was conducted in a Chinese-Swedish co-operation during the years 1986 and 1992 [6, 13, 39]. The blade had a radius of 2.648 m and was equipped with NACA 44tt [1] with $tt = 14$ to 22%. Reynolds

number in general was about 500k. A main finding was the confirmation of the stall
delay at sections closer to the hub, see Fig. 8.19.

The difference is particularly large when comparing the rotating and non-rotating
cases, but the stall behavior seems to be very different from what is expected by older
measurements [1] and 2D Boundary-Layer-Panel codes (3) [15].[2]

8.5 NREL NASA-Ames Wind Tunnel Experiment

The *National Renewable Energy Laboratory (NREL)*, located in the vicinity of Den-
ver, Colorado, performed field measurements from the early 90ies on [11, 58]. As
mentioned earlier, the idea was to have more control on the inflow conditions, so
a larger wind tunnel campaign was performed in the world largest wind tunnel
$(24.4 \times 36.6\,\mathrm{m}^2)$, at NASA-Ames, see Fig. 8.21.

The blade was designed from only one profile, the S809 [60]. A complete description of the turbine is included in [30]. Other main parameters of the machine are (Fig.
8.20):

- rotor diameter 10.0 to 11.06 m,
- hub height 12.2 m,
- cone angle: 0, 3.4 and 18°,
- rated power 19.8 kW,
- 2-bladed up- and down-wind configuration.

As an example of the again confirmed stall delay for the S809 profile, see Fig. 8.22.
Predicting the stall behavior is very complicated and requires a sophisticated CFD
model as discussed in Chap. 7, Sect. 7.8.1 in detail. The most important aspect of
the experiment was the so-called *blind comparison* [57]. Here participants had to
calculate power and load data before the experimental data was compared to the
measurements. A sample graph is shown in Fig. 8.23. The spread is about 10% in the
attached case and diverges to a factor of 7 in the stalled case. Clearly, modeling of
stall behavior is of utmost importance as was emphasized by Jim Tangler [63, 64].
For the first time [61], CFD gave more accurate results and details of the flow and
loads than ordinary BEM and vortex methods. A detailed investigation of findings
was conducted between 2003 to 2008 for IEAwind task XX, see (7) [52].

---

[2] This code is comparable to the so-called Eppler code (3) [16], developed somewhat earlier and
was extended to include 3D-effect from stall delay for rotation in [71].

Fig. 8.20 Outline of the NASA-AMES wind tunnel, the open inlet has dimensions of $108 \times 39\,\text{m}^2$ size *Source* NASA

Fig. 8.21 Outline of the NASA-AMES (NREL Phase VI) 2-blades rotor experiment (the cross-section for testing is $24.4 \times 36.6\,\text{m}^2$. With Permission of NWTC/NREL, Golden, CO, USA

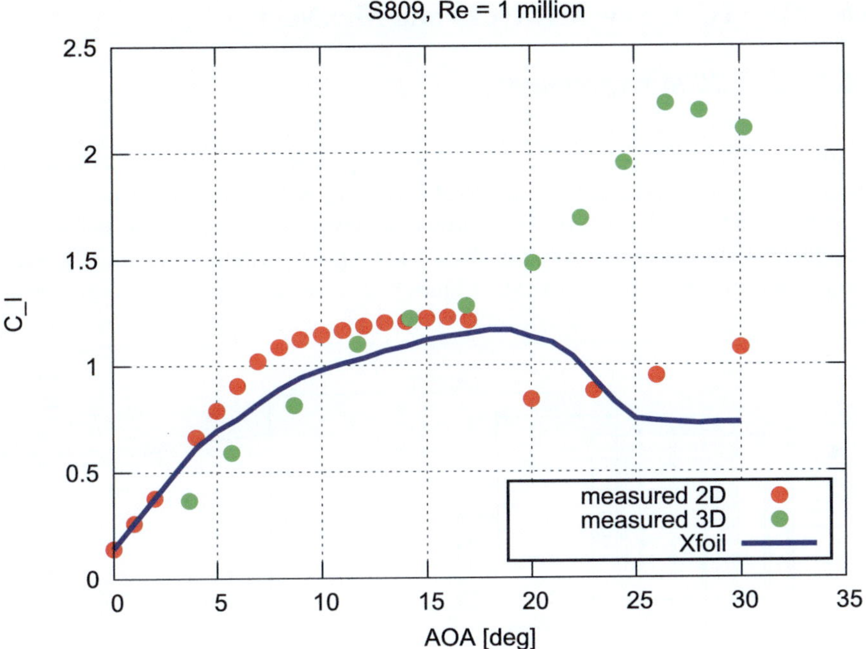

Fig. 8.22 Example of stall delay on a rotating blade from the NASA Ames Phase VI experiment, data from [64]

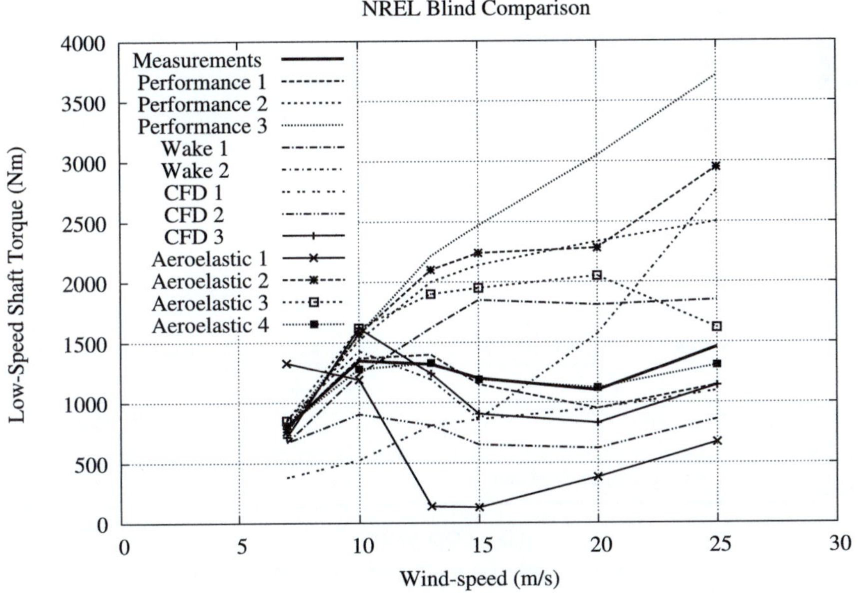

Fig. 8.23 Blind Comparison of computed and measured low-speed shaft-torque for the NASA-AMES (NREL unsteady experiment phase VI) 2-bladed rotor

8.6 MEXICO, New MEXICO and MexNext

8.6.1 The 2006 Experiment

Shortly after the NASA-Ames experiment a European-Israeli project was planned in Europe's largest wind tunnel the *DNW-LLF* see. Fig. 8.24. It has an open nozzle and a cross-section of $9.5 \times 9.5 m^2$. The turbulence level was measured using pressure transducers (Kulites, used resolution about 5 kHz) giving a range of about 3‰. Much more details are described in [8, 51]. Figure 8.25 gives an impression of the set-up.

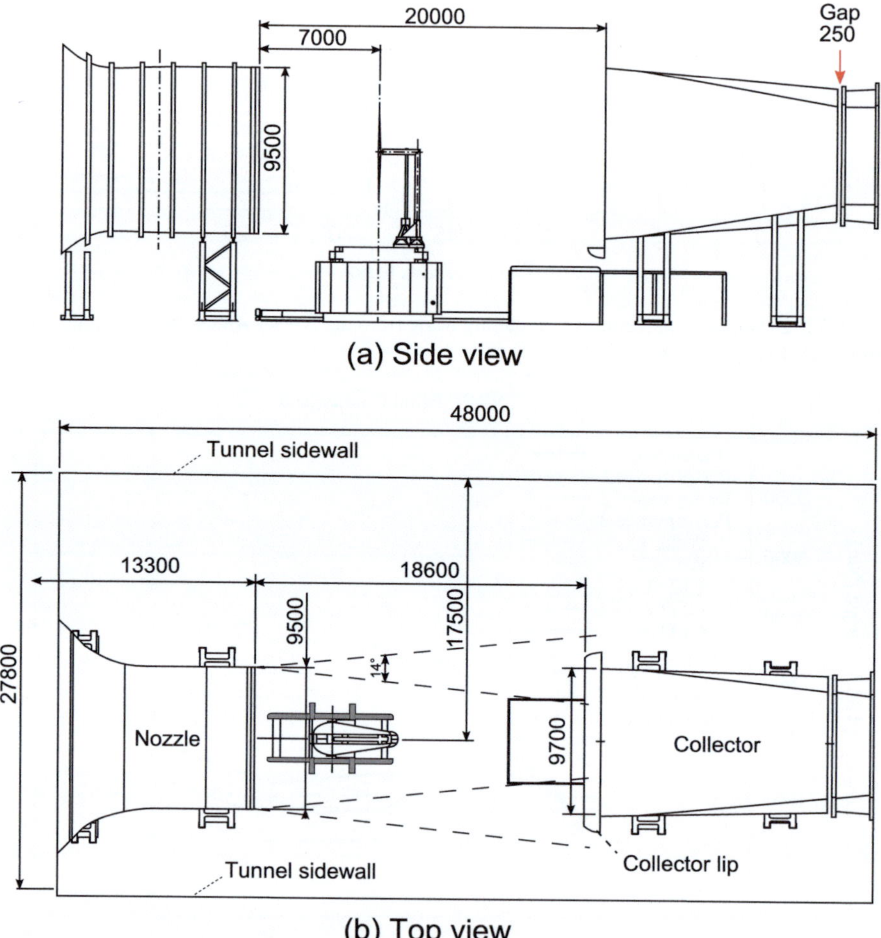

Fig. 8.24 Outline of the DNW-LLF wind tunnel

Fig. 8.25 Outline of the Mexico Experiment in December 2006. With Permission of ECN, The Netherlands

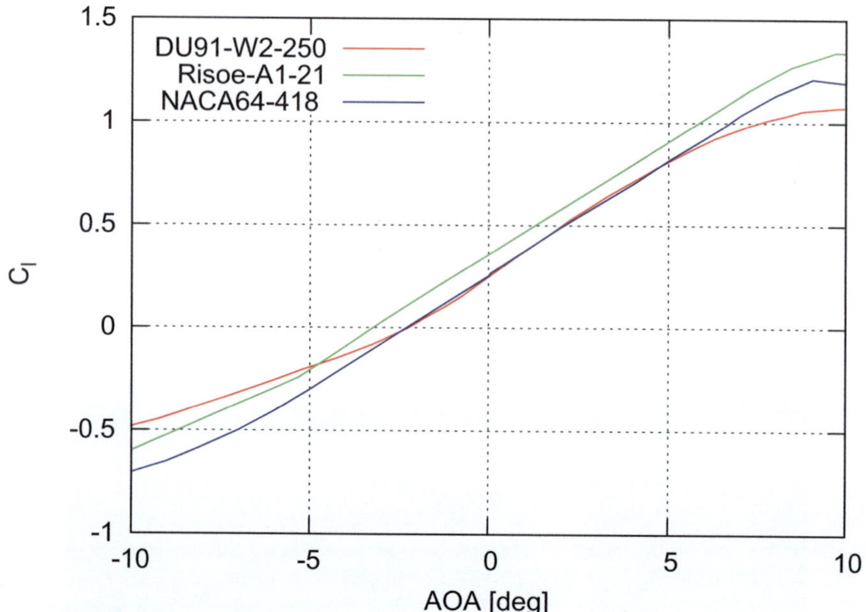

Fig. 8.26 Lift data of used profiles

The turbine has a diameter of 4.5 m and is three-bladed. As it was a co-operative European project, three profiles (see Fig. 8.26) were used:

- DU91-W2-250 from $r/R_{tip} = 0.11$ to 0.36,
- RISØ A1-21 from $r/R_{tip} = 0.45$ to 0.56,
- NACA64-418 from $r/R_{tip} = 0.65$ to 0.91.

At about 5% chord, a zig-zag tape was glued on the surface to avoid laminar separation bubbles. The chord-based Reynolds number again was in the order of 500 k for a rotational speed of 425 RPM. Unlike the NASA-Ames experiment, the velocity field was also measured by *particle image velocimetry (PIV)*. This was especially useful when comparing with CFD models, see Sect. 8.6.3 and Fig. 8.27. In the upper right corner, a tip vortex seemed to emerge.

8.6.2 The 2014 Experiment

During the first two phases of MexNext (IEAwind task 29), it became clear that the agreement between CFD and experiment and accuracy (in terms of accuracy) was less than perfect. Therefore, fortunately, a 2nd tunnel entry was possible in 2014 [9, 35]. Table 8.1 compares four arbitrarily chosen values.

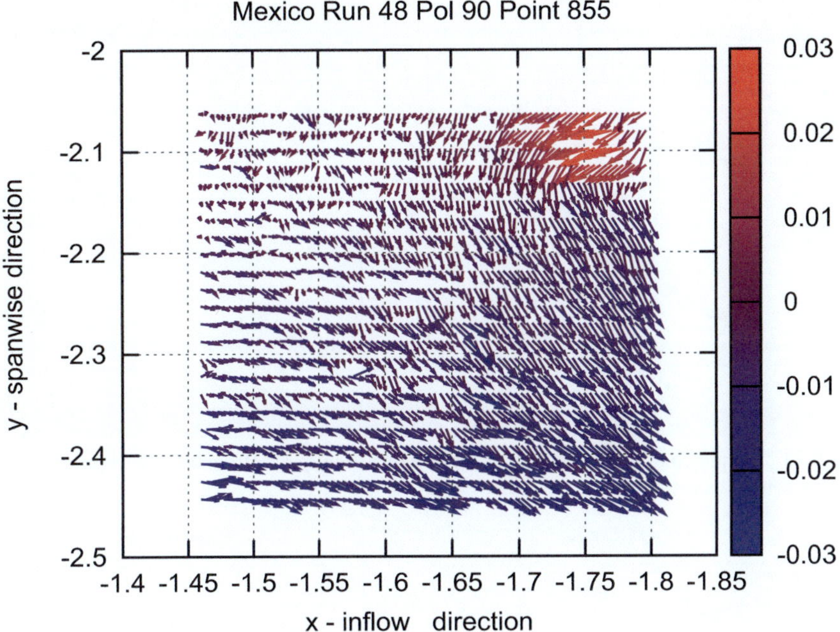

Fig. 8.27 Sample PIV data: axial traverse, pitch = 5.3 deg and inflow velocity of 24 m/s, in the upper right corner, a tip vortex seems to emerge

Table 8.1 Comparison of arbitrarily chosen values. Differences are in the order of 4%. See [35] for more information

Name	10 m/s Thrust/N	Torque/Nm	15 m/s Thrust/N	Torque/Nm
Mexico	854	61	1517	285
New Mexico	974	68	1163	317

8.6.3 *Using the Data: MexNext*

As it was the case for the NASA-Ames wind tunnel experiment (7) [52], an IEAwind task (20 and 29, respectively) was established to analyze the data in detail. The final report of the first phase (2009–2011) (5), [49] gives an impressive view of the gained result. As may be seen from Figs. 8.28 and 8.29, many more groups used CFD than 10 years before for Annex 20. Unfortunately, the spread is rather large and it can hardly be concluded that a firm improvement has been reached. A more comprehensive analysis (7) [52] shows that most of the theoretical (BEM-like as well as CFD) models overpredict the loads when compared to the measurement. Therefore, the uncertainty (systematic as well as statistical) of the experimental data should be discussed. Due to the fact of the special arrangement of the inflow (nozzle) and outflow (collector), it seems difficult to fit the exact prescribed inflow conditions

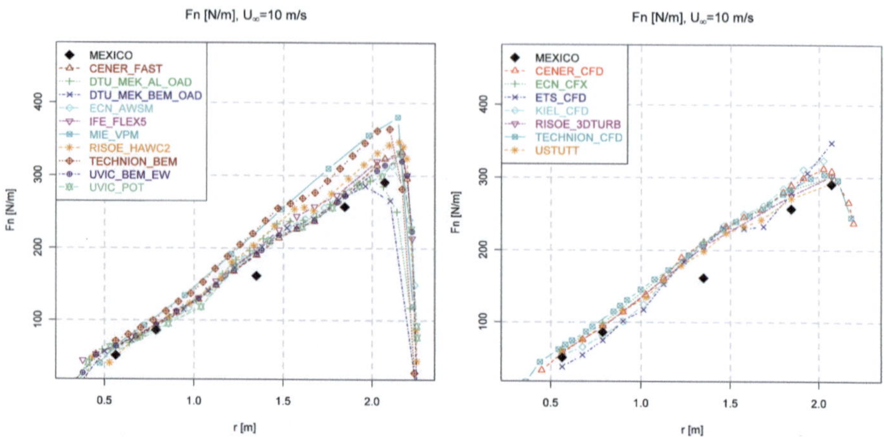

Fig. 8.28 Sample comparison (at 10 m/s inflow) of lifting line (left) and CFD (right) models in comparison to the experimental data—normal forces

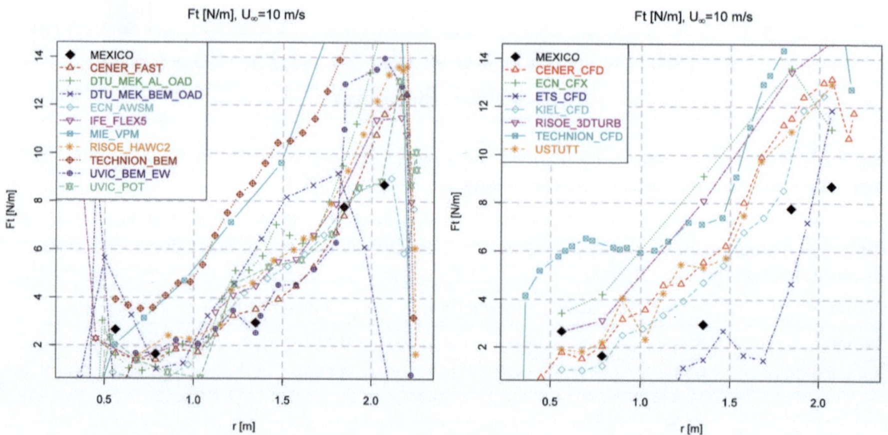

Fig. 8.29 Same as in Fig. 8.28 but for tangential forces

(see Fig. 8.30). Therefore, several authors have investigated the whole wind tunnel including the rotor modeled as an actuator disk [38, 56] and (7) [28].

Figures 8.30 and 8.31 show some sample results. It is clearly seen that the differences for the low-speed case (high tip-speed-ratio λ) are most pronounced.

8.6.4 Wind Turbines in Yaw

A subject of continuous interest is the investigation of so-called *yawed* flow. Here wind direction and axis of rotation do not coincide. It has to be noted that yaw

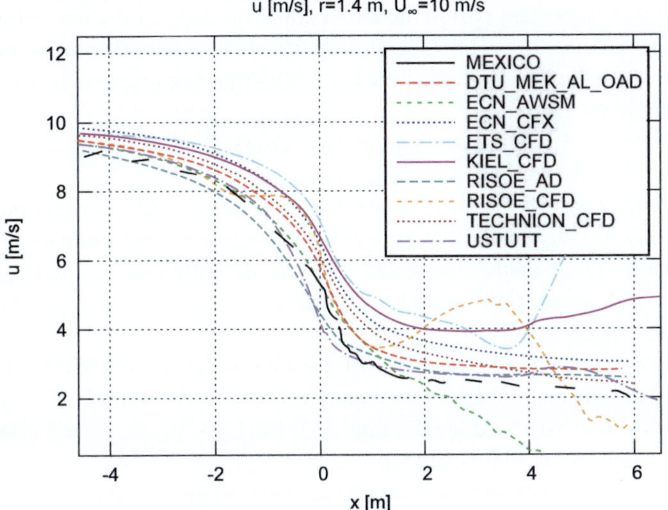

Fig. 8.30 Sample CFD calculations of various authors of the velocity field behind the Mexico-Rotor, $u_{in} = 10\,\mathrm{m/s}$, $\lambda = 10$. Data from (5) [49]

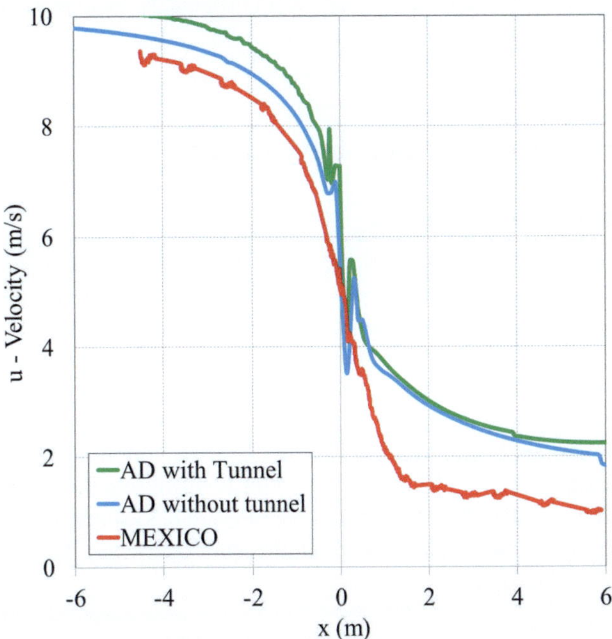

Fig. 8.31 CFD calculation of the velocity field behind the Mexico-Rotor comparing open flow and wind tunnel from [28]. Same data as if Fig. 8.30

control plays an important part in the total control strategy of a wind turbine. A very frequently asked question is how fast the control system should compensate for such misalignments. It is clear from the very beginning that because power roughly varies with yaw angle as

$$c_{P,yawed} = c_{P0} \cdot cos(\gamma)^n \tag{8.1}$$

with some exponent $n \in \{2, 3\}$ (5) [10] or even $n \in \{0 \cdots 6\}$ [7] more power could be extracted if the alignment is very strict. Unfortunately, the price to pay is in the form of much higher loads introduced into the azimuthal drive due to gyroscopic moments, which seem to be so high [42] that a reasonable compromise consists of very slow yaw control only.

In addition, each revolution of the rotor now produces cyclic inflow conditions. For this reason, Haans, Schepers and many others [7, 19, 42, 47, 48, 50, 59], and more recently, the MexNext consortium (5), [49] discussed yawed conditions to improve engineering guidelines [45].

In this case, mesh generation is much more elaborate. Necessarily, transient calculations make the turn-around time approximately one order of magnitude higher when compared to the stationary non-yawed case. A careful analysis by Schepers [48], supported by an equally careful analysis of measurements on 2 MW state-of-the-art turbine, indicates that the tip speed ratio influences the exponent γ in Eq. 8.1 strongly up to a TSR where c_P is maximum.

8.7 Experiments in the Boundary Layer

8.7.1 Introductory Remarks

As we have seen in Chap. 5, the high ($c_P \approx 0.5$) values are only possible with a high lift-to-drag ratio (>100). Using standard analyzing tools like XFoil or RANS-CFD, it becomes clear that both laminar (low shear stress) and turbulent (high shear stress) areas on the wing's surface must be present. Wind tunnel measurements (see Fig. 8.32) prove this, including the dependence on the turbulent inflow condition. From the computational side, we have to note that most simplified transition models rely on the use of correlation [23][3] ($N \sim Ti$) for determining the location of a fully developed turbulent state of the boundary layer. Figure 8.33 shows the measurements of transition locations on flat plates (without pressure gradients). These findings may be summarized in Eq. 8.3:

$$N = -8.43 - 2.4 \cdot ln(Ti) , \tag{8.2}$$
$$0.1 < Ti < 2\% . \tag{8.3}$$

---

[3] Recently, Fava and Lobo [17] used LES and stability investigations to reach for a proposal of a modified Mack correlation.

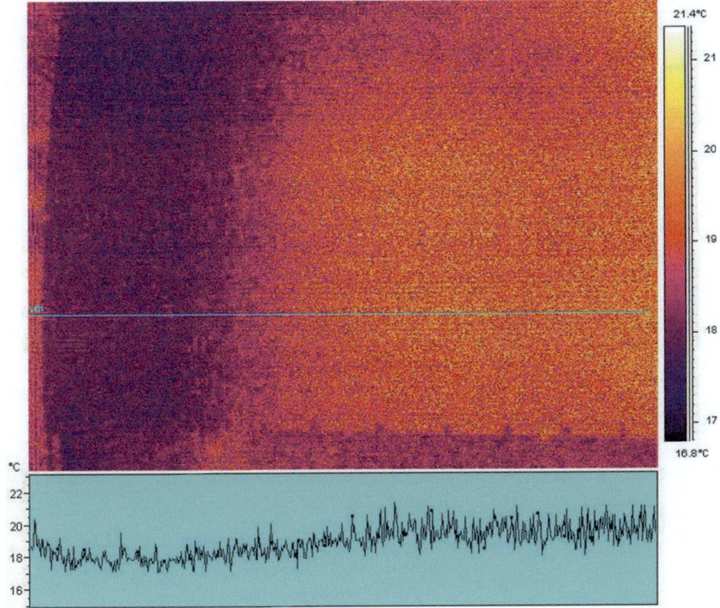

Fig. 8.32 Example of thermogrammetry in a wind tunnel experiment for a 2D airfoil. Flow from left to right. Dark areas indicate laminar parts and lighter colors turbulent flow. From [68]

It shows that the amplification factor N^4 varies from $N = 9$ (at 0.1% TI) up to $N = 1$ (at 2%) turbulence intensity. This lowest possible N of 1 means that the distance between the first instability of laminar flow and fully developed flow (compare with Fig. 3.20) is zero. It has to be noted that this scenario only refers to the so-called *Tollmien-Schlichting*-process where special wave packets occur and may be amplified. In a higher turbulent environment, this transition to turbulence may occur via *by-passes* [62].

Turbulence intensities used in wind energy are mostly determined for (fatigue) load estimation and are derived by averaging over a time period of 600 sec with sampling frequency of a few Hertz. They span a much wider range from 5% at minimum up to about 30% in the so-called complex terrain. Clearly, there should be a low-frequency cut-off in the whole energy spectrum which is aerodynamically active. Empirically, this value is guessed to be on the order of 10 Hz.[5] A comparison with Fig. 3.27 shows that only a very small fraction of the total energy content is located in this frequency range. Reference [31] investigated the combined receptivity and tran.

---

[4] Roughly speaking this is the ratio where a small disturbance has grown to a relevant size.

[5] In [69, 70], even a much lager range of $\nu < 500$ Hz is excluded.

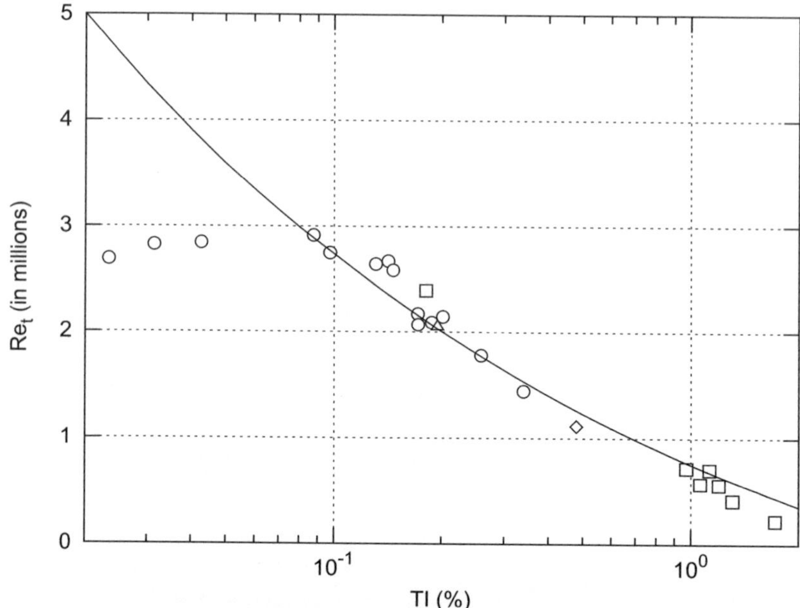

Fig. 8.33 Mack's empirical correlation for the dependency of transition-location to free-stream inflow turbulence. Data from [23]

8.7.2   Laminar Parts on the Wind Turbine Blades

In [14] the thermographic technique used in wind tunnel environments was extended for use on operating wind turbines, see Fig. 8.34. Meanwhile [37, 44], the capabilities and accuracy of thermography were investigated in much more detail. Figures 8.35 and 8.36 show the same operational state of a Multi-MW turbine thermographic as well as microphone date on the suction side of a wind turbine blade. The first one detects transition at $x_{tr}^{Th}/c = 0.255 \pm 0.003$ and the second method $x_{tr}^{Mi}/c = 0.26 \pm 0.01$. The larger inaccuracy is mainly due to the spreading of the microphones within the array.

8.7.3   The DAN-AERO MW Experiments

A very large field experiment was undertaken during the DAN-Aero MW Experiment on state-of-the-art commercial turbines (NecMicon 2MW and Siemens SWT-2.3 both equipped with LM38.8 blades) [3, 25]. Among others, investigation of boundary layer transition was possible by the use of a dense distribution of 60 microphones around a test cross-section. Figure 8.37 gives an example. It has to be noted that comparable measurements on much smaller rotors (D = 25 m) have been undertaken, but were not

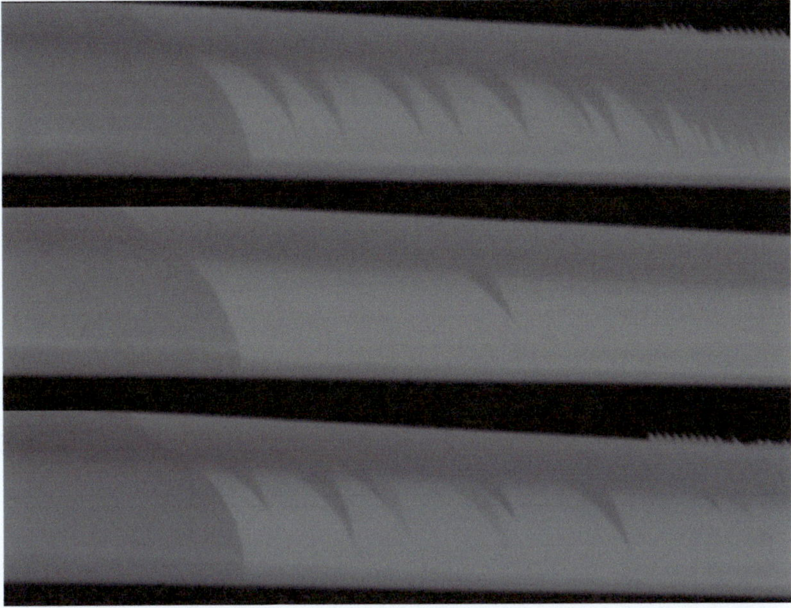

Fig. 8.34 Example of a thermographic photograph of a rotating blade of a commercial wind turbine. Light colors indicate laminar flow and dark ones indicate turbulent flow. Used by Permission of Deutsche WindGuard Engineering GmbH, Bremerhaven, Germany

Fig. 8.35 Thermography

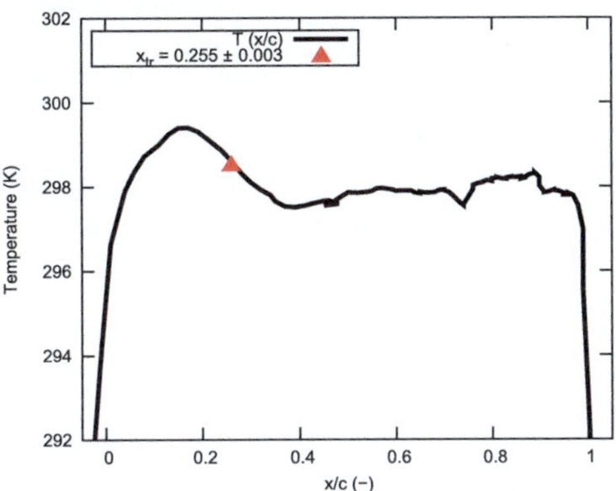

published [69, 70]. The sudden increase from $\nu > 500$ Hz between POS. 1.7% and POS 2.3% may indicate transition to a turbulent flow state. Recently [26, 32], Madsen re-examined the DAN-AEro data with refined data-processing methods; confirming most of the conclusions concerning the transition mechanism: Bypass seems more likely than Tollmien-Schlichting.

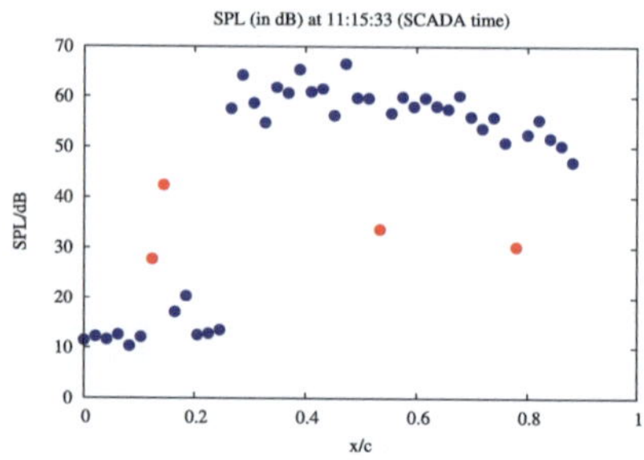

Fig. 8.36 Microphone data

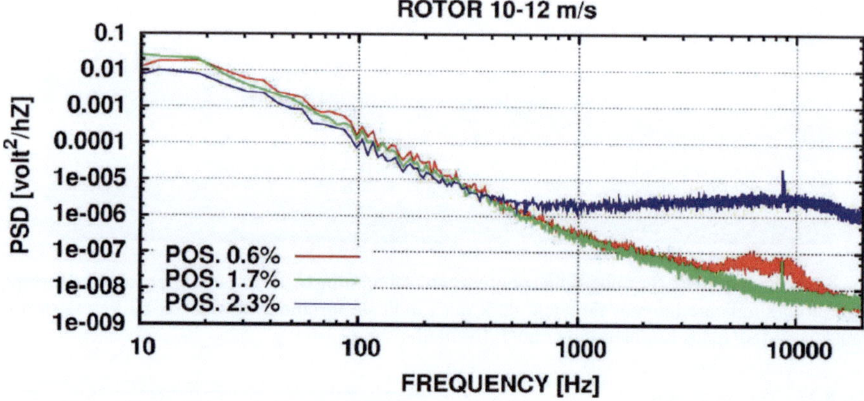

Fig. 8.37 Microphone measurements on a 38.8 m blade, data from [24, 25], Used with Permission of DTU Wind energy

8.7.4 *Hot Film Measurements*

Similar experiments on a much smaller wind turbine than used in Sect. 8.7.3 [53] were performed on an ENERCON E-30 wing. For investigating the boundary layer state, a technique comparable to airplane wings was used. Those *aerodynamic gloves* experiments are common there [34, 36, 54]. Figure 8.38 shows high-frequency time-series of a *hot film*. The active area was much smaller than in the experiment described in Sect. 8.7.3 and reaches from $x/c = 0.26$ (bottom) to $x/c = 0.35$ (top) only. The similarity with Fig. 3.22 (Schubauer's experiments on a flat plate) seems striking on a first glance but genuine *Tollmien-Schlichting-Waves* are lacking. Figures 8.39, 8.40 and 8.41 give show corresponding spectra. Recently [22] a thorough discussion, comparison and interpretation of both the DanAero and the Aerodynamic Glove experiments have been given.

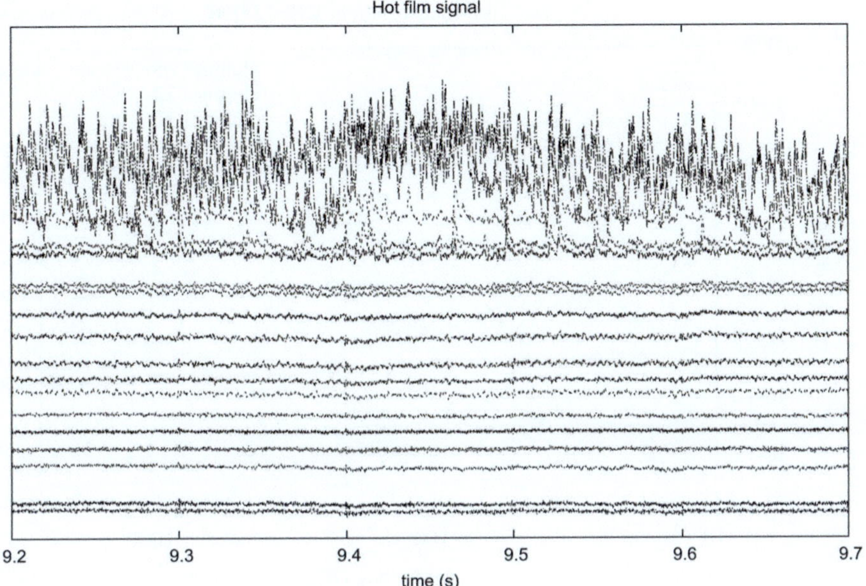

Fig. 8.38 Sample time-series of a hot film located on a wind turbine blade from x/c = 0.25 (bottom) to x/c = 0.35 (top) for RN = 1 M, TI = 3%. From [53]

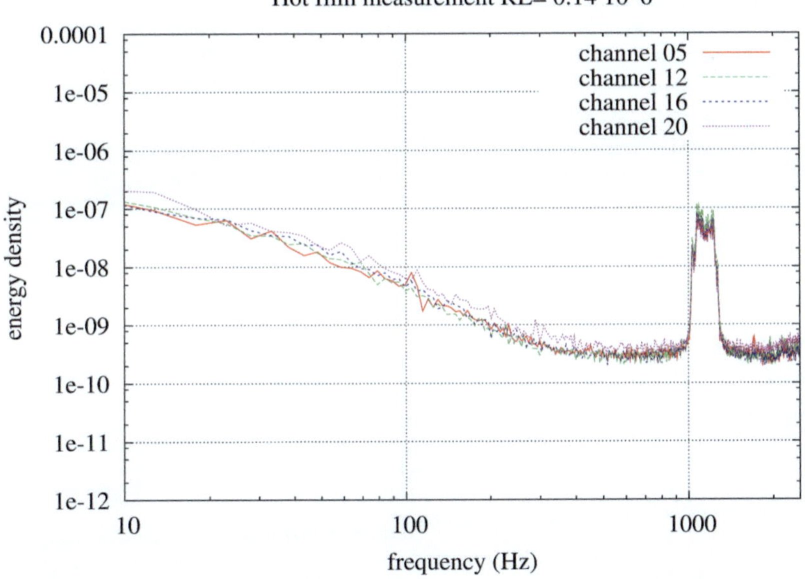

Fig. 8.39 Energy spectra for RN = 0.2 M [53]

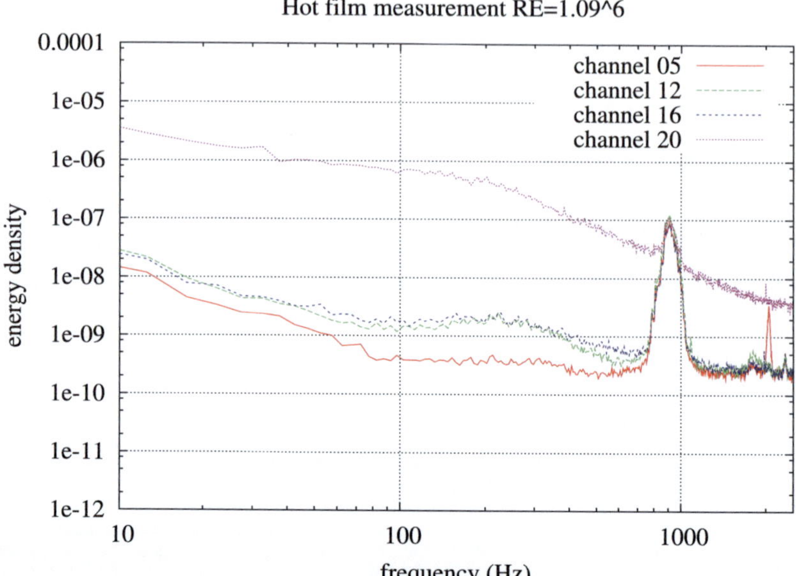

Fig. 8.40 Re = 1.1 M [53]

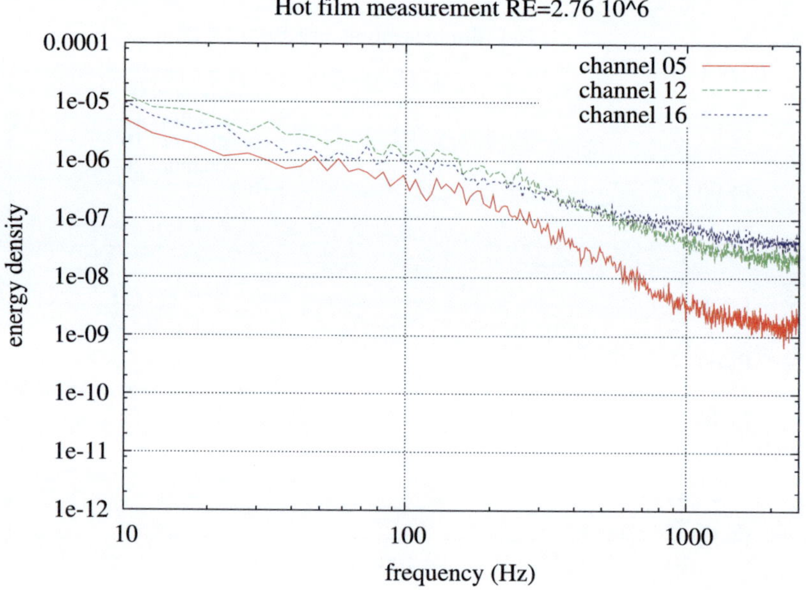

Fig. 8.41 Re = 2.8 M [53]

8.8 Summary of Experiments in Wind Turbine Aerodynamics

It has been shown that experiments even only on small parts of a wind turbine are extensive and complicated. Also, it has been stated many times that—compared to the aerospace industry—wind turbine research is comparatively young. Therefore, it may be easy to predict that new experiments, both in wind tunnels and in the field, will be conducted in the near future and will bring much progress in scientific and practical engineering.

8.9 Problems

Problem 8.1 Use the c_p data given in Table 8.2 to integrate for lift.

Problem 8.2 Download data for the NTk Wind turbine from **wind data.com** to estimate a power curve.

Problem 8.3 Use measured data of Problem 8.1 to prepare plots of lift-to-drag ratio.

Table 8.2 Measured pressure coefficients for DU-W-300-mod

x/c	cp-0	cp-8
0.950000	0.070000	0.040000
0.900000	0.000000	–0.020000
0.850000	–0.070000	–0.110000
0.800000	–0.150000	–0.240000
0.750000	–0.220000	–0.340000
0.700000	–0.300000	–0.450000
0.650000	–0.380000	–0.560000
0.600000	–0.470000	–0.700000
0.550000	–0.550000	–0.830000
0.490000	–0.650000	–1.010000
0.440000	–0.810000	–1.170000
0.390000	–0.920000	–1.380000
0.340000	–1.020000	–1.540000
0.290000	–1.140000	–2.000000
0.240000	–1.210000	–2.140000
0.210000	–1.190000	–2.230000
0.180000	–1.140000	–2.270000
0.160000	–1.090000	–2.300000
0.140000	–1.030000	–2.300000
0.130000	–0.930000	–2.270000

(continued)

Table 8.2 (continued)

x/c	cp-0	cp-8
0.110000	−0.860000	−2.280000
0.100000	−0.800000	−2.290000
0.090000	−0.700000	−2.260000
0.080000	−0.640000	−2.250000
0.070000	−0.570000	−2.230000
0.060000	−0.480000	−2.210000
0.050000	−0.400000	−2.190000
0.040000	−0.320000	−2.160000
0.040000	−0.210000	−2.140000
0.030000	−0.110000	−2.120000
0.020000	−0.040000	−2.110000
0.020000	0.070000	−2.080000
0.010000	0.180000	−2.030000
0.010000	0.320000	−1.970000
0.010000	0.440000	−1.880000
0.000000	0.570000	−1.780000
0.000000	0.740000	−1.540000
0.000000	0.870000	−1.190000
0.000000	0.980000	−0.660000
0.000000	1.000000	−0.350000
0.000000	0.990000	0.280000
0.010000	0.810000	0.810000
0.020000	0.600000	0.980000
0.030000	0.370000	0.970000
0.050000	0.130000	0.880000
0.070000	−0.140000	0.710000
0.100000	−0.460000	0.450000
0.150000	−0.840000	0.090000
0.240000	−1.520000	−0.550000
0.310000	−1.450000	−0.700000
0.380000	−1.090000	−0.500000
0.450000	−0.570000	−0.180000
0.520000	−0.260000	−0.020000
0.580000	−0.060000	0.120000
0.650000	0.130000	0.250000
0.720000	0.250000	0.330000
0.860000	0.370000	0.430000
0.930000	0.410000	0.460000
1.000000	0.130000	0.060000
0.000000	0.000000	0.000000

References

1. Abbott I, von Doenhoff A (1959) Theory of wing sections. In: Including a summary of airfoil data. Dover
2. Ahmad MM (2014) Cfd investigations of the flow over flat back airfoils using openfoam and different turbulence models. Master's thesis, Kiel University of Applied Sciences, Kiel, Germany
3. Aagaard Madsen H et al (2009) The dan-aero mw experiment final report. Technical Report Risø-R-1726(EN), Risø National Lab, Roskilde, Denmark
4. Schepers JG et al. (1997) Final report of iea annex xiv, field rotor aerodynamics. Technical Report ECN-C–97-027, ECN, Petten, The Netherlands
5. E. A/s (1992) The tjaæreborg wind turbine. Technical Report contract EN3W.0048, ELSAM, CEC, DG XII, DK, Fredericia, Denmark
6. Björck A, Ronsten G, Montgomerie B (1995) Aerodynamic section characteristics of a rotating and non-rotation 2.375 m wind turbine blade. Technical Report FFA TN 1995-03, FFA, Bromma, Sweden
7. Boorsma K (2012) Power and loads for wind turbines in yawed conditions. Technical Report ECN-E–12-047, ECN, Petten, The Netherlands
8. Boorsma K, Schepers JG (2011) Description of experimental setup - mexico measurements. Technical Report ECN-X–11-120, ECN, Petten, The Netherlands
9. Boorsma K, Schepers JG (2018) Description of experimental setup, new mexico experiment, version 3. Technical Report ECN-X-15-093, ECN, Petten, The Netherlands
10. Burton T, Sharpe D, Jenkins N, Bossanyi E (2011) Wind energy handbook, 2nd edn. Wiley, Chichester
11. Butterfield CP, Musial WP, Scott GN, Simms DA (1992) Nrel combined experimental final report - phase ii. Technical Report NREL/TP-442-4807, NREL, Golden, CO, USA
12. Commission IE (2013) Iec 61400–12-2, wind turbines - part 12–2: Power performance of electricity producing wind turbines based on nacelle anemometry. Technical report, Geneva, Switzerland
13. Dexin H, Thor S-E (1993) The execution of wind energy projects 1986–1992. FFA TN, FFA TN 1993(19), 1993–19, (1993) FFA. Bromma, Sweden
14. Dollinger C, Balaresque N (2013) Messverfahren zur akustisch-aerodynamischen optierung von rotorblättern im winkanal. Private communication
15. Drela M (1990) XFOIL: an analysis and design system for low reynolds number airfoils. Springer lecture notes in engineerings, vol 54. Springer, Berlin
16. Eppler R (1990) Airfoil design and data. Springer, Berlin
17. Fava T, Lobo B, Schaffarczyk A, Breuer M, Henningson D, Hanifi A (2023) Numerical investigation of transition on a wind turbine blade under free-stream turbulence at $re_c = 10^6$. Turbulence and Combustion, page to be published, Flow
18. Freudenreich K, Kaiser K, Schaffarczyk AP, Winkler H, Stahl B (2004) Reynolds number and roughness effects on thick airfoils for wind turbines. Wind Eng 28(5):529–546
19. Haans W (2011) Wind turbine aerodynamics in yaw. PhD thesis, TU Delft, Delft, The Netherlands
20. Himmelskamp H (1950) Profiluntersuchungen an einem umlaufenden propeller. Technical report, Mitt. Max-Planck-Inst. f. Strömungsforschung, Nr. 2, Göttingen, Germany (in German)
21. Lissaman PBS (1983) Low-reynolds-number airfoils. Ann Rev Fl Mech 15:223–239
22. Lobo BA, Özçakmak OS, Madsen HA, Schaffarczyk AP, Breuer M, Sørensen NN (2023) On the laminar-turbulent transition mechanism on megawatt wind turbine blades operating in atmospheric flow. Wind Energy Sci 8(3):303–326
23. Mack LM (1977) Transition and laminar instability. Technical Report JPL Publication, 77-15, Jet Propulsion Lab, Pasadena, CA, USA
24. Madsen HA, Bak C (2012) The dan-aero mw experiment. Technical Report IEA wind Annex 29 (MeNext) annual meeting, NREL, Golden, CO, USA

25. Madsen HÅ et al (2010) The dan-aero mw experiment. Technical Report AIAA-2010-645, AIAA, Orlando, FL, USA

26. Madsen HÅ et al (2019) Transition charactereistics measured on a 2 mw 80m diameter wind turbine rotor in comparison with transition data from wind tunnel measurement. AIAA- Sci Tech Forum. San Diego, CA, USA. AIAA, pp 2019–0801

27. JMadsen J, Lenz K, Dynampally P, Sudhakar P (2009) Investigation of grid resolution requirements for detached eddy simulation of flow around thick airfoil sections. In: Proceedings of the EWEC 2009, Marseille, France

28. Mahmoodi E, Schaffarczyk AP (2012) Actuator disc modeling of the mexico rotor experiment. In: Oldenburg U (ed) Proceedings of the Euromech Coll.: wind energy and the impact of turbulence on the conversion process, Oldenburg, Germany, 2012. University of Oldenburg

29. Menter F (1992) Improved two-equation k-ω turbulence models for aerodynamical flows. Technical Report 103975, NASA Technical Memorandum, Moffett Field, CA, USA

30. NN. Basic machine parameters. Paper circulated during NASA Ames blind comparison panel, NREL, Golden, USA, 2000, 2000

31. Ohno D, Romblad J, Rist U (2020) Laminar to turbulent transition at unsteady inflow conditions: numerical simulations with small scale free-stream turbulence. In: DGLR 2018, pp 214–225. DGLR 2018

32. Özçakmak S, Sœrensen N, Madsen HA, Sœrensen J (2019) Laminar-turbulent transition detection on airfoils by high-frequency microphone measurements. Wind Energy 22(10)

33. Özlem CY, Pires O, Munduate X, Sørensen N, Reichstein T, Schaffarczyk AP, Diakakis K, Papadakis G, Daniele E, Schwarz M, Lutz T, Prieto R (2017) Summary of the blind test campaign to predict high reynolds number performance of du00-w-210 airfoil. AIAA 915:2017–0915

34. Peltzer I et al (2009) In flight experiments for delaying laminar-turbulent transition on a laminar wing glove. Proc IMechE 223:619–626

35. Phisipsen I, Heinrich S, Pengel K, Holthusen H (2015) Test report for measurements on the mexico wind turbine model in dnw-llf. Technical Report LLF-2014-19, DNW, Marnesse, The Netherlands

36. Reeh AD, Weissmüller M, Tropea C (2013) Free-flight investigations of transition under turbulent conditions on a laminar wing glove. Technical Report AIAA-2013-0994, AIAA, Grapevine, TX, USA

37. Reichstein T, Schaffarczyk AP, Dollinger C, Balaresque N, Schülein E, Jauch C, Fischer A (2019) Investigation of laminar-turbulent transition on a rotating wind-turbine blade of multimegawatt class with thermography and microphone array. Energies 12(11):2102

38. Réthoré P-E et al (2011) Mexico wind tunnel and wind turbine modeled in cfd. In: AIAA-3373, Orlando

39. Ronsten G (1992) Static pressure measurements on a rotating and a non-rotating 2.375 m wind turbine blade. comparison with 2d calculations. J Wind Eng 39:105–118

40. Schaffarczyk A et al (2024) Aerodynamic design and performance of a 60%-thick aerofoil for use in inboard sections of wind-turbine blades. In: Proceedings of the the Science of Making Torque from Wind, Firenze

41. Schaffarczyk AP (2008) Numerische polare eines 46% dicken aerodynamischen profils. Technical Report 58, Kiel University of Applied Sciences, Kiel, Germany

42. Schaffarczyk AP (2011) Expertise zum einsatz eines lasermesssystems zur verbesserung des energieertrages und reduzierung der lasten mittels genauerer windnachführung einer windenergieanlage (use of a laser system for increased energy yield and load reduction by improved yaw control). Technical Report 83, Kiel University of Applied Sciences, Kiel, Germany

43. Schaffarczyk AP, Arakawa C (2021) A thick aerodynamic profile with regions of negative lift slope and possible implications on profiles for wind turbine blades. Wind Energy 24(2):162–173

44. Schaffarczyk AP, Schwab D, Breuer M (2016) Experimental detection of laminar-turbulent transition on a rotating wind turbine blade in the free atmosphere. Wind Energy 20:2

45. Schepers JG (1999) An engineering model for yawed conditions, developed on the basis of wind tunnel measurements. Technical Report AiAA-paper, 1999-0039, ECN, Petten, The Netherlands

46. Schepers JG (2002) Verification of European wind turbine design codes, vewtdc; final report. Technical report, ECN-C-01-055, ECN, The Netherlands

47. Schepers JG (2004) Annexlyse: validation of yaw models, on basis of detailed aerodynamic measurements on wind turbine blades. Technical Report ECN-C–04-097, ECN, Petten, The Netherlands

48. Schepers JG (2012) Engineering models in wind energy aerodynamics. PhD thesis, TU Delft, Delft, The Netherlands

49. Schepers JG et al (2012) Final report of iea task 29, mexnex (phase 1): Analysis of Mexico wind tunnel measurements. Technical report, ECN-E-01–12-004, ECN, The Netherlands

50. Schepers JG, Snel H (1995) Dynamic inflow: yawed conditions and partial span pitch control. Technical Report ECN-C–95-056, ECN, Petten, The Netherlands

51. Schepers JG, Snel H (2007) Model experiments in controlled conditions, final report. Technical report, ECN-E-07-042, ECN, The Netherlands

52. Schreck S (2008) Iea wind annex xx: Hawt aerodynamics and models from wind tunnel measurements. Technical Report NREL/TP-500-43508, NREL, Golden, CO, USA

53. Schwab D, Ingwersen S, Schaffarczyk AP, Breuer M (2012) Pressure and hot film measurements on a wind turbine blade operating in the atmosphere. In: Proceedings of the science of making torque from wind, Oldenburg, Germany

54. Seitz A (2007) Freiflug-Experimente zum Übergang laminar-turbulent in einer Tragflügelgrenzschicht. PhD thesis, TU Braunschweig, Germany. DLR-FB-2007-01

55. Shen WZ, Hansen MOL, Sœrensen JN (2009) Determination of the angle of attack on rotor blades. Wind Energy 12:91–98

56. Shen WZ, Zhu WJ, Sørensen JN (2012) Actuator line/navier-stokes computations for the Mexico rotor: comparison with detailed measurements. Wind Energy 15(5):811–825

57. Simms D, Schreck S, Hand M, Fingersh LJ (2001) Nrel unsteady aerodynamics experiment in the nasa-ames wind tunnel: a comparison of predictions to measurements. Technical Report NREL/TP-500-29494, NREL, Golden, CO, USA

58. Simms DA, Hand MM, Fingersh LJ, Jager DW (1999) Unsteady aerodynamics experiment phases ii–iv, test configurations and available data campaigns. Technical Report NREL/TP-500-25950, NREL, Golden, CO, USA

59. Snel H, Schepers JG (1995) Joint investigation of dynamic inflow effects and implementation of an engineering method. Technical Report ECN-C–94-056, ECN, Petten, The Netherlands

60. Somers D (1997) Design and experimental results for the s809 airfoil. Technical Report NREL/SR-440-6918, NREL, Golden, CO, USA

61. Sørensen NN, Michelsen JA, Schreck S (2002) Navier-stokes prediction of the nrel phase vi rotor in the nasa ames 80 ft x 120 ft wind tunnel. Wind Energy 5:151–168

62. Suder KL, OBrian JE, Roschko E (1988) Experimental study of bypass transition in a boundary layer. Technical Report NASA, TM 100913, NASA, Cleveland, Ohio, USA

63. Tangler JL (2004) The nebulous art of using wind-tunnel airfoil data for predicting rotor performance. Technical Report NREL/CP-500-31243, NREL, Golden, Co, USA

64. Tangler JL, Kocurek JD (2005) Wind turbine post-stall airfoil performance characteristics guidelines for blade-element momentum methods. Technical report, NREL/CP-500-36900 and 43rd AIAA Aerospace Meeting and Exhibition, Reno, Nevada, USA

65. Timmer WA (2007) Wind turbine airfoil design and testing. Wind turbine aerodynamics: a state-of-the-art. von Karman Institute for Fluid Dynamics, Rhode Saint Genese, Belgium, pp 2005–2007

66. Timmer WA, Schaffarczyk AP (2004) The effect of roughness at high reynolds numbers on the performance of airfoil du9 97-w-300mod. Wind Energy 7(4):295–307

67. Stahl B, Zhai J (2003) Experimentelle untersuchung an einem 2d-windkraftprofil im dnw-kryo kanal. Technical Report DNW-GUK-2003 C 02, Köln, Germany

68. Stahl B, Zhai J (2004) Experimentelle untersuchung an einem 2d-windkraftprofil bei hohen reynoldszahlen im dnw-kryo kanal. Technical Report DNW-GUK-2004 C 01, Köln, Germany(in German)
69. van Groenwoud GJH, Boermans LMM, van Ingen JL (1983) Onderzoek naar de omslag laminair-turbulent van de grenslaag op de rotorbladen vand de 25 m hat windturbine. Technical Report Rapport LR-390, Techische Hogeschool Delft, Delft, The Netherlands
70. van Ingen JL, Schepers JG (2012) Prediction of boundary layer transition on wind turbine blades using e^n-method and a comparison with measurements. Private communication, Technical report, ECN, Petten, The Netherlands
71. van Rooij RPJOM (1996) Modifications of the boundary layer calculation in rfoil for improved airfoil stall prediction. Technical Report report IW-96087R, TU Delft, Delft, The Netherlands
72. van Rooij RPJOM (2007) Open air experiments on rotors, a state-of-the-art. In: Lecture series, von Karman Institute for fluid dynamics, Rhode Saint Genese, Belgium, pp 2005–2007. VKI
73. Wolf M, Jeromin A, Schaffarczyk AP (2010) Numerical prediction of airfoil aerodynamics for thick profiles applied to wind turbine blade roots. In Proceedings of the DEWEK, Bremen, Germany

Chapter 9
Impact of Aerodynamics on Blade Design

So wie die Sache steht, ist das Beste, auf das zu hoffen ist, ein Geschlecht erfinderischer Zwerge, das für alles zu mieten ist (B. Brecht, Life of Galileo, 1941) [9].[1]

9.1 The Task of Blade Design

Now having introduced all knowledge from fluid mechanics, it is high-time to try to give an overview on what is really used in practical wind turbine blade design. Referring to Chap. 10 and especially Fig. 10.1 we see that with the development of huge off-shore wind turbines up to 170 m rotor diameter, a period of exponential growth has started again after some years of dormancy. An European Project *UpWind* [29] (compare to [21]) forecasts feasible blades for wind turbines up to 20 MW with blade lengths up to 120 m, while currently (2014) turbines reaching rotor diameters of more than 230 m and rated power of 16 MW have been constructed. Recently [5] a 50 MW wind turbine with rotor diameter of 500 m and estimated blade mass of 400 tons has been proposed. Design of a real commercial wind turbine blade is a complicated process and involves (at least) the following:

- aerodynamics,
- structural design
- and manufacturing.

Aerodynamics therefore only touches the outer skin of the blade giving an idea of how the shape should be. Manufacturing relies very much on the structural design, because this more or less defines the amount of material used. On the other hand, the amount of material used determines mass and this is related by engineering rules to

---

[1] As things are, the best that can be hoped for is a generation of inventive dwarfs who can be hired for any purpose.

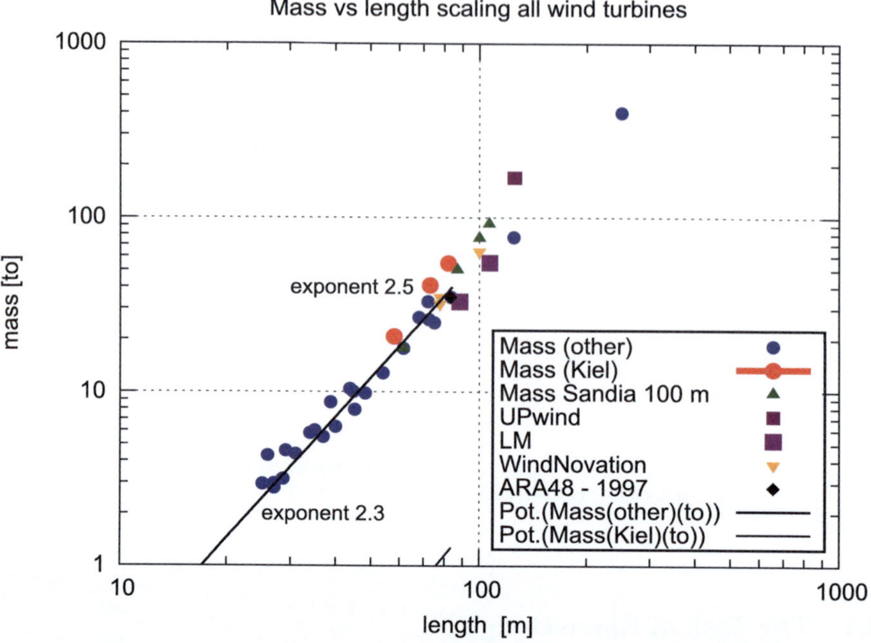

Fig. 9.1 Scaling of wind turbine blade masses in metric tons

costs and prices [2]. To give an order of magnitude, € 10 per kg have not be exceeded to be competitive on the market place. To see how masses evolve or *scale* we may have a look on Figs. 9.1, 10.2 and 2.6 of Chap. 10.

A double-logarithmic plot reveals the scaling exponent n from $m \sim D^n$. It must clearly lie between 2 and 3: The first dimension, the length of the blade, is determined by the gross design of the rotor. Then the chord c follows from $\Gamma \sim c \cdot c_L$ and yields—by an appropriate airfoil choice—(see Sect. 9.2) the second dimension. The remaining third dimension then is left to the structural engineer. Scaling only length and chord without augmenting the internal structure then results in $n = 2$, whereas simply increasing **all three** dimensions results in $n = 3$ [20]. From Fig. 9.1 and (5) [22] we may see that smaller ($R_{tip} < 80$ m) blades have had $n \leq 2.4$ where larger blades seem to need so much more material resulting in $n > 3$ [2, 21]. The reason for this is rather simple: structural design according to the international standard of [30] designated loads that the structure has to withstand for a design lifetime of 20 years, for example. We will come back to these items in Sect. 9.4.

9.2 Airfoils for Wind Turbine Blades

Choosing the right airfoil from an *airfoil catalog*, see (8) [1] or [4, 27], comes next after having chosen blade length and having identified lift c_L and chord c. Numerous profiles exist but only some measured aerodynamic lift data have even been published. As a general introduction, the reader may consult [25]. To minimize losses, lift-to-drag ratio has to be as high as possible, as we saw in Chap. 5. Nevertheless the highest forces are associated with c_L^{max} so this has to be controlled during the design phase and confirmed by measurements.

The inverse problem, designing the shape of a profile when the (inviscid) pressure distribution around the profile is given, can be performed by *conformal mapping* (3) [15] and was used by Tangler and Somers (NREL) [32, 33] for designing the Sxxx family of airfoils. One of them (S809) has been used for the NREL NASA-Ames experiment as already described in Sect. 8.5.

Somewhat earlier (in the late seventies) during the *GROWIAN* program [34] at the University of Stuttgart, Wortmann(FX) and Althaus(AH) [4] profiles were investigated.

When XFoil (3) [12] become popular in the late nineties at the Aeronautical Research Association in Sweden (FFA, [3, 7, 8]) and at the Technical University Delft in the Netherlands [35, 36], special profiles were investigated carefully. Risø (now DTU Wind energy) [17] did the same. Only very limited information is freely available how much these genuine wind turbine blade-profiles are used in commercial blades.

Sometimes very slim blades [11, 16] are considered for blade design in combination with *high-lift airfoils*.

Some thick ($\geq$ 30% thickness, see Figs. 8.4, 8.5 and 8.6) airfoils for root sections have already been presented in Chap. 8. There seems to be still room for further improvement [19].

So-called *flat-black airfoils* [6, 31] have been re-investigated again (Fig. 9.2).

9.3 Aerodynamic Devices

Aerodynamic devices are small devices which may easily added or removed from the wing. Among them two play a noteworthy role:

- Vortex generators and
- Gurney flaps.

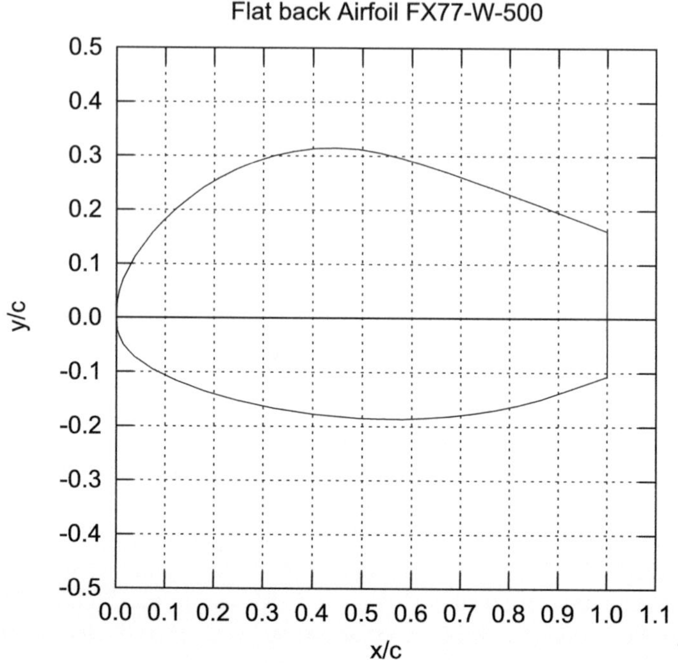

Fig. 9.2 Flat-back Airfoil from University of Stuttgart [4]

9.3.1 Vortex Generators

Vortex Generators (VGs) are common in the aerospace industry. Their use and theoretical understanding is quite extensive, so that a meaningful use is possible. Figure 9.3 shows a typical overview of an arrangement used in medium-sized wind turbines. The triangle extensions are $40 \times 10\,\text{mm}^2$. They are used mainly to introduce (additional) lift without regard to the additional drag they produce. This is clearly seen in Fig. 9.4: c_L^{max} increases by about 20% but the drag as well, so that L2D decreases from about 80 (clean blade)^2 to about 45 only. VGs have been used from the beginning of modern wind turbine applications on blades, mainly to prevent the part close to the hub from early separation. Details of the flow pattern behind VGs have been investigated recently by Velte [39] and others.

---

^{2} This open-jet wind tunnel has a comparably high turbulence intensity of more than 1%.

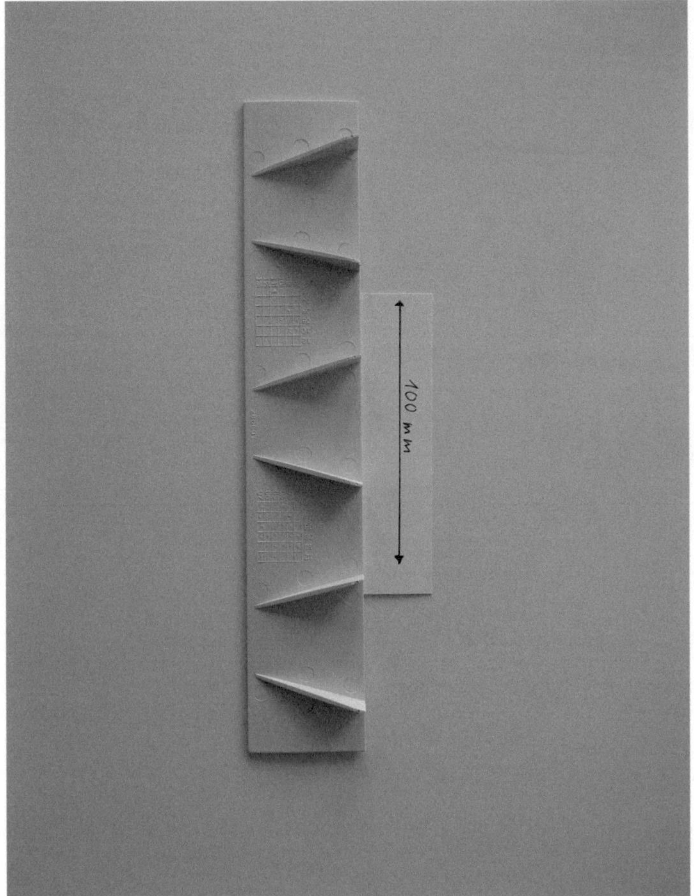

Fig. 9.3 Example of vortex generators used on wind turbine blades. Flow from left. Photo: Schaffarczyk

9.3.2 Gurney Flaps

This device—apparently invented by accident [24]—consists of a small trailing edge flap; see Fig. 9.5. Because of its simplicity to increase lift, it may be used to compensate fouled profiles. In Fig. 9.6 results are shown where the fouling (by insects, sand or other influences) was modeled by some defined surface roughness (Karborundum). Figure 9.6 shows the results of attempts to improve for higher lift. The impact on the lift-to-drag ratio is summarized in Fig. 9.7. Only in combination with Gurney flaps, a small improvement in terms of lift-to-drag ratio seems to be possible when a severe roughness close to the nose is present.

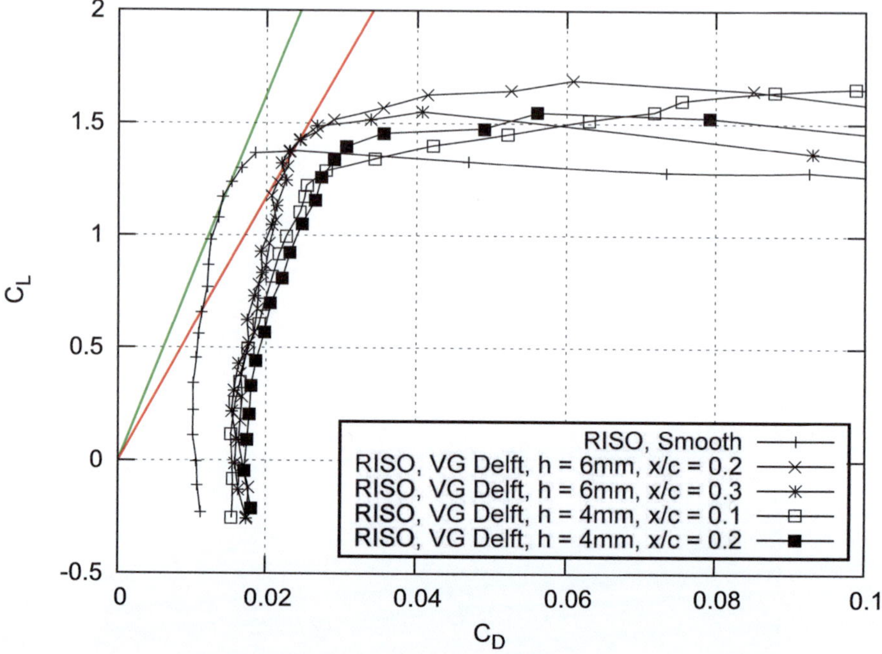

Fig. 9.4 Influence of VGs on the performance on the FFA-W3-211 [3]

9.4 Structural Design and Manufacturing

9.4.1 Structural Design

Wind turbines are used in most cases to produce electricity. The first part of a design therefore is to ensure the predicted power performance. Next to this—and probably much more important—is to ensure *structural integrity* during the foreseen lifetime which—in most cases—is at least 20 years. As we have seen the primary load is the thrust of a wind turbine, Eq. 3.20. Guidelines and regulations for a safe structural design have evolved gradually [18, 30, 40]. Apart from aerodynamics, knowledge from rotor-dynamics [41] and structural design [13][3] is necessary. Some of the tasks are to

- ensure safety against aero-elastic instabilities such as divergence/flutter,
- investigate loads from extreme events expected only a few times within the estimated lifetime and
- sum the effects of loads from rapidly changing operating conditions during normal or electricity generation.

---

[3] This textbook from the late 1980s is surly out-dated but may help to bridge the gap between engineering mechanics and black-box tools like FLEX5 or BLADED.

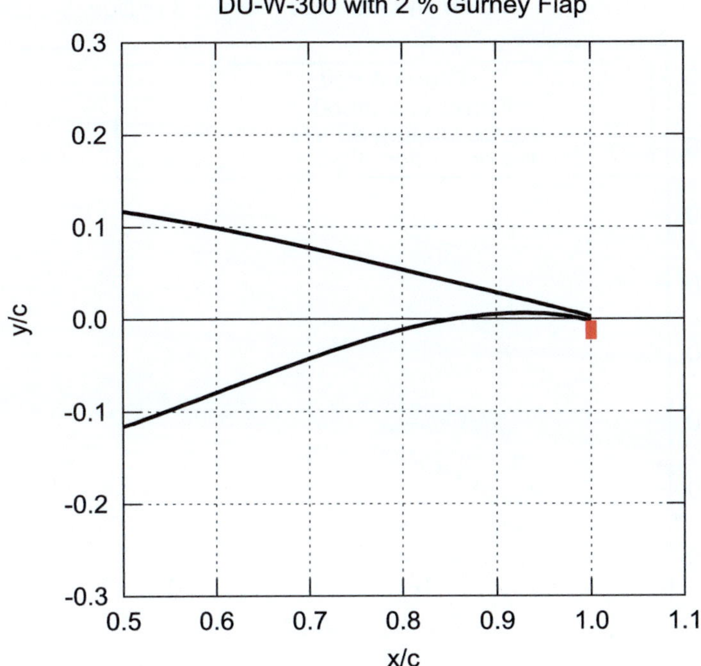

Fig. 9.5 Sketch of a Gurney flap used in the 2002/2003 Kölner KK experiment on high Reynolds number flow on the DU-W-300-mod profile (8) [37, 38]

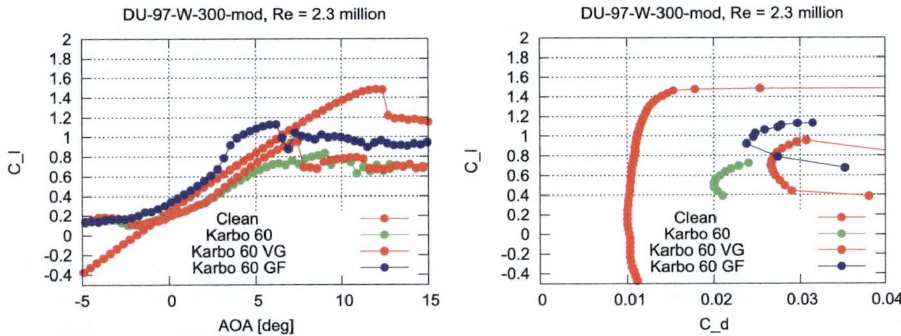

Fig. 9.6 Influence of various devices on DU97-W-300-mod measured at Kölner KK, Re = 2.3 M

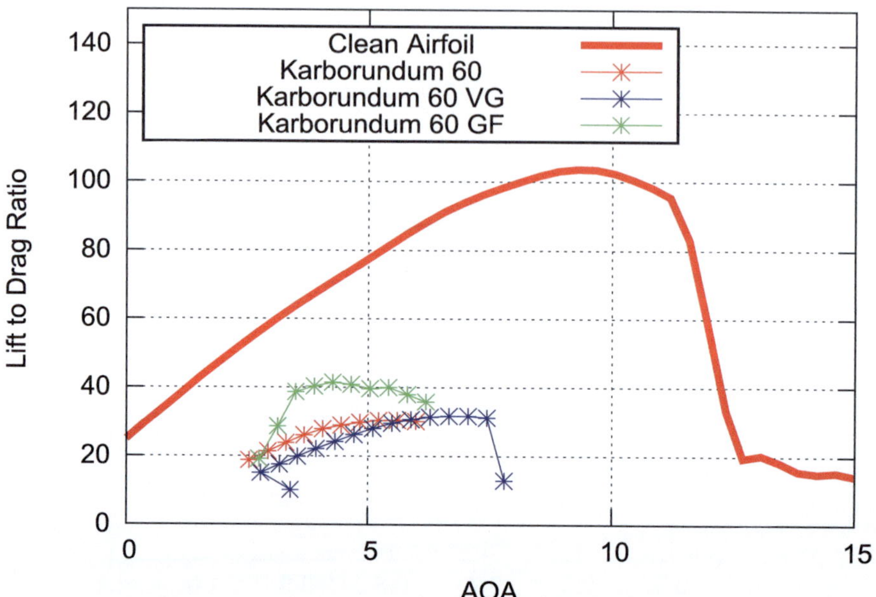

Fig. 9.7 Influence of various measures on lift-to-drag ratio. Karbo = Karborundum = severe roughness in form of sand. VG = Vortex generators, GF = Gurney flaps

A typical iterative process is shown in Fig. 9.8. A turbine may be called optimized if it can resist equally extreme winds and fatigue loads. The IEC standard [30] proposes three possibilities for proof of safety:

- a so-called *simplified load model,*
- simulation modeling (by aero-elastic codes) and
- load measurement.

To compare the amount of work required, it may be interesting to note that the first model is available as an Excel spreadsheet [42], and a safety-check may be finished in hours. A full aero-elastic load model takes much more time (many months), and clearly for measurements, a full prototype must be available.

Aero-elastic modeling may be identified at different stages of sophistication:

I Aerodynamic

1. Blade Element Momentum Code (BEM).
2. Wake-Codes.
3. Full 3D CFD including turbulence modeling.

Blade structural design process

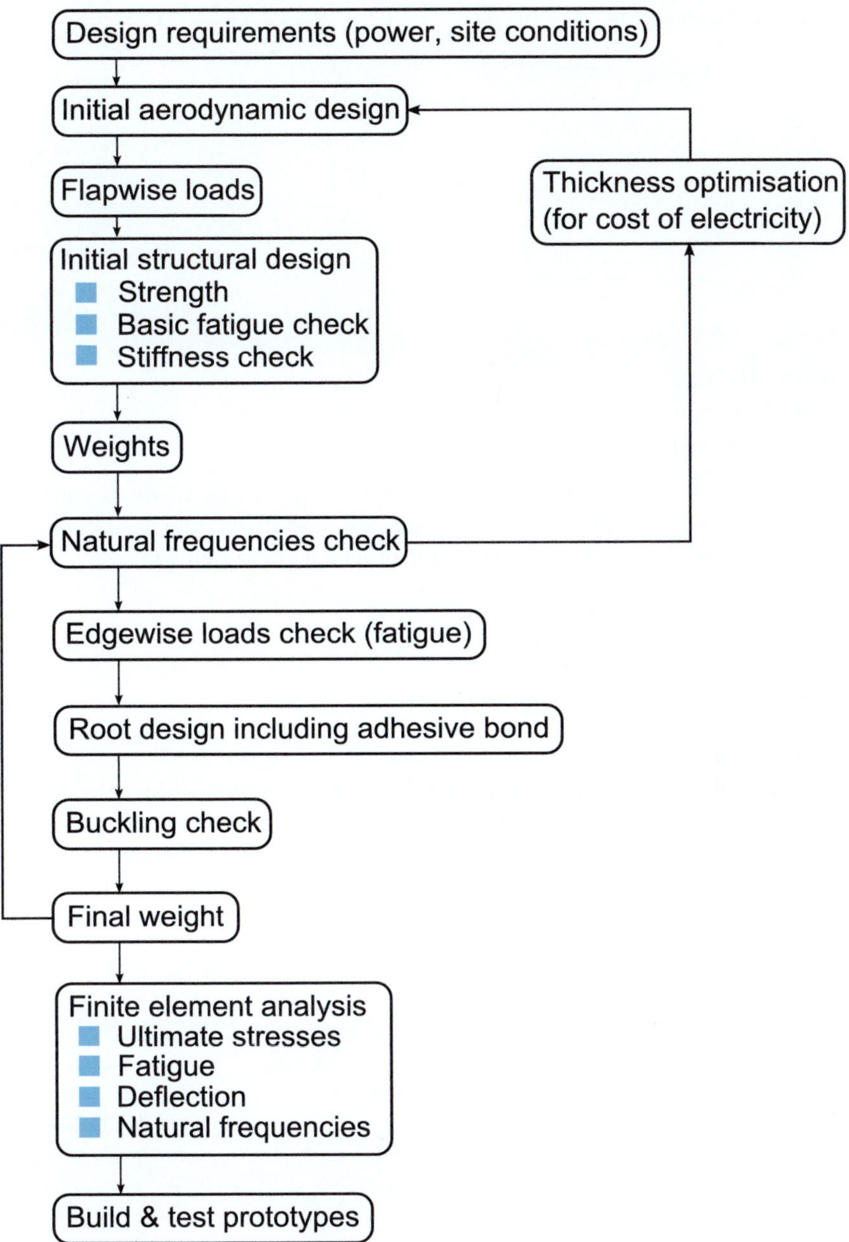

Fig. 9.8 Simplified chart of a structural design

Fig. 9.9 Drag coefficient of
cylindrical shells. Measured
data from Wieselsberger.
Turbulent and transitional
CFD simulations

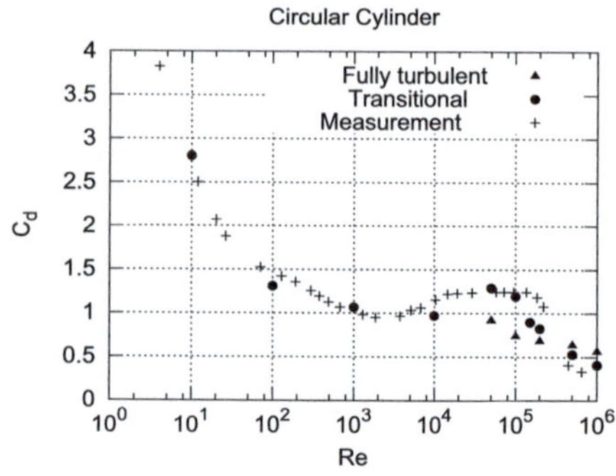

II Structural Mechanics

4. Beam (1D) model.
5. Shell (2D) model.
6. Solid (3D) model.

In principle $3 \cdot 3 = 9$ possibilities for coupling the various methods can occur.
At the present time (1 with 4) coupling is the one most commonly used in industry.
Several industrial codes are available, some of which are as follows:

- FLEX, from Stig Øye, DTU;
- BLADED, by Garrad Hassan;
- GAROS, by Arne Vollan, FEM;
- Phatas, by ECN

and many more.

One special item of structural design in connection with ultimate loads [30]
and more directly related to aerodynamics may be presented in more detail here.
Figure 9.10 shows radially resolved drag coefficients which are important in stand-
still conditions during severe storms. It can be seen that—compared to older blades—
c_D is decreased in the inner parts(closer to hub) and increased in the thinner outer

Fig. 9.10 Radial resolved
drag coefficient of two wind
turbine blades: LM 19—an
old stall blade and more
modern blades

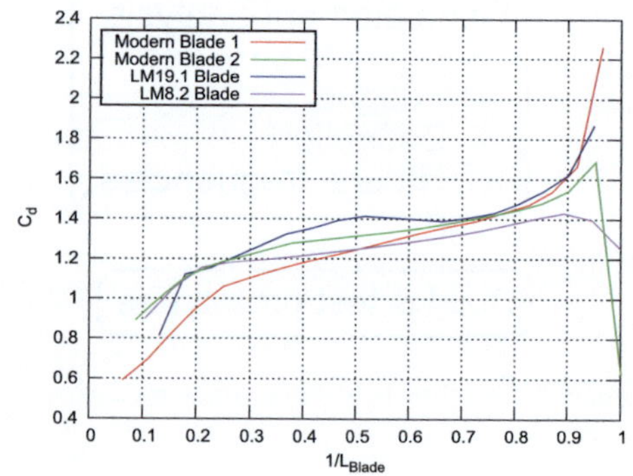

parts (closer to tip). As Reynolds numbers then vary between 500 k and 5 M, predictions must be capable to model the so-called *drag crisis*. Modern CFD codes with transition modeling included seem to be able to do so; see Fig. 9.9. It may be seen from Fig. 9.10 that newer blades try to reduce these loads from the inner part of the blade to the tip.

9.4.2 Manufacturing

Wind turbine blades are manufactured mostly from *glass (or carbon) fiber reinforced plastics (GFRP)* [23, 26]. A typical process is to use moulds which are separated into a lower and upper part. The price for such a mould-system varies based on the number of actually manufactured blades between one and several million Euros. [26] gives a detailed sample-calculation for a 55 m blade of about 18 metric to mass. Material and labor costs (€ 45 per hour) are of almost equal share to give a total production price of about € 160 T corresponding to € 8.9 per kilogram. This price is much lower than typical prices for comparable parts used in commercial airplanes. Nevertheless, and due to the fact that the price per produced kWh has to be decreased further, blades have to become even cheaper. Large-scale production and automation of manufacturing are the key issues for achieving these goals. The normalized price of wind turbines, see Fig. 10.3, will be presented in more detail in Chap. 10.

9.5 Examples of Modern Blade Shapes

Figure 9.11 shows cross-sections for an active stall blade from the late 1990s producing about 0.5 MW rated power. It was presented in some detail in Chap. 7. The aerodynamic experts used profiles from (8) [1] and BEM methods improved by empirical and so-called 3D corrections. This rated power was state of the art back then. Approximately 20 years later, two to three MW now is state of the art for on-shore turbines. Figures 9.12 and 9.13 give some impression for blade shapes from German manufacturer ENERCON. As is well known, ENERCON uses direct-driven (no gear box) turbines which require larger nacelle diameters. It is therefore pertinent to design the inner part of the rotor interlocking with the nacelle. This program already started in 2002 [28] and a new type E-70 (rated power 2.0–2.3 MW) was released in 2004. Unfortunately, very limited information on the aerodynamic properties is available for these kinds of manufacturer-owned blade designs.

In 2014 the largest manufactured blades [14] had a length of 83 m and a mass of less than 35 metric tons. They belong to a 7 MW (S7.0-171) rated power off-shore wind turbine from Samsung Heavy Industries; see Fig. 9.15. Partially flat-back airfoils are used; see Fig. 9.14.

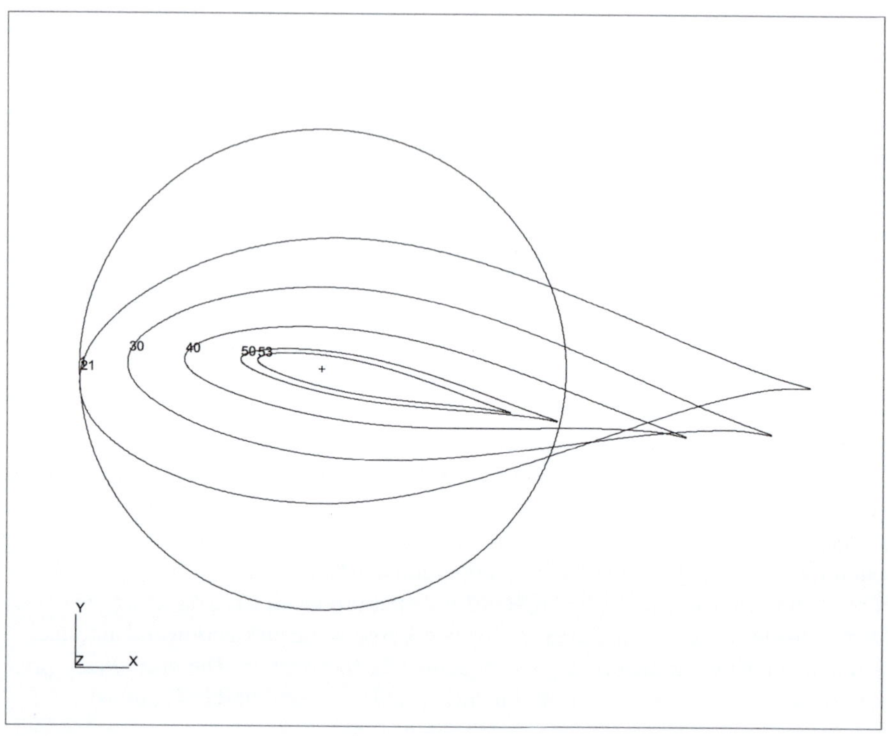

Fig. 9.11 Cross section of an active stall blade. Rated power around 0.5 MW, length 25 m

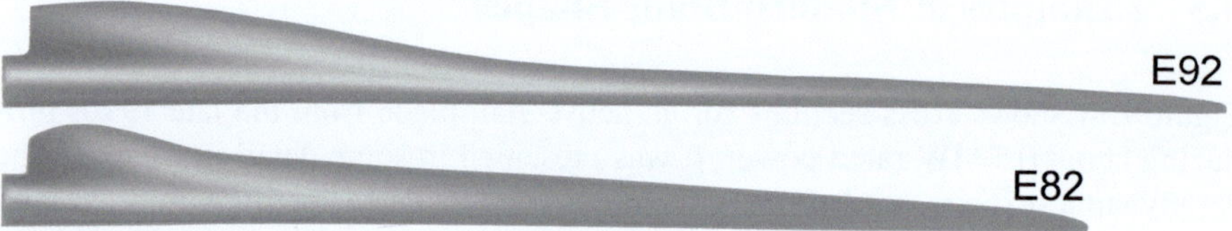

Fig. 9.12 Rated Power around 2 MW, diameter 82 m and 92 m, respectively. Reproduced with permission of ENERCON GmbH, Aurich, Germany

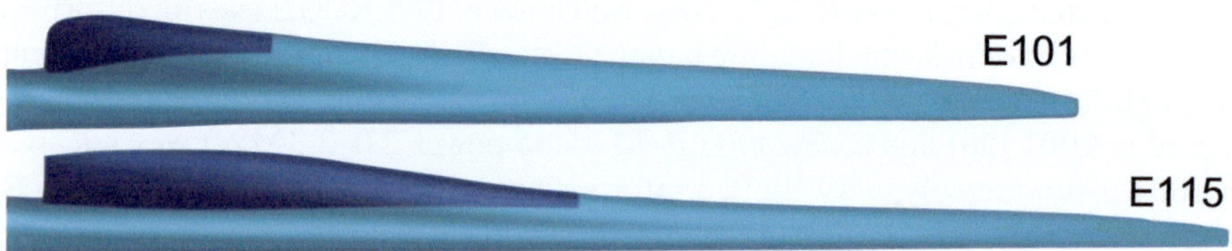

Fig. 9.13 Rated Power around 3 MW, diameter 101 m and 115 m, respectively. Reproduced with permission of ENERCON GmbH, Aurich, Germany

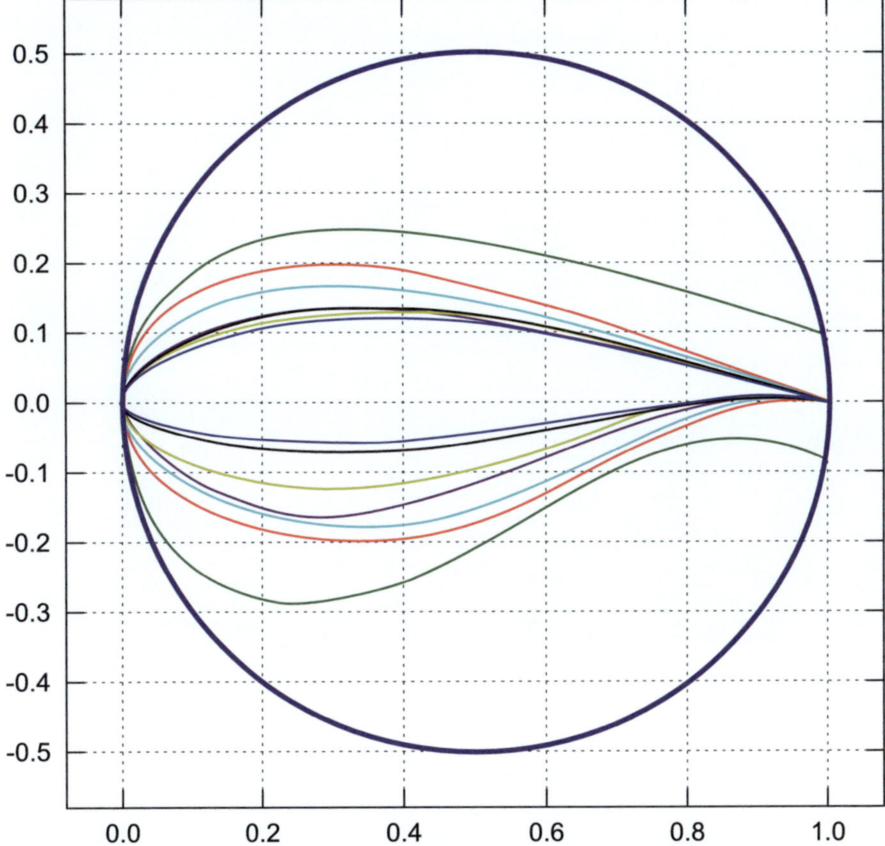

Fig. 9.14 Cross section of a large blade which uses partially flat-black profiles

The largest manufactured blades in Summer 2019 had a length of 107 m (see Fig. 9.16) and an estimated mass between 55 and 63 metric tons. They belong to a 12 MW (GE Haliade-X 12 MW) rated power off-shore wind turbine from GE.

Maximum blade length increased further to about 122 m (see Fig. 9.19). They belong to a 16 MW (GW242-16MW) rated power off-shore wind turbine from GOLDWIND.

There are further projects in research (Avatar, INNwind); see Chap. 10. Industry certainly will bring additional ongoing progress (Figs. 9.17 and 9.18).

Fig. 9.15 Largest (R = 83 m) wind turbine blade in Spring 2014 (1st edition. Reproduced with permission of SSP Technology A/S, Stenstrup, Denmark

Fig. 9.16 Largest (R = 107 m) wind turbine blade in Summer 2019. Reproduced with permission of LM

Fig. 9.17 Largest (D = 252 m) wind turbine blade (December 2023) thus far. Reproduced with permission of GOLDWIND

9.6 Problems

Problem 9.1 Design blades for a wind-car (see problem 5.5)[4] which is specified as follows: $B = 3$, TSR $= \lambda = 5.5$, L2D-ratio $= 80$, $R_{tip} = 0.9$ m, $c_{L,des} = 1.0$ at AOA $= \alpha_{des} = 6° = 6$ deg, $\tilde{w}/u_1 = 0.5$ and drive-train efficiency $\eta_{DrTr} = 0.7$.

---

[4] Much of the underlying principle (low induction) has recently been re-introduced for the design of 10+ MW ordinary wind turbines (10) [10].

Fig. 9.18 Largest (D = 260 m) wind turbine blade (December 2023) thus far. Reproduced with permission of MingYang

Fig. 9.19 Largest blades in 2023. GOLDWIND (left, R = 122) and Ming Yang (right, R = 123)

References

1. Abbott I, von Doenhoff A (1959) Theory of wing sections. Dover, Including a summary of airfoil data
2. Sieros G et al (2012) Upscaling wind turbines: theoretical and practical aspects and their impact on the cost of energy. Wind Energy 15(1):3–17
3. Fuglsang P et al (1998) Wind tunnel tests of the ffa-w3-241, ffa-w3-301 and naca 63-430 airfoils. Technical Report Risø-R-1041(EN), Risø National Lab, Roskilde, Denmark
4. Althaus D (1996) Niedriggeschwindigkeitsprofile: Profilentwicklungen und Polarenmessungen im Laminarwindkanal des Institutes für Aerodynamik und Gasdynamik der Universität Stuttgart. Vieweg, Braunschweig, Germany
5. Escalera Mendoza AS, Shulong Y, Chetan M, Griffith DT (2022) Design and analysis of a segmented blade for a 50mw wind turbine rotor. Wind Eng 46(4):1146–1172
6. Baker JP, van Dam CP, Gilbert BL (2008) Flat-back airfoil wind tunnel experiment. Technical Report SAND2008-2008, Sandia National Laboratories, Albuquerque, NM, USA
7. Björk A (1990) Coordinates and calculations for the ffa-w1-xxx, ffa-w2-xxx and ffa-w3-xxx series of airfoils for horizontal axis wind turbines. Technical Report FFA TN 1990-15, Stockholm, Sweden
8. Björk A (1996) A guide to data files from wind tunnel test of a ffa-w3-211 airfoil at ffa. Technical Report FFAP-V-019, FFA, Bromma, Sweden
9. Brecht B (2008) Life of Galileo. Penguin Classics, London, UK, Reprint
10. Chaviaropoulos P, Siros G (2014) Design of low induction rotors for use in large offshore wind farms. In: Proc. TORQUE 2010, Crete, Greece
11. Corten G (2007) Vortex blades - proposal to decrease turbine loads by 5 %. In WindPower, Los Angeles. CA, USA, p 2007
12. Drela M (1990) XFOIL: an analysis and design system for low Reynolds number airfoils, vol 54. Springer lecture notes in engineering. Springer-Verlag, Berlin, Heidelberg, Germany

13. Eggleston DM, Stoddard FS (1987) Wind turbine engineering design. van Nordstand, New York, USA

14. Eichler K (2013) SSP technology - blade design. In: VDI-conference. VDI Verlag

15. Eppler R (1990) Airfoil design and data. Springer-Verlag, Berlin, Heidelberg, Germany

16. Fuglsang P (2004) Aero-elastic blade design - slender blades with high lift airfoils compared to traditional blades. In: Wind turbine blade workshop, Albuquerque, NM, USA

17. Fuglsang P, Bak C (2004) Development of the risø wind turbine airfoils. Wind Energy 7(2):145–162

18. GmbH GLW (2010) Guidelines for the certification of wind turbines. Technical report, Germanischer Lloyd Windenergie GmbH. Hamburg, Germany

19. Grasso F (2012) Design of thick airfoils for wind turbines. In: Wind turbine blade workshop, Albuquerque, NM, USA

20. Griffith DT, Ashwill D (2011) The Sandia 100-meter all-glass baseline wind turbine blade: SNL 100-00. Technical Report SAND2011-3779, Sandia National Laboratories, Albuquerque, NM, USA

21. Hillmer B et al (2007) Aerodynamic and structural design of MultiMW wind turbine blades beyond 5 mw. In: The science of making torque from wind. J Phys Conf Ser 75:01202

22. Jamieson P (2018) Innovation in wind energy, 2nd edn. Wiley & Sons, Chichester, UK

23. Jaquemotte P (2012) Fertigung von rotorblättern, einsatz von carbonfasern (manufacturing of rotor blades - use of carbon fibers). private communication

24. Katz J (2006) Aerodynamics of race cars. Annu Rev Fluid Mech 38:27–63

25. Lissaman PBS (2009) Wind turbine airfoils and rotor wake. In: Spera A (ed) Wind turbine technology, 2nd edn. ASME Press, New York, USA

26. Ludwig N (2013) Automated processes and cost reductions in rotor blade manufacturing. In: VDI-conference, rotor blades of wind turbines, Hamburg, Germany. VDI

27. Miley SJ (1982) A catalog of low Reynolds number airfoil data for wind turbine applications. Technical Report RFP-3387 UC-60, Golden, CO, USA

28. NN. Considerably higher yields - revolutionary rotor blade design. Technical Report 3, Wind Blatt, Aurich, Germany, 2004

29. NN. Upwind - design limits and solutions for very large wind turbines. Technical report, EAWE, 2011. Brussels, Belgium

30. NN. Iec 61400, wind turbines, design requirement. Technical report, International Electrotechnic Commision, 2019

31. Schaffarczyk A et al (2024) Aerodynamic design and performance of a 60%-thick aerofoil for use in inboard sections of wind-turbine blades. In: Proc the science of making torque from wind, Firenze, Italy

32. Tangler J, Smith B, Jager D (1992) Seri advanced wind turbine blades. Technical Report NREL/TP-257-4492, Golden, CO, USA

33. Tangler J, Somers DM (1995) NREL airfoil families for haw turbine dedicated airfoils. Technical report. AWEA

34. Thiele HM (1983) Growian-rotorblätter: Fertigungsentwicklung, bau und test (growian's rotor-blades: development, construction and testing). Technical Report BMFT-FB-T 83-11, BMFT, Bonn, Germany

35. Timmer WA (2007) Wind turbine airfoil design and testing. In: Wind turbine aerodynamics: a state-of-the-art. von Karman Institute for Fluid Dynamics, Rhode Saint Genese, Belgium, pages 2007–05

36. Timmer WA, van Rooij RPJOM (2003) Summary of the delft university wind turbine dedicated airfoils. J Sol Energy Eng 125(4):488–496

37. und BS, Zhai J (2003) Experimentelle untersuchung an einem 2d-windkraftprofil im dnw-kryo kanal. Technical Report DNW-GUK-2003 C 02, Köln, Germany

38. und BS, Zhai J (2004) Experimentelle untersuchung an einem 2d-windkraftprofil bei hohen reynoldszahlen im dnw-kryo kanal. Technical Report DNW-GUK-2004 C 01, Köln, Germany(in German)

39. Velte CM (2009) Characterization of vortex generator induced flow. PhD thesis, Technical University of Denmark, Lyngby, Denmark
40. Veritas DN (2002) (DNV)/Riso/o. Guidelines for design of wind turbines, 2nd ed. Technical report, Roskilde, Denmark
41. Vollan A, Komzsik L (2012) Computational techniques of rotor dynamics with the finite element method. CRC Press, Boca Ratton, Fl, USA
42. Wood D (2011) Small wind turbines. Springer-Verlag, London, UK

Chapter 10
Concluding Remarks on Further Developments

10.1 State of the Art

In this book we have tried to give an overview of aerodynamics of wind turbines and its application to blade design. During the last few decades rotor-size (see Fig. 10.1) increased more than one order of magnitude (exponentially). After a five-year period of stagnation, starting in 2010 there seems to be exponential rotor-size growth occurring again culminating so far with the design and construction of 252/260 m diameter 16 MW off-shore wind turbine.

Nevertheless design methods developed in the first third of the last century (BEM) are still state of the art for industry. Progress to more sophisticated fluid mechanics (CFD) is slow partly due to the high amount of computational effort and partly due to the limited gain in accuracy. It has to be noted that blade masses have shown to be much smaller than originally estimated; see [10] and Fig. 9.1 from Chap. 9. Advanced control plays a key role but any details are out of the scope of this textbook.

10.2 Upscaling

In Chap. 9, Fig. 9.1 we have already seen a general trend for upscaling when including all sizes of wind turbines. In Fig. 10.2 only *big* (larger than state of the art, larger than 2–3 MW rated power, $R > 60$ m) turbines are shown. Early attempts of UAS Kiel (9) [5] for a feasible 10 MW blade design resulted in a *scaling exponent $m \sim R^n$* of $n \approx 2.5$. Sandia (9) [4] simply **used** $n = 3$ for their approach. Very recent design (9) [3] and theoretical investigations [1, 2], see Fig. 10.2, show that a even smaller ($n \approx 2.2$) should be possible. They based their estimates on glass-fiber blade data for lengths between 40 and 80 m.

Since 2006 many projects were started for investing in turbines up to 20 MW rated power. The first of them called *UPwind* lasted from March 2006 to February 2011 (9) [7]. A blade length of about 126 m was estimated accompanied with a blade

© The Author(s), under exclusive license to Springer Nature Switzerland AG 2024
A. P. Schaffarczyk, *Introduction to Wind Turbine Aerodynamics*, Green Energy
and Technology, https://doi.org/10.1007/978-3-031-56924-1_10

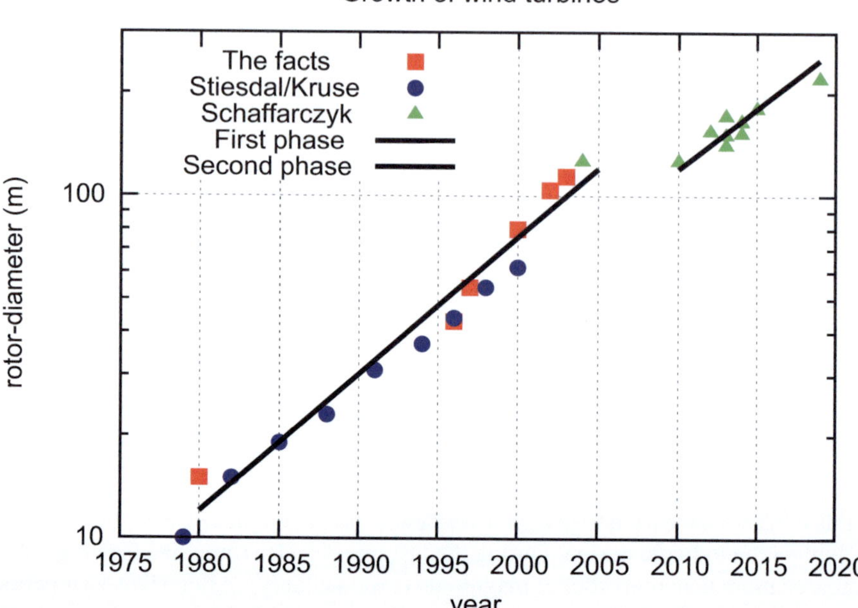

Fig. 10.1 Increase of sizes of wind turbine rotor diameters, exponential versus non-exponential size growth

mass of about 150 metric tons. One lesson learned is that standard aerodynamics seems to be insufficient for further upscaling. Therefore two further projects have been launched:

1. **INN**ovative **WIND** conversion systems (10–20 MW) for off-shore applications (duration: November 2012–October 2017).
2. **Ad**V**anced **A**erodynamic **T**ools for l**A**rge **R**otors (duration: November 2013–October 2017). This project is administrated by the *European Energy Research Alliance*.

Somewhat different but nevertheless closely related to improve rotor performance, a German project *Smart Blades* (duration: December 2012 to February 2016) which concentrates on active flow control is underway.

10.3   Outlook

Future developments of wind turbine aerodynamics are currently discussed more in terms of upscaling and less in terms of new aerodynamic concepts. Increasing the size immediately gives increased masses and one begins to wonder if (normalized)

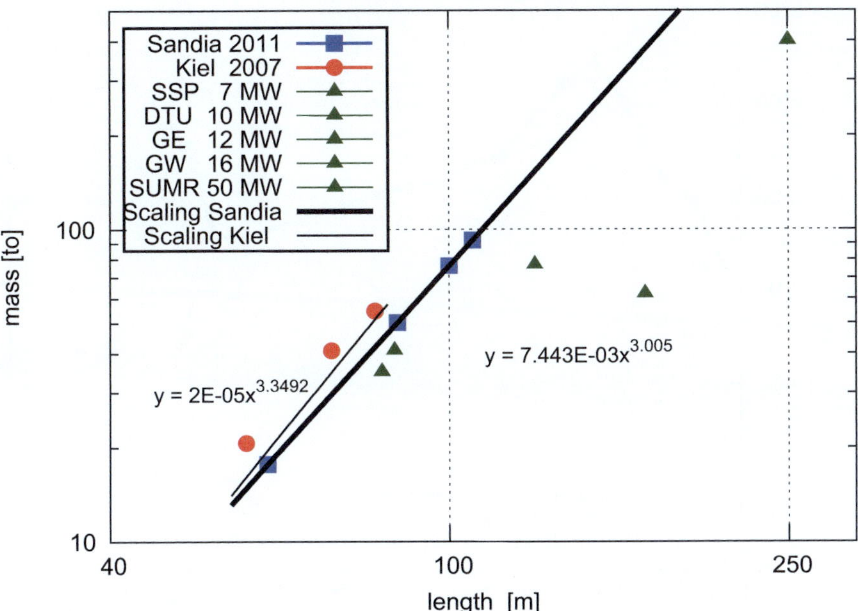

Fig. 10.2 Scaling of wind turbine blade masses (in metric tons)—Large Turbines only. Mass of GE's Haliade XL is estimated and shows a big technological leap to much smaller masses than expected

prices will change. We have emphasized this in the last sections. As energy economics is closely connected to energy policy [9], we may see the driving force for further aerodynamic research from this point of view: the price per kWh of produced electricity [6][1] must decrease further.

One important part of the overall cost comes from investment in the turbine. Figure 10.3 shows the development for turbines in general and Chinese [8] turbines in particular. It is interesting to see that in 2004 both prices were equal and reached a maximum (at different levels) in 2008. IN 2019 a trend toward convergence (in 2025) was seen. However, due to several events of global impact since 2020 a lot of additional variabilities were added; see Fig. 10.3.[2] It is interesting to see what it will mean to the international market if the differences in price remain so large.

---

[1] It is interesting to observe [9] the difference for residential electricity prices around the world from 2012 to 2035 (in US$ per MWh): Japan: 240, European Union 200, USA 120–130 and 70–90 for China. Only in China, prices for industrial consumers are higher than for residential consumers.

[2] In 2022 and 2023 two MoUs were signed around the EU prime-ministers with the result that ambitious plans for off-shore wind in the North Sea (and Baltic Sea) were formulated: 2030: 110 GW, 2040: 230 GW and finally in 2050 around 300 GW.

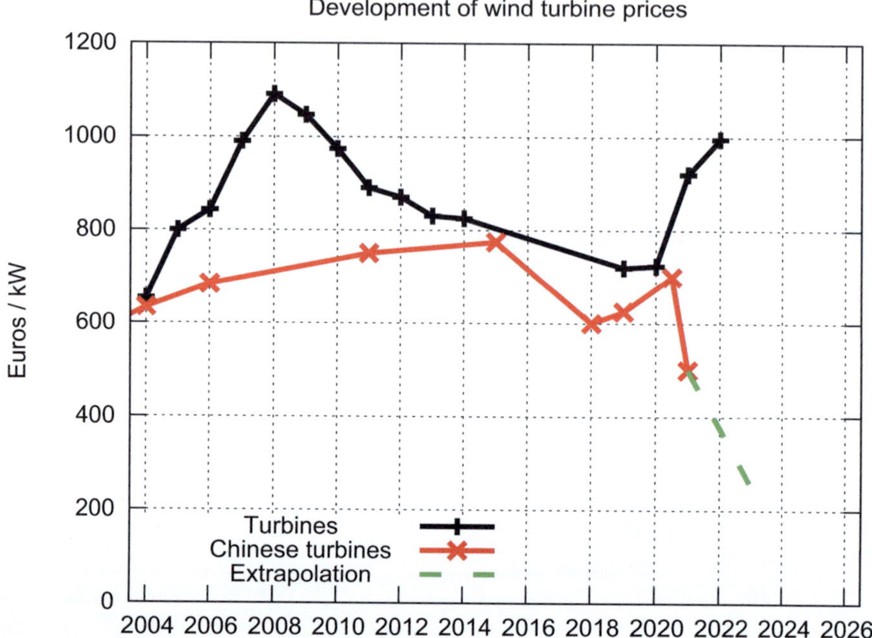

Fig. 10.3 Development of wind turbine prices, *Sources* Wind Power Monthly, IEA Technology Road map Wind Energy, [8] CWEA, Wang ZhongXia [11] and others

References

1. Bak C (2013) Aerodynamic design of wind turbine rotors. In: Advances in wind turbine blade desing and materials, pp 59–108
2. Chaviaropoulos P, Siros G (2014) Design of low induction rotors for use in large offshore wind farms. In: Proceedings of the TORQUE 2010, Crete, Greece
3. Eichler K (2013) Ssp technology - blade design. In: VDI-Conference. VDI Verlag
4. Griffith DT, Ashwill D (2011) The sandia 100-meter all-glass baseline wind turbine blade: Snl 100-00. Technical Report SAND2011-3779, Sandia National Laboratories, Albuquerque, NM, USA
5. Hillmer B, et al (2007) Aerodynamic and structural design of multimw wind turbine blades beyond 5 mw. In: The science of making torque from wind, pp 01202. J Phys Conf S. 75
6. Lantz E, et al (2012) Iea wind task 26 the past and future cost of wind energy. Technical Report NREL/TP-6A20-53510, IEAwind TCP, Golden, CO, USA
7. NN. Upwind - design limits and solutions for very large wind turbines. Technical report, EAWE (2011). Brussels, Belgium
8. NN. Iea wind (2012) Annual report. Technical report, IEAwind TCP, p 2013
9. Energy Outlook NNWORLD (2013) IEA/OECD. France, Paris, p 2013
10. Peeringa J, Brood R, Ceyhan O, Engles W, de Winkel G (2011) Upwind 20mw wind turbine pre-design. Technical Report ECN-E–11-017, ECN, Petten, The Netherlands
11. Schepers G (2023) Private communication. Oktober

Chapter 11
Basics of Wind Farm Aerodynamics

11.1 Introduction

Since more than two decades [10] wind wake aerodynamics is an established part
wind wind energy aerodynamics [7, 9]. The main objectives are to understand and
model the interaction of large wind farms with the atmosphere. By *Large Wind Farm*
we may refer to *Hornsea 2* which has 1.6 GW rated power and covers an area of
$462 \, \text{km}^2$. As can be easily seen the power per unit area is only $P_{wf} = 3.46 \, \text{W/m}^2$
which is obviously small (compared to the solar constant of $340 \, \text{W/m}^2$, when averaged over the whole earth and day) so improvements are highly desirable.

Another example which highlights the importance of wind farm aerodynamics
(WFA, for short) is seen from [3] where it is reported that the average **over**-prediction
of annual yield from 783 selected US wind farms is $5.5\% \pm 0.7\%$. We will not enter
into a discussion about the economical impact.

We start by defining a wind farm efficiency:

$$\eta_{WF} := \frac{\sum_{i=1}^{N} P_i}{N \cdot P_{i,ref}} \, , \tag{11.1}$$

in other words, it is the output of the wind farm divided by the reference values
of all individual wind farms. Early reference values (alpha ventus, Germany, 2011,
[12]) indicate a value of 95%. This wind farm consists of 12 turbines only with an
approximate distance of 6.5 D. Barthelmie et al. [1] investigated wind farms Nysted
and Horns Rev (both DK) in more detail, especially directional changes. Although
no overall efficiency is given, it seems that these values here are somewhat smaller
than for alpha ventus, although the average distance seems to be somewhat larger
(approx. 7D).

© The Author(s), under exclusive license to Springer Nature Switzerland AG 2024
A. P. Schaffarczyk, *Introduction to Wind Turbine Aerodynamics*, Green Energy
and Technology, https://doi.org/10.1007/978-3-031-56924-1_11

11.2 Selected Results

11.2.1 Wake of a Single Wind Turbine

The main quantities to be investigated for comparisons are as follows:

1. Mean stream-wise velocity profiles for different stream-wise locations;
2. Same as above but for (added) turbulence intensity.

Stevens and Menneveau [8] as well as Porté-Agel [7] review several models and compare them to wind tunnel measurements. It seems that the rather strong impact of wind turbines to generate a long-lasting wake (more than 10D upstream) is well reproduced. This may help to answer (positively) Menneveau's first question; see [5].

11.2.2 Wind Farms

The situation becomes much more complicated when changing to wind farms, as they are embedded in the Atmospheric Boundary Layer (ABL). At this point wind energy meets meteorology [2].

In case of interest, we recommend to become familiar with SOWFA from https://github.com/NREL/SOWFA-6. This open-source package combines the LES approach (see Sect. 7.5.12) for ABL and the Actuator Line approach (see Sect. 7.7) for the turbines. Apart from its resource-consuming (mostly for CPU-time) behavior it has proven to be rather flexible and easy to use. An example is shown in Fig. 11.1 [11]. Upper row shows the development of the velocity deficit up to 10 diameters upstream. A remarkable change in the shape is seen by comparing $D = 7$ and $D \geq 8$. The row below shows $< u'u' >$ the velocity autocorrelation in wind direction which constitutes the main component of the turbulence intensity.

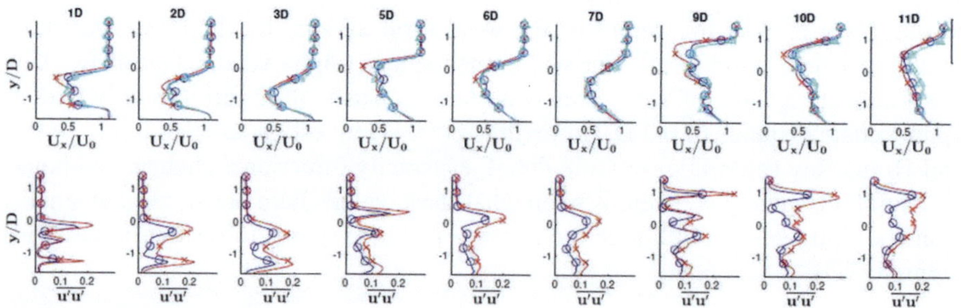

Fig. 11.1 Wake development, reproduced with permission of Prof. Botasso, Technical University of Munich, Germany, from [11]

Luzatto-Fegis et al. [4] try to model the interaction of a wind farm with the lower atmosphere with the goal for a Betz'-type upper limit of wind farm efficiency. In accordance with [5] it is found that $P_{wf}^{max} \approx 30\,\mathrm{W/m^2}$ seems to be possible as an upper limit.

Many attempts have been made to reach for an optimized spacing within wind farms. These questions can only be answered if some cost-model is included. It is interesting to note that [6] found by very few general assumptions that this optimized spacing lies in a range $4 < s < 8$. Limits correspond to situations where either turbine- (4) or land-(8) prices go to zero.

References

1. Barthelmie R, Pryor S, Frandsen S, Hansen K, Schepers J-G, Rados K, Schlez W, Neubert A, Jensen L, Neckelmann S (2010) Quantifying the impact of wind turbine wakes on power output at offshore wind farms. J Atmos Oceanic Tech 27:1302–1317
2. Emeis S (2013) Wind energy meteorology. Springer, Berlin, Heidelberg, Germany
3. Fields MJ, Optis M, Perr-Sauer J, Todd A, Lee JC, Meissner J, Simley E, Bodini N, Williams L, Sheng S, Hammond R (2021) Wind plant performance prediction benchmark phase 1 technical report. Technical Report NREL/TP-5000-78715, NREL, Golden, CO, USA
4. Luzzatto-Fegiz P, Colm-cille PC (2018) Entrainment model for fully-developed wind farms: effects of atmospheric stability and an ideal limit for wind farm performance. Phys Rev Fluids 3(9)
5. Meneveau C (2019) Big wind power: seven questions for turbulence research. J Turbul 20(1):2–20
6. Meyers J, Meneveau C (2012) Optimal turbine spacing in fully developed wind farm boundary layers. Wind Energy 15(2):305–317
7. Porté-Agel F, Bastankhah M, Shamsoddin S (2020) Wind-turbine and wind-farm flows: a review. Bound-Layer Meteorol 174:1–59
8. Stevens RJ, Meneveau C (2017) Flow structure and turbulence in wind farms. Annu Rev Fluid Mech 49(1):311–339
9. Stoevesandt B et al (2022) Handbook of wind energy aerodynamics. Springer, Cham, Switzerland
10. Vermeer L, Sœrensen J, Crespo A (2003) Wind turbine wake aerodynamics. Prog Aerosp Sci 39(6):467–510
11. Wang C, Muz-Simon A, Deskos G, Laizet S, Palacios R, Campagnolo F, Bottasso CL (2020) Code-to-code-to-experiment validation of les-alm wind farm simulators. J Phys: Conf Ser 1618(6):062041
12. Westerhellweg A, Cañadillas B, Kinder F, Neumann T (2014) Wake measurements at alpha ventus - dependency of stability and turbulence intensity. J Phys: Conf Ser 555:012106

Solutions

A.1 Solutions for Problems of Sect. 1.2

1.1 Power density in the wind:

    6 m/s : 136 W/m^2

    8 m/s : 323 W/m^2

    10 m/s : 630 W/m^2

    12 m/s : 1089 W/m^2.

    Annual averaged values:

    4 m/s : 46 W/m^2

    5 m/s : 89 W/m^2

    6 m/s : 154 W/m^2

    7 m/s : 245 W/m^2.

1.2 According to [7] Eq. (7), we have (*1-2-3-formula*)

$$\bar{P} = \rho^1 \left(\frac{2}{3}D\right)^2 \cdot \bar{v}^3 = 1.13 \cdot \frac{\rho}{2}\frac{\pi}{4}\bar{v} \tag{A.1}$$

for a Betz turbine ($c_P = 16/27 = 0.596$) (Table A.1).

1.3 Let m = 0. Then

Table A.1 Numerical values

c_P	$\bar{c_P}$
16/27	1.13
0.58	1.10
0.5	0.948
0.4	0.758

$$p(y, \sigma) = \frac{1}{2\pi\sigma^2} \int du \int dv \cdot exp(-u^2/2\sigma^2) \cdot exp(-v^2/2\sigma^2) \cdot \delta(y - \sqrt{u^2 + v^2}).$$

$$(A.2)$$

Introducing 2D polar coordinates $u = r \cdot cos(\varphi)$ and $v = r \cdot sin(\varphi)$, it follows:

$$p(y, \sigma) = \qquad (A.3)$$

$$\frac{1}{2\pi\sigma^2} \int_0^{2\pi} d\phi \int_0^{\infty} dr \cdot r \cdot exp(-r^2/2\sigma^2)\delta(r - y) == \qquad (A.4)$$

$$\frac{y}{\sigma^2} \cdot exp(-y^2/2\sigma^2). \qquad (A.5)$$

1.4 $\eta = 6/30 = 0.186$.

A.2 Solutions for Problems of Sect. 2.10

2.1 Calculating $\frac{\partial c_P}{\partial \lambda} = 0$, we arrive at

$$\lambda_m^2 - \frac{2}{3}\frac{c_2}{c_1} \cdot \lambda - \frac{1}{3} = 0 \qquad (A.6)$$

with solution

$$\lambda_m = \frac{1}{3}\left(\frac{c_2}{c1} \pm \sqrt{\left(\frac{c_2}{c_1}\right)^2 - 3}\right). \qquad (A.7)$$

2.2 Power increases with swept area which is $\sim D^2$. Therefore

$$P = P_= \cdot \left(\frac{A_D}{A_R}\right) \cdot \eta = 1.16. \qquad (A.8)$$

In practice and especially for a small wind turbine losses have to be included. They may vary between 20 and 50%.

2.3 Clearly two rotors have to be present, but they are extracting energy from the same stream tube. In the best case then these two rotors feed in torque on one shaft. See Fig. A.1 as an example.

A.3 Solutions for Problems of Sect. 3.11

3.1 The Euler equation for steady flow reads as

$$\rho(\mathbf{u} \cdot \nabla)\mathbf{u} = -\nabla p \mathbf{f} . \qquad (A.9)$$

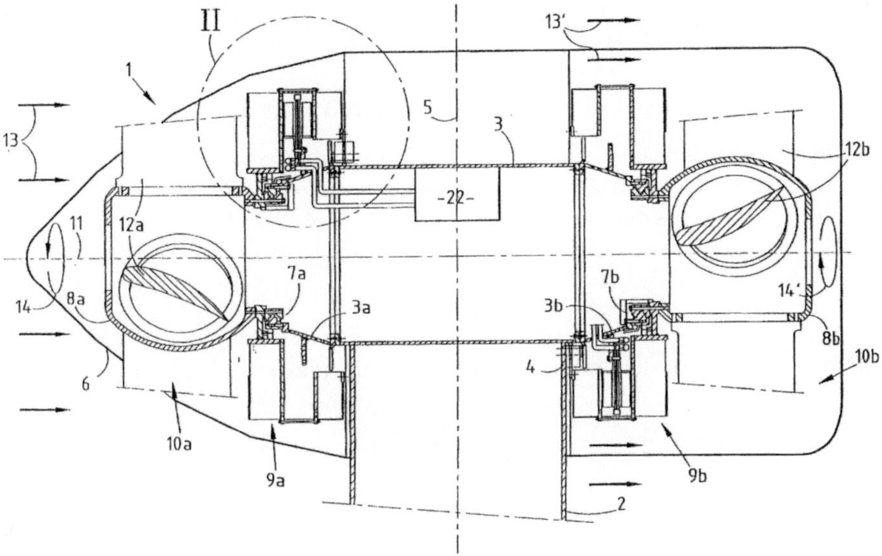

Fig. A.1 Example of a drive train for a counter-rotating wind turbine

Assuming that $\mathbf{f} = \nabla\Phi$[1] and using $\mathbf{u} \cdot \nabla\mathbf{u} = \nabla(\frac{1}{2}\mathbf{u} \cdot \mathbf{u}) - \mathbf{u} \times \omega$, we obtain

$$-\mathbf{u} \times \omega = -\nabla\left(\frac{p}{\rho} + \Phi + \frac{1}{2}\mathbf{u}^2\right). \tag{A.10}$$

If there is no vorticity $\omega = 0$ that may be interpreted that *along* a streamline $\mathbf{u} = \frac{d\mathbf{r}}{dt}$, the quantity H

$$H := \frac{p}{\rho} + \Phi + \frac{1}{2}\mathbf{u}^2 \tag{A.11}$$

remains constant.

Remark: Read (3) [1] for a careful discussion on the abuse of *Bernoulli's* equation for explaining lift.

3.2 A line vortex with concentrated constant circulation along the z-axis (x = y = 0) induces a purely tangential velocity field $\mathbf{u}_\phi$

$$\mathbf{u}_\phi = \frac{\Gamma}{2\pi r}. \tag{A.12}$$

As kinetic energy in a given volume V is $E = \rho \int_V v^2 \cdot dV$, we have per unit length of vortex line:

---

[1] Van Kuik [33] argues that for an actuator disk, this may not be generally true.

$$E = \int_0^{2\pi} d\phi \cdot \int_\epsilon^R \frac{r \cdot dr}{r^2} = const \cdot (log(R) - log(\epsilon)) \ . \tag{A.13}$$

This expression clearly is singular for $\epsilon \to 0$ as well as for $R \to \infty$.

3.3 Remember that 2D potential flow is most easily discussed with use of complex analysis $z = x + i \cdot y$ and the complex velocity $w = u - i \cdot v$. The complex potential $F(z)$ from which w is determined by

$$w(z) = \frac{dw}{dz} \tag{A.14}$$

reads (with U the constant inflow velocity along the real axis [2]):

$$\frac{1}{U} F(z) = -i \left(\frac{kU}{w}\right)^{\frac{1}{2}} + \left(1 - \frac{kU}{w}\right)^{\frac{1}{2}} . \tag{A.15}$$

On integrating:

$$2i \left(\frac{w}{kU}\right)^{\frac{1}{2}} + \left(\frac{w}{kU}\right)^{\frac{1}{2}} \left(\frac{w}{kU} - 1\right)^{\frac{1}{2}} + \frac{1}{2}\pi i - log\left\{\left(\frac{w}{kU}\right)^{\frac{1}{2}} + \left(\frac{w}{kU} - 1\right)^{\frac{1}{2}}\right\} = \frac{z}{k} , \tag{A.16}$$

with

$$k = \frac{2a}{\pi + 4} \quad \text{and } 2a = \text{plate breadth.} \tag{A.17}$$

The flow is symmetrical along the real axis $y \to -y$. At point B $(z_B = a \cdot i)\, w = kU$. Let s being the distance from B on the boundary streamline [2] find

$$x(s) = (s^2 + sk)^{\frac{1}{2}} - k \cdot log\left\{\left(\frac{s}{k} + 1\right)^{\frac{1}{2}} + \left(\frac{s}{k}\right)^{\frac{1}{2}}\right\} , \tag{A.18}$$

$$y(s) = 2(sk + k^2)^{\frac{1}{2}} + \frac{1}{2}\pi k. \tag{A.19}$$

For $s \to \infty$, the last terms in Eqs. (A.18) and (A.19) may be dropped and s can be eliminated so that the asymptotic shape

$$y^2 = 4kx = \frac{8a}{4 + \pi} x \tag{A.20}$$

is an quadratic parabola.

3.4 The (conformal) mapping between the $\zeta = \xi + i \cdot \eta$ and $z = x + i \cdot y$ planes is given by

$$z = \zeta + \frac{\lambda^2}{\zeta} \tag{A.21}$$

and the sub-set (a circle) is due to the equation

$$\zeta = \zeta_m + R \cdot e^{i \cdot \varphi} \quad 0 \le \varphi < 2\pi \tag{A.22}$$

$$z_m = x_m + y_m. \tag{A.23}$$

R is determined by the constraint that the circle must pass through $z = 1$. If $z_M = 0$ and $\lambda = 1$ Eq. (A.21) maps to a strip on the real axis with $-2 < x < 2$ so that the *chord* is $c = 4$. For other values for λ and z_M, we have [2]

$$c \approx 4\lambda \left\{ 1 + \left(\frac{x_M - \lambda}{\lambda} \right) \right\}. \tag{A.24}$$

The (relative) thickness is approximately given by

$$t \approx \frac{3}{4}\sqrt{3} \cdot \left(\frac{x_M}{\lambda} - 1 \right). \tag{A.25}$$

Shifting z_M off the real axis introduces *camber* expressed through an angle β

$$y_M = \lambda \cdot \tan(\beta) \tag{A.26}$$

so that for zero angle of attack a finite lift is generated:

$$c_{L,0} = 2\pi \cdot \tan(\beta) . \tag{A.27}$$

3.5 (a) Kolmogorov and Taylor scales (Table A.2).

 Note that all values have an estimated uncertainty of 10% at least.

 (b) Spatial average and resolution of sensors

 It is clear from its definition, Eq. (3.109), that the spatial correlation (or if we use Taylor's *frozen turbulence* hypothesis, temporal as well) is estimated by λ_T. If we use of an unavoidable finite sensor (of characteristic length scale ℓ) typical correction factors must be of order

$$constant \cdot \left(\frac{\ell}{\lambda_T} \right) . \tag{A.28}$$

Table A.2 Results

ϵ	$u'^2 (m^2/s^2)$	$\eta (mm)$	Taylor microscale (mm)
$5.0 \cdot 10^{-3}$	1.00	0.743	34.5
$4.5 \cdot 10^{-2}$	2.25	0.429	8.6
$3.0 \cdot 10^{-2}$	4.00	0.475	23.0

It can be estimated that if the sensor length is of the same size as Taylor's length, then an error of about 10% may be introduced. See more, for example, in [23] for the case of hot-wire sensor in boundary layer flow.

A.4   Solutions for Problems of Sect. 4.7

4.1 Influence of Wind Shear

(a) $z_0 = 1$ mm: $v_{avr} = 11.87$ m/s, $z_0 = 10$ mm: $v_{avr} = 11.83$ m/s.

(b) $z_0 = 1$ mm: $P_{avr}/P_=0.974$, $z_0 = 10$ mm: $P_{avr}/P_=0.97$.

4.2 TI correction for power:

To start we present a solution for cubic dependence $P(v) \sim v^3$:

$$P(v) = \frac{\rho}{2} c_P \frac{\pi}{4} v^3 \tag{A.29}$$

$$v(t) = \bar{v} + v', \ \bar{v} = <v>, \ <v'> = 0 \tag{A.30}$$

$$<(\bar{v} + v')^3> = \bar{v}^3 + 3\bar{v}^2 \cdot <v'> + 3\bar{v} \cdot <v'^2> + <v'^3> = \tag{A.31}$$

$$= \bar{v}^3 + 3\bar{v}\sigma. \tag{A.32}$$

For more general dependencies expand P(v) into a power series:

$$f(x - x_0) = \sum_{n=1}^{\infty} \frac{f^{(n)}(x_0)}{n!} (x - x_0)^n. \tag{A.33}$$

4.3 Analysis of Wind Data (Fig. A.2)

(a) Gaussian Distribution (Fig. A.3)

(b) Energy spectra (Fig. A.4)

(c) Correlation time.

Correlations often decrease exponentially:

$$c(t) = exp-(t/T). \tag{A.34}$$

This cannot be seen in Fig. A.5, even not the decrease to $1/e = 0.368$.

According to Eq. 3.109 and [24], the part of c(t) close to $t \approx 0$ can be deduced from a quadratic fit: $c(t) = a \cdot t^2 + b \cdot t + c$ (Fig. A.6). From this $\lambda_T \approx \sqrt{c}$ a value may be deduced.

4.4 Model of gust return time

(a) Because [3] assumes Gaussian distributions for U and its derivative the values are comparable small (Table A.3):

(b) A Rayleigh distribution which is often used for a first guess gives (Table A.4).

4.5 For simplicity we assume that a gust may be described by a constant wind speed of value u_g during a period T_g. If there is only one such gust during the measurement

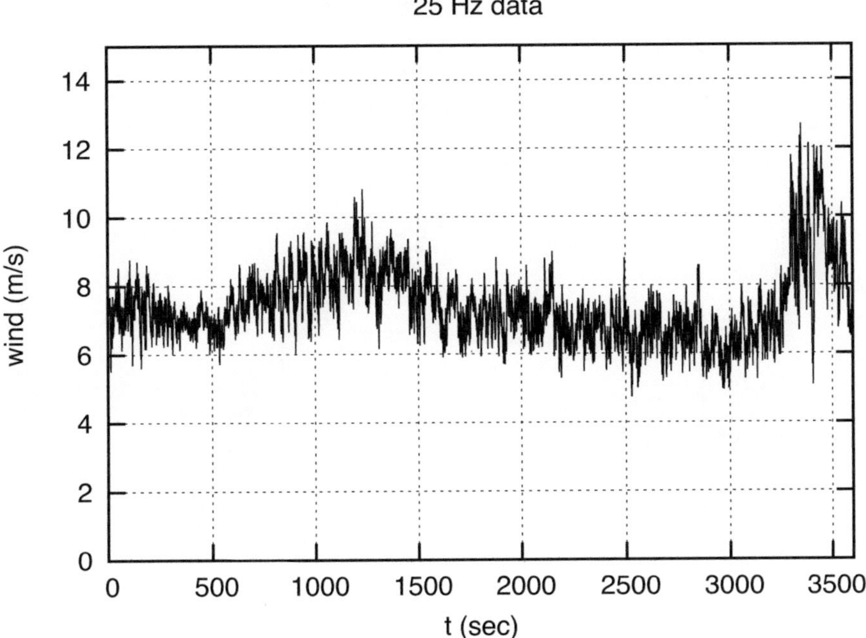

Fig. A.2 Time series of 1 h sample wind data

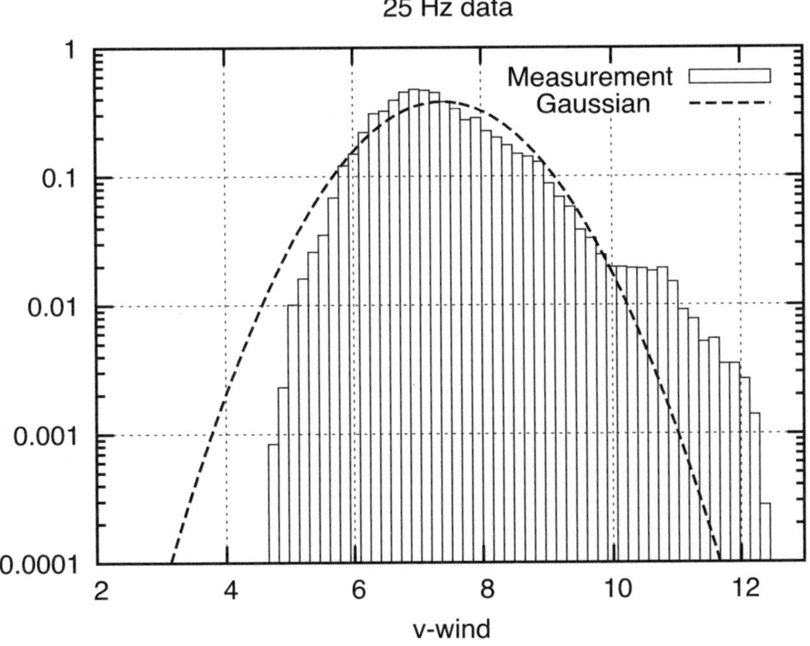

Fig. A.3 Histogram of 1 h wind data equivalent to 90 k data

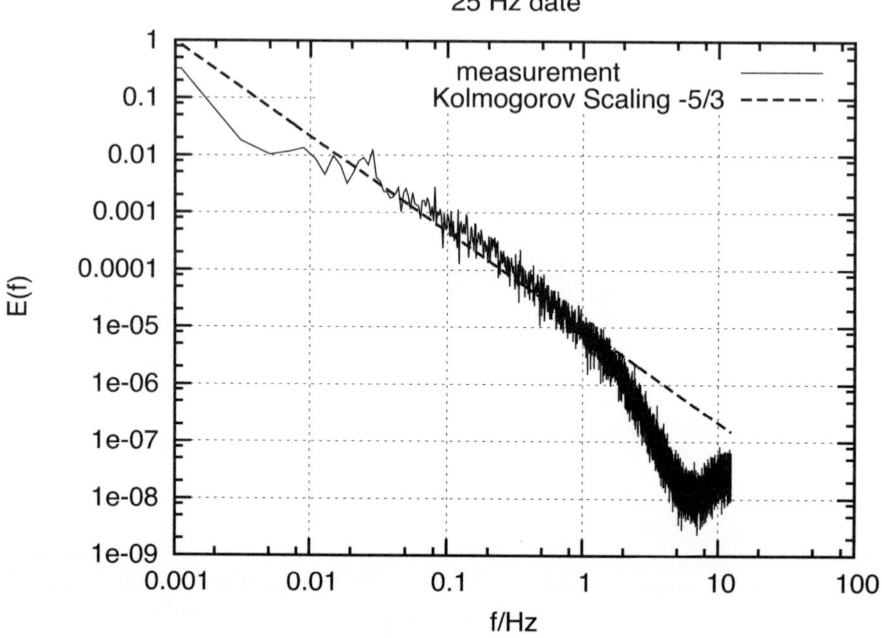

Fig. A.4 Energy content

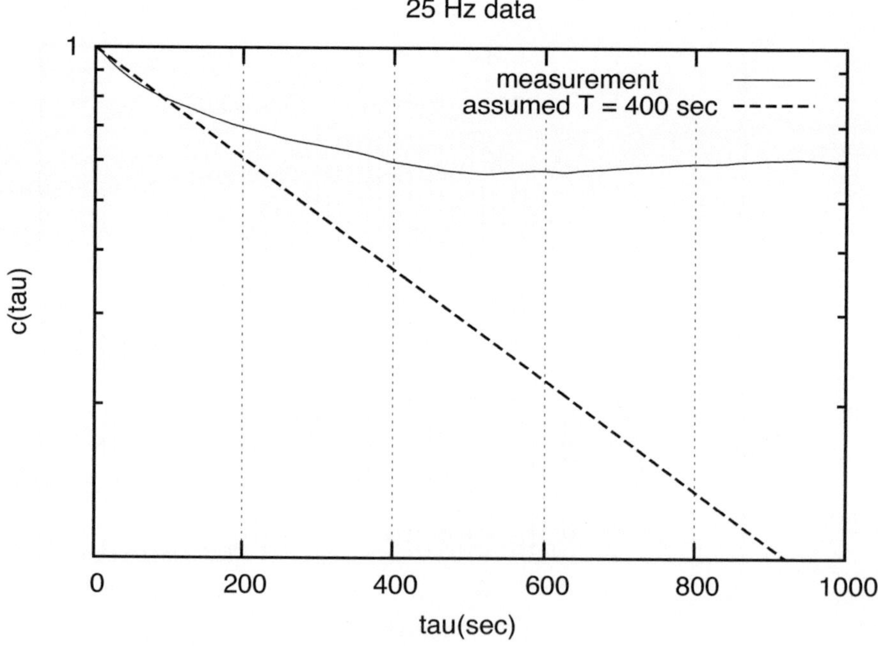

Fig. A.5 Correlation function and exponential with T = 400 s

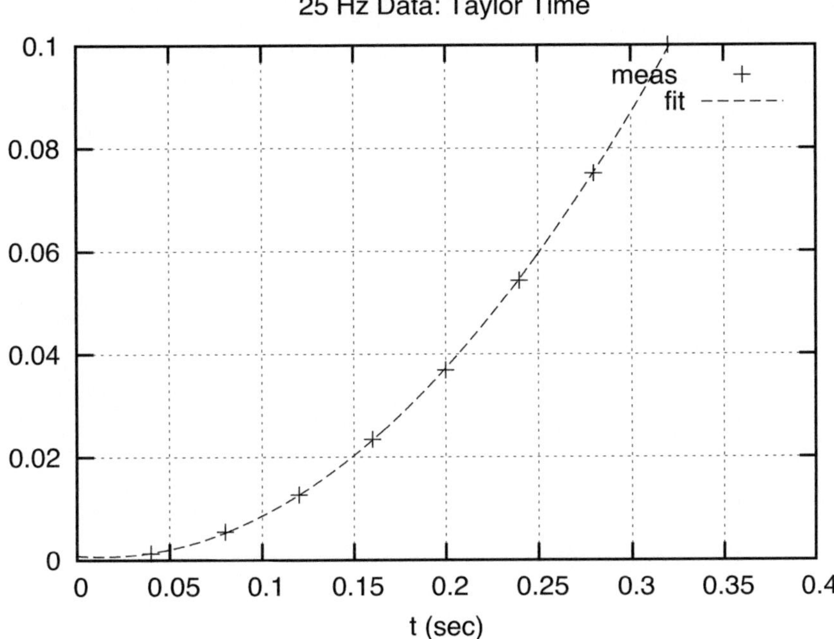

Fig. A.6 Quadratic fit according to [24] gives a Taylor time of about 30 ms

Table A.3 Results for Gaussian distribution

T in years	U_{max}
1	12.3
10	12.6
50	12.9
100	13.0

Table A.4 Results for a Rayleigh distribution

v (m/s)	$10^6 \cdot r(v)$	Return time (years)
22	24.9	0.76
23	9.76	1.95
24	3.65	5.21
25	1.31	14.6
26	0.45	42.6
27	0.15	130

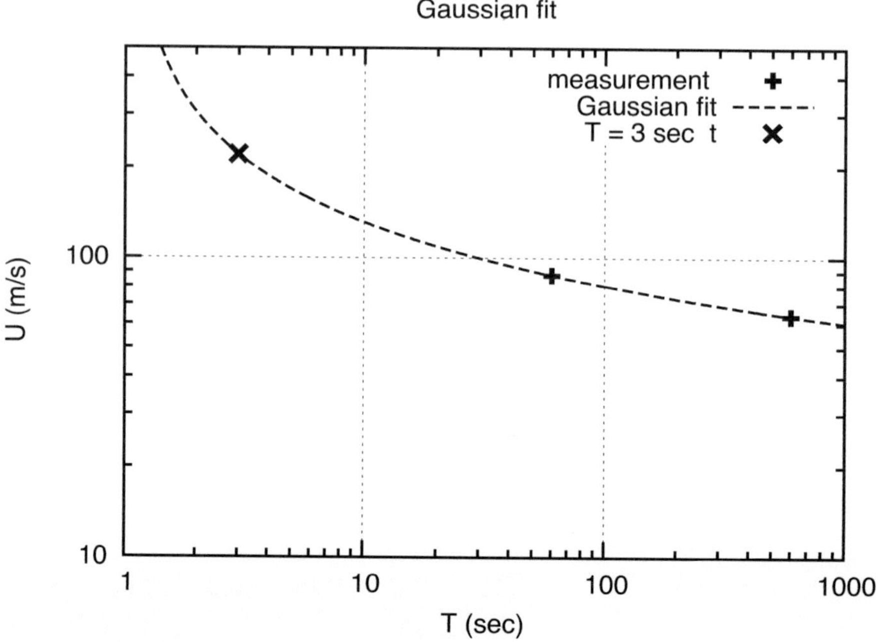

Fig. A.7 Extrapolation of 3 s gust wind speed assuming Gaussian distribution

(and averaging) time T_m, the averaged wind speed should vary only linearity with T. This then gives $v_{3s} \approx 90$ m/s. An extrapolation along the lines of [3] gives a much higher value of 220 m/s; see Fig. A.7 (Fig. A.8).

4.6 (a) TurbSim: The results were generated with the standard Kaimal.inp file but with a somewhat higher resolution: $\Delta t = 0.01$ s and $T = 600$ s (Figs. A.9, A.10 and A.11).

(b) Mann Model

These results come from a time series of the main-flow component (Figs. A.12, A.13 and A.14).

(*c) CTRW Here we present a sample (Fig. A.15).

A.5 Solutions for Problems of Sect. 5.8

5.1 Wind turbine used as a fan

(a) Simple momentum theory applied to a fan (5) [10]:

$$T_{ideal} = \sqrt[3]{2\rho \cdot A_r P^2}. \tag{A.35}$$

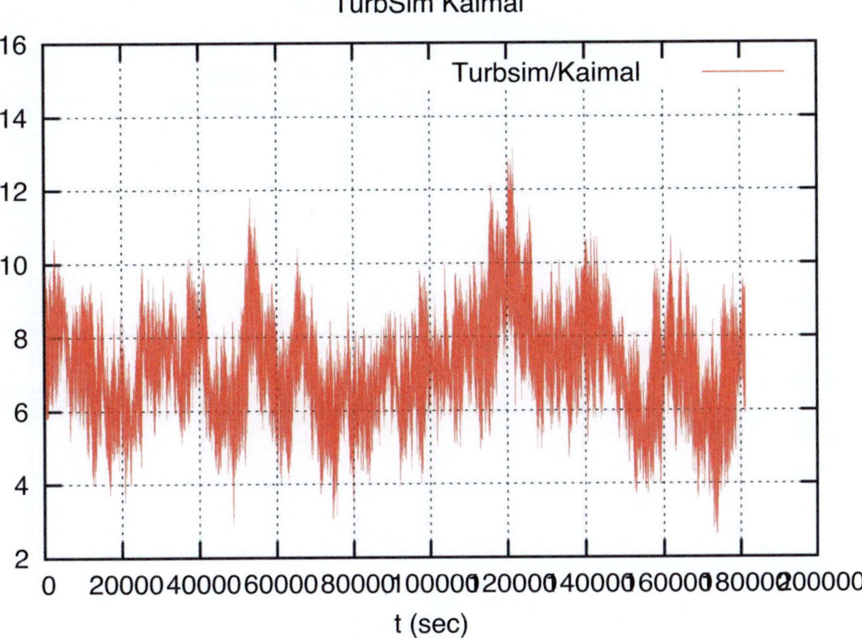

Fig. A.8 Sample time series from TurbSim

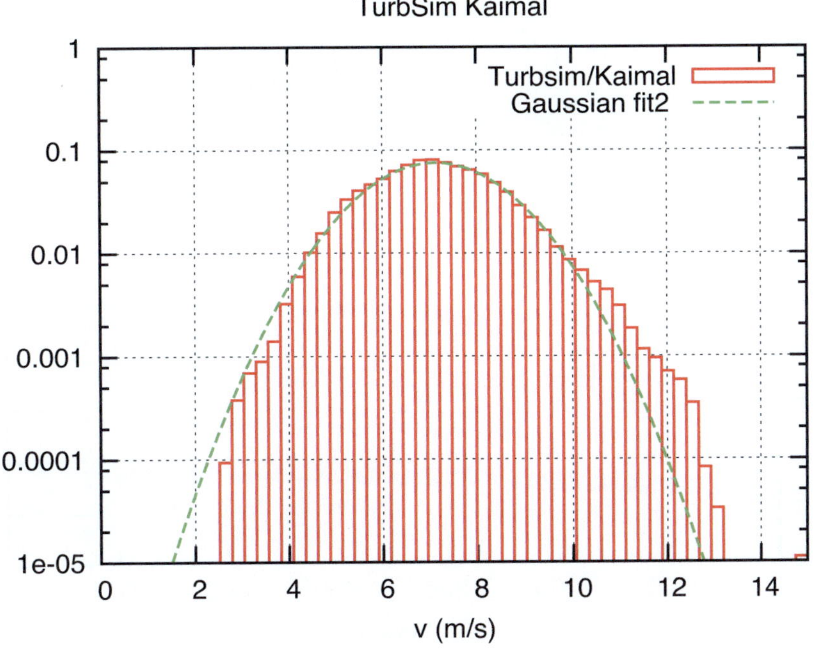

Fig. A.9 Sample histogram for time series from Fig. A.8

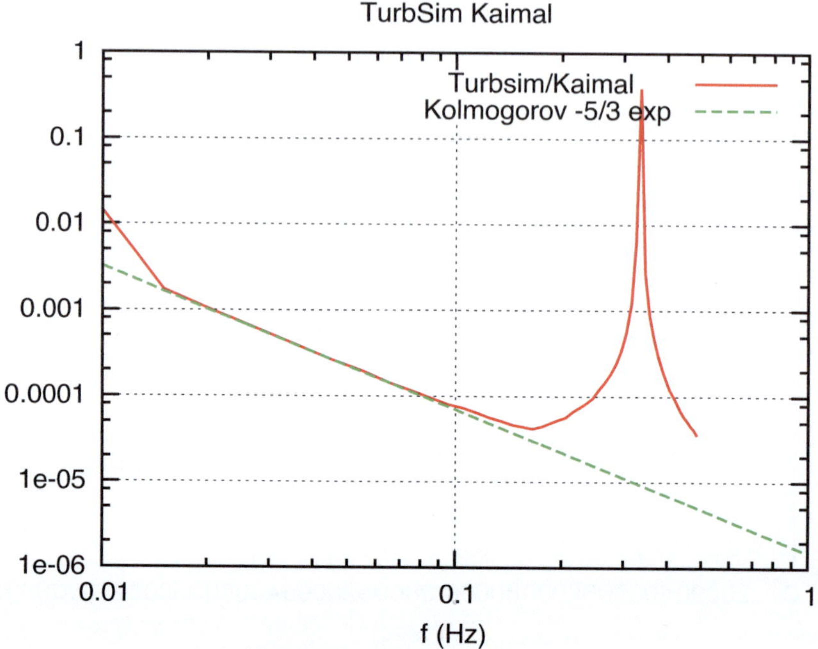

Fig. A.10   Sample energy spectrum for time series from Fig. A.8

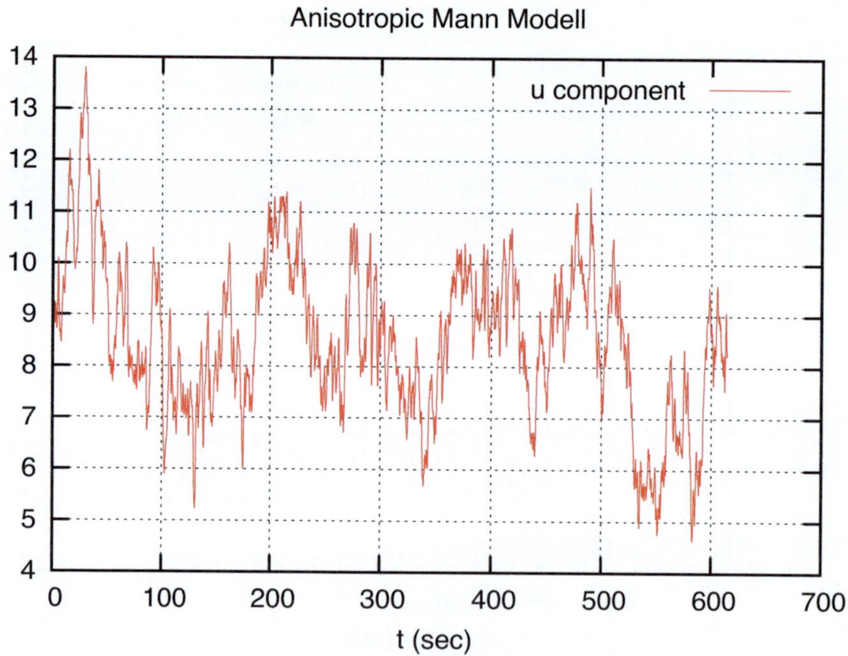

Fig. A.11   Sample time series from anisotropic Mann model [11]

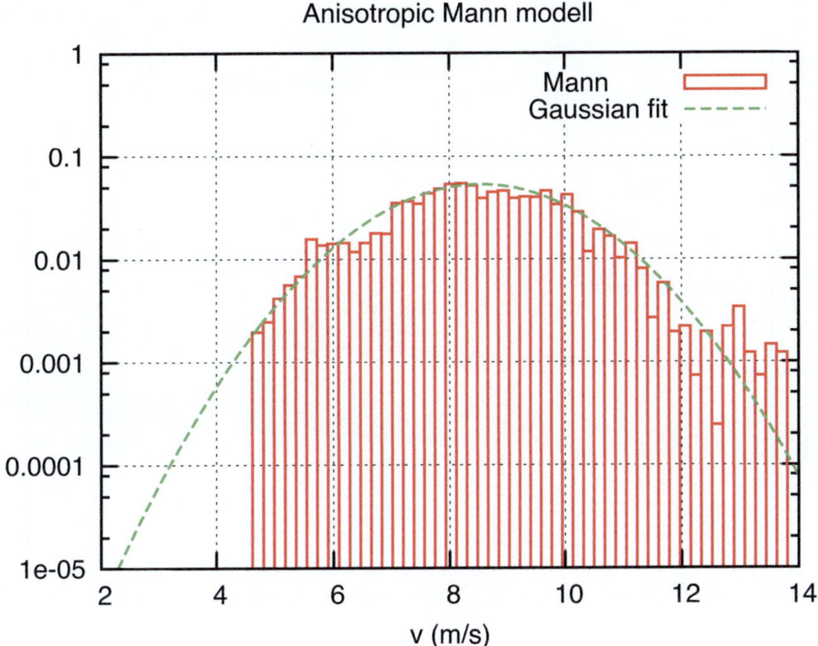

Fig. A.12 Sample histogram for time series from Fig. A.11

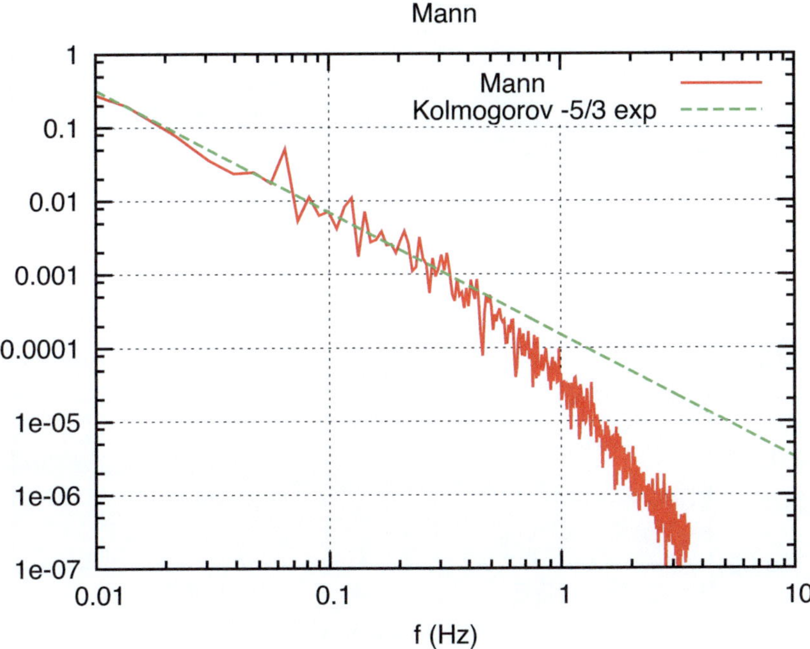

Fig. A.13 Sample energy spectrum for time series from Fig. A.11

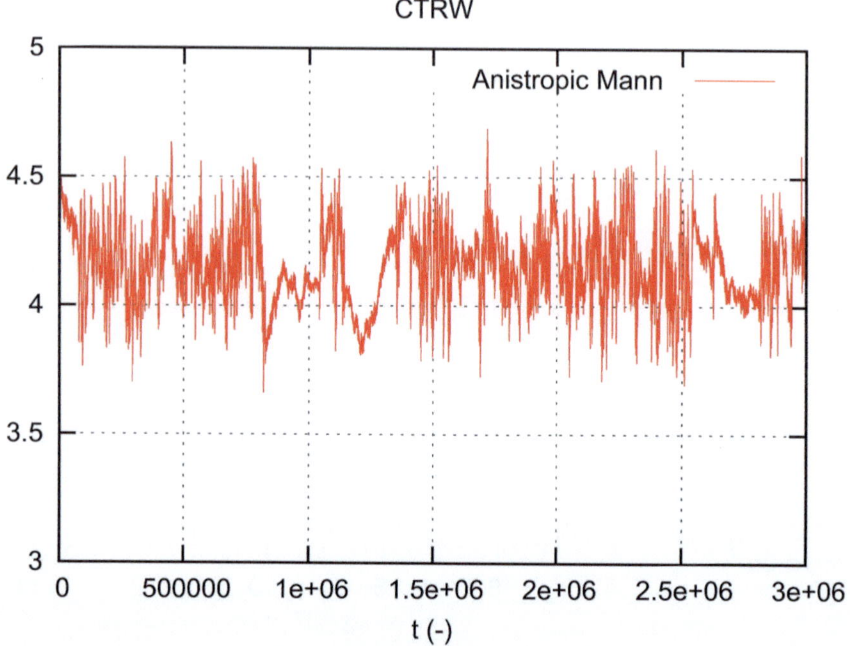

Fig. A.14 Sample time series from Kleinhans continuous time random walk model (4) [15]

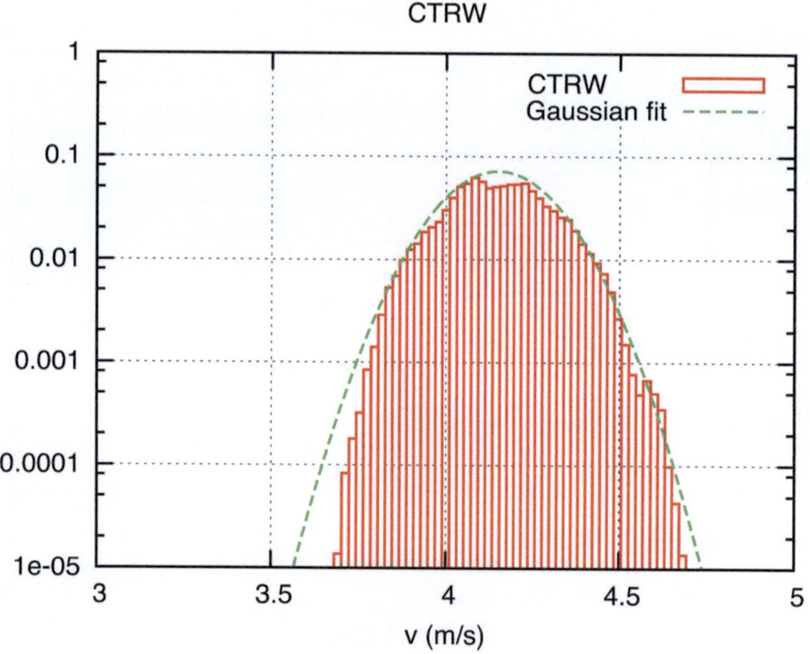

Fig. A.15 Sample histogram for time series from Fig. A.14

Table A.5 Summary of theoretical results for a wind turbine to be used a fan

Model	Thrust/kN	Power/KW	v_3/m/s
Momentum theory	61	330	11
Momentum theory	41	200	10
Wt-perf	22	199	~7
Own code	40	185	~9

As first input we use our rated values of $P = 330$ kW and 45 RPM.
Then $T_{fan} = 61$ kN, $v_2 = 5.4$ m/s and $v_3 = 2 \cdot v_2 = 11$ m/s.
Now lowering P down to 200 kW the corresponding numbers become
$T_{fan} = 40.7$ kN, $v_2 = 4.9$ m/s and $v_3 = 2 \cdot v_2 = 9.8$ m/s.
(b) BEM model
If we assume that NREL's wt-perf is reliable for the fan mode as well, we get
for the same RPM and a negligible inflow velocity of $v_1 = 0.01$ m/s the following
values:
$T_{fan} = 22$ kN, $P = 199$ kW and $v_3 \approx 7$ m/s.
Pitch was varied until (negative !) thrust was minimal.
Compared to pure momentum theory this thrust is more than a factor of two
smaller. Therefore one may write his/her own code to have a third approach:
If we use the same RPM of 45 and the same pitch, we now have
$T_{fan} = 38$ kN, $P = 175$ kW and $v_3 = 8.5$ m/s (Table A.5).
Pure momentum theory neglects all losses so that we may expect less thrust for
the BEM case. The final decision can only be drawn from the experiment.

5.2 Multiple actuator disks (5) [18]
(a) Equations 5.74 and 5.75 are simply the BEM equations for two disks if the
inflow for the second disk is the wind far downstream for the first disk.
(b) Using Eq. (5.73) we arrive at

$$c_{P2} = (1 - 2a)^3 \cdot 4b(1 - 2b)^3. \tag{A.36}$$

As there is no back-reaction from disk 2 to disk 1, one finds for a maximum of
disk 2:

$$b = 1/3 \tag{A.37}$$
$$c_{P2} = (1 - 2a)^3 (16/27) . \tag{A.38}$$

The total power of both disks then is a function of a only:

$$c_{P,totaL} = 4a(1 - a)^2 + (1 - 2a)^3 (16/27) . \tag{A.39}$$

By simple algebra one finds the results from Eqs. (5.76) and (5.79).
(c) A formal proof that Eq. (5.80) is correct is possible by induction.

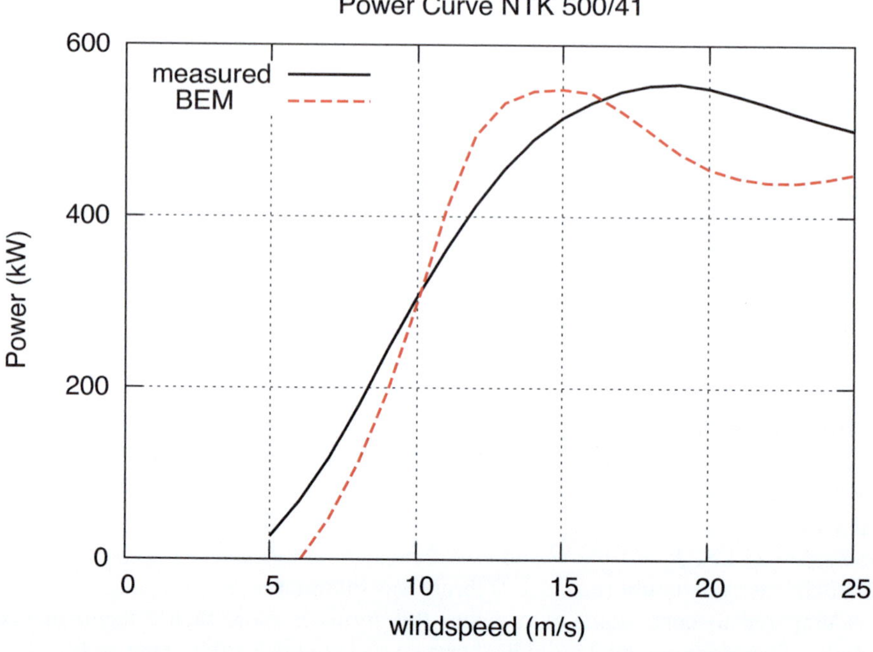

Fig. A.16 Comparison of measured data and a simple BEM model with global pitch of $-6°$

The article further discusses different shapes for an array of two disks.

5.3 This turbine is discussed in more detail in (3) [13] (p. 62 ff) and (7) [22]. Profile data has to be added. If we choose from Abbot-Doenhof data from NACA 63-series, we get a picture as in Fig. A.16. Obviously this does not fit.

5.4 Wilson (Chap. 5a from [31]) uses *along* streamlines conservation laws of

$$\text{mass}: u_1 r_2 dr = u_3 r_3 dr \tag{A.40}$$

$$\text{angular momentum}: r_2^2 \omega = r_3^2 \omega_1 \tag{A.41}$$

$$\text{and energy } \frac{1}{2}(u_1 - u_2)^2 = \left(\frac{\Omega + \omega_1/2}{u_3} - \frac{\Omega + \omega/2}{u_2} \right) u_2 \omega_1 r_3^2. \tag{A.42}$$

In addition an equation *across* the streamlines is necessary which may be obtained from Euler's equation (3) 3.31:

$$\frac{d}{dr_3}\left(\frac{1}{2}(u_1^2 - u_3^2) \right) = (\Omega + \omega_1)\frac{d}{dr_3}(\omega_1 r_3^2). \tag{A.43}$$

Axial inductions at the disk and far downstream are different:

$$u_2 = u_1 \cdot (1 - a) \, , \tag{A.44}$$

$$u_3 = u_1 \cdot (1 - b) \, . \tag{A.45}$$

Using this and the energy Eq. (A.42) one gets

$$a = \frac{b}{2} \left[1 - \frac{(1-1)b^2}{4\lambda^2(b-a)} \right] . \tag{A.46}$$

The interested reader should also compare this to [26, 30].

5.5 Wind Vehicle Part 1: Momentum theory.

Estimation of Gross Parameters

Figure A.17 shows the principle of a wind vehicle. Before we proceed to the formal analysis, it may be useful to present some typical numbers. At the races between university wind-car teams, the rotor area is limited to $2 \times 2 = 4$ m^2.

Using a HAWT the reduction goes further to $\pi = 3.14$ m^2. With a wind speed of 7 m/s and an assumed car speed of 3.5 m/s this gives—using an ordinary wind-turbine—$P = 1320$ W and $T = 188$ N. If no other forces have to be overcome, we need for translating the thrust alone $P_{thrust} = 660$ W. This is a highly ideal case. In reality a well-designed (SWT) rotor has only $c_P = 0.35$ which is only 60% of the power from above $= 780$ W. If the car has a roller friction of (only) 50 N, we have to have a net power of $50 \times 3.5 = 175$ W. Therefore this car does **not** drive against the wind.

By analyzing many races (see Fig. A.18), it is clear that $\tilde{w} = v_{car}/v_{wind} > 0.5$ was frequently reached (the actual *official* world record being 72%), so a design goal may be $\tilde{w} \geq 0.5$.

Momentum Theory

Now, to derive a momentum formulation we follow again (5) [28]. To drive the car we need power of

$$P = (D + T) \cdot w \, . \tag{A.47}$$

On the other side, wind contains a power of

$$P = T \cdot u_1 \cdot (1 - a) \tag{A.48}$$

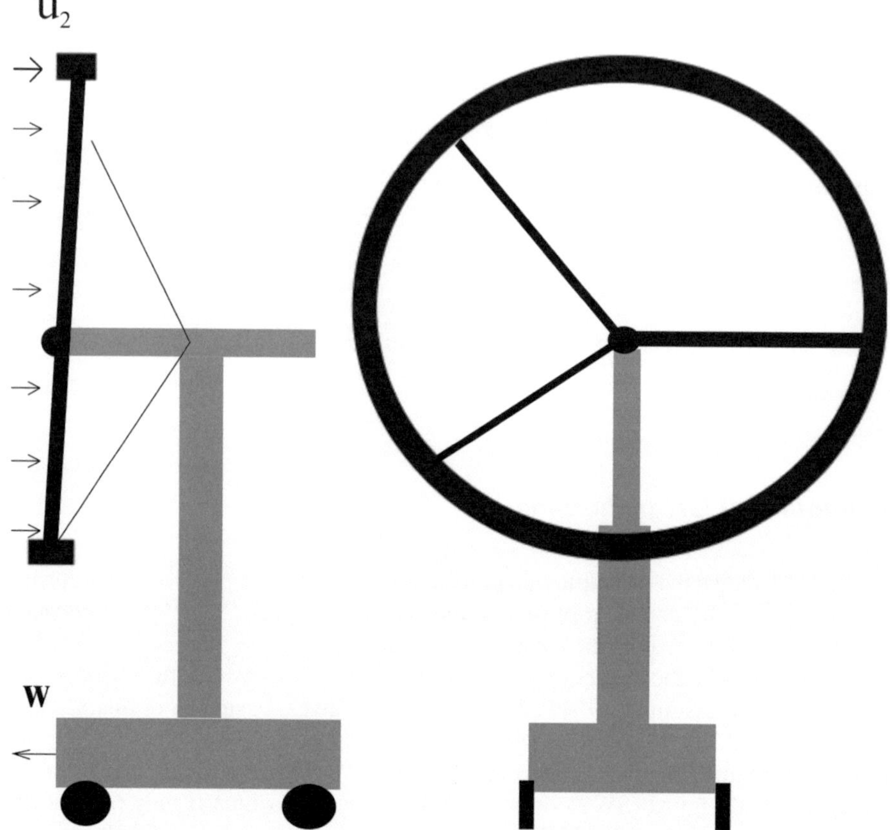

Fig. A.17 Principles of wind car

as usual. All other drag forces on the car are assumed to be summarized in

$$D = \frac{1}{2}\rho A_V(u_1 + w) \cdot c_D \ . \tag{A.49}$$

Equalizing Eqs. (A.47) and (A.48) we have

$$2\rho A_R(u_1 + w - u_2)u_2^2 = 2\rho A_R(u_1 + w - u_2)u_2 w + \frac{1}{2}\rho A_V c_D(u_1 + w)^2 w \ . \tag{A.50}$$

Note that now $u_2 = (u_1 + w) \cdot (1 - a)$. The non-dimensional car velocity (the *ratio*) may be defined as $\tilde{w} := w/u_1$, so Eq. (A.50) reads as

$$a(1 - a)^2 \cdot (1 + \tilde{w})^3 - a(1 - a) \cdot \tilde{w}(1 + \tilde{w})^2 - K \cdot \tilde{w}(1 + \tilde{w})^2 = 0 \tag{A.51}$$

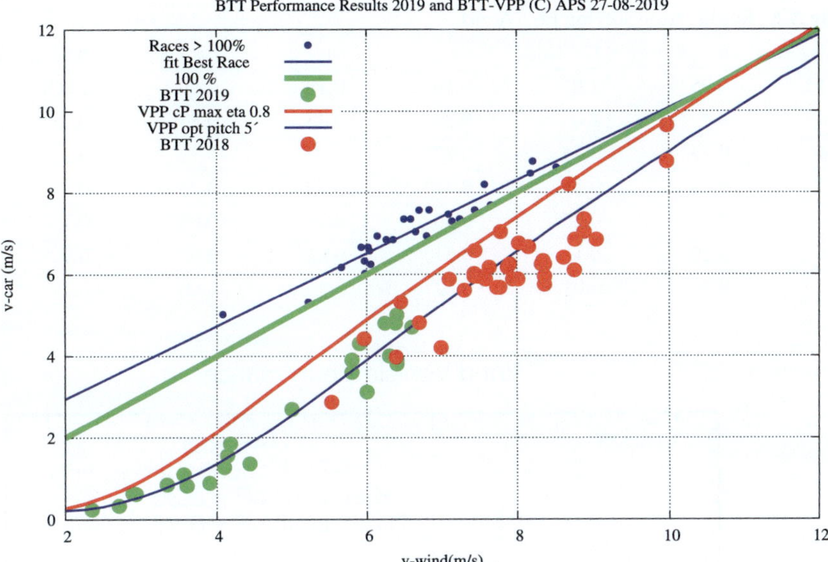

Fig. A.18 Results for the wind vehicle Baltic Thunder from 2010 to 2019. Compared to the first edition, an almost doubling of performance ($50 \rightarrow 100\%$ has been achieved)

with K from Eq. (5.87) and

$$\tilde{w} = \frac{a(1-a)^2}{K + a^2(1-a)} .$$
(A.52)

The car's speed is at a maximum if

$$\frac{d\tilde{w}}{da} = 0 ,$$
(A.53)

which gives after some simple algebra the cubic equation:

$$a^3 - a^2 - 3ka + K = 0 .$$
(A.54)

Results

Equation A.54 may be solved by either using a numerical (Newton-Raphson) algorithm or directly by using Cardano's method. Results are shown in Table A.6 (Fig. A.19).

5.6 Derivation of Prandtl's tip correction.

Table A.6 Results from solving Eq. (A.54)

K	a	v_c/v_{wind}	c_P	c_T	c_P/c_T	$\eta_{Drivetrain}$
0.013	0.05	2.0	0.16	0.18	0.89	>0.8
0.06	0.1	1.0	0.29	0.35	0.83	0.7
0.14	0.13	0.75	0.37	0.47	0.79	0.7
0.33	0.18	0.5	0.44	0.60	0.73	0.7
1.0	0.24	0.25	0.50	0.74	0.68	0.7
3.1	0.29	0.1	0.51	0.82	0.62	0.7
36	0.31	0.01	0.52	0.86	0.60	0.7

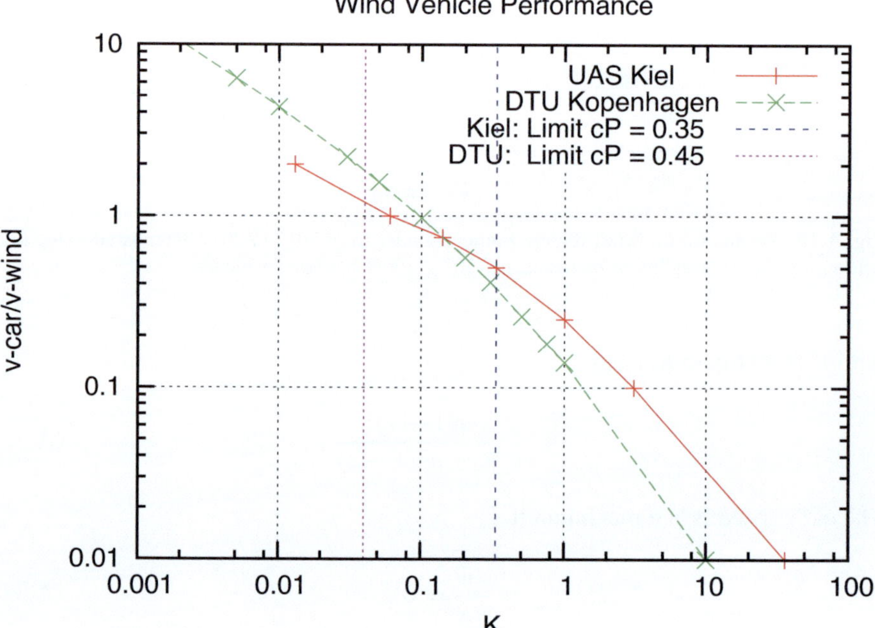

Fig. A.19 Plot of results for different rotor optimization

2D Fluid Mechanics and Complex Analysis

2D potential flow is most elegantly discussed with the use of complex analysis. Let $\mathbb{C} \ni z := x + i \cdot y, i^2 = -1$ and $(x, y) \in \mathbb{R}$. Any inviscid irrotational flow field then may be described by a stream function Φ (to obey the continuity equation) and a velocity potential Ψ $((u, v) = \nabla \cdot \Psi)$ which are combined into one complex quantity $w = u - i \cdot v$ and $F(z) = \Phi + i \cdot \Psi.$[2] F is called the *complex (velocity) potential.*

---

[2] Note that the complex potential and the tip-correction factor have the same symbol F.

Velocity is related to F via complex derivation like

$$w = \frac{dF}{dz} \, .$$

(A.55)

One special technique possible only in complex analysis is the use of *conformal mapping* [5] $f : \mathbb{C} \to \mathbb{C}$ of which the Joukovsky map

$$f(z) = z + \frac{c^2}{z}$$

(A.56)

is the most well known. It maps a circle into a shape which has become famous as the *Joukovsky profile* [2]. As a result of conformal mapping, the simpler flow field around a cylinder is mapped by the same function to the more complicated flow around this profile.

Further, complex analysis shows that **real** multi-valued functions like $\sqrt{(x)}$ or $\log(x)$ may be transformed to a one-to-one complex-valued function by introducing the concepts of *analytical continuation* and *Riemannian surfaces*. As an example, Fig. A.20 shows the complex logarithm but a proper picture in 3D is not possible because of Nash's embedding theorem. In general all manifolds are smooth and there is no intersection between the different *branches*. For a more mathematical introduction to this **extremely beautiful and easy** subject, see [16].

Model Description and Derivation

Prandtl starts with a simple circulatory flow on $\zeta \ni G \subset \mathbb{C}$; $|\zeta| > 1$ whose potential is **assumed** to be

$$F(\zeta) = \Phi + i \cdot \Psi := ic \cdot log(\zeta)$$

(A.57)

with $\zeta = \xi + i \cdot \eta$ and $\xi, \eta \in \mathbb{R}$.

By using the map $z : \zeta \mapsto z$

$$z(\zeta) := \frac{d}{\pi} \cdot log\left(\frac{1}{2}\left(\zeta + \frac{1}{\zeta}\right)\right) .$$

(A.58)

G is mapped into a sliced complex plane; see Fig. A.21. Each circle in the ζ-plane $\zeta = R \cdot e^{i \cdot \phi}$ with $R = R_0 + dr$ with $R_0 = 0.01$; $dr = 1.0$; $0 \leq \phi < 2 \cdot \pi$ corresponds to a wavy line and seems to be sliced by strips of distance $2 \cdot d$ in the imaginary direction in the z-plane.

Using the definition of the complex cosine:

$$cos(z) := \frac{e^{iz} + e^{-iz}}{2}$$

(A.59)

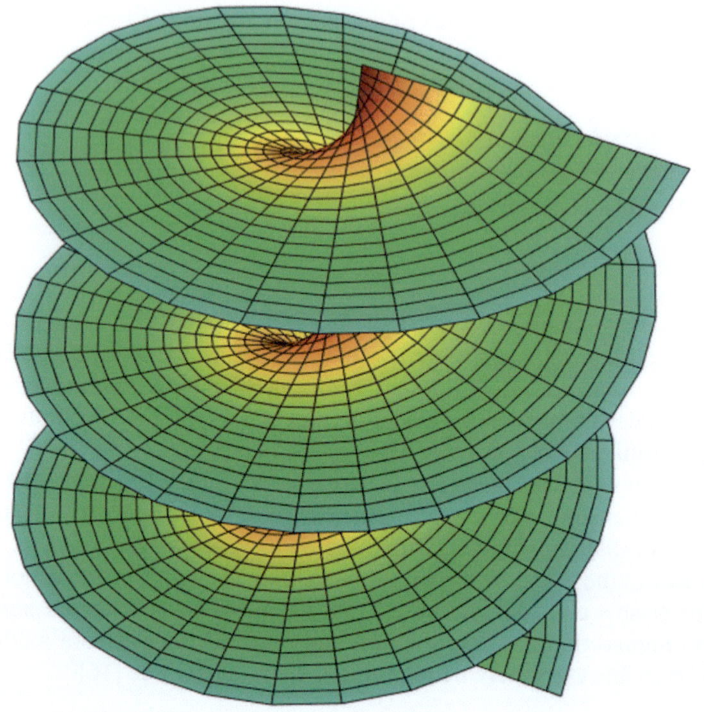

Fig. A.20 Example of a Riemannian surface: the complex logarithm

we have

$$z(F) = x + i \cdot y = \frac{d}{\pi} \cdot log\ cos\left(\frac{F}{c}\right) . \tag{A.60}$$

Discussing the steam line $\Psi = 0$, F becomes real and y has to vanish, so

$$x = \frac{d}{\pi} \cdot log\ cos\left(\frac{\Phi}{c}\right) , \tag{A.61}$$

or

$$\pm\,\Phi = c \cdot \mathrm{arccos}(\exp(\pi x/d)) , \tag{A.62}$$

because $cos(\pm z) = cos(z), \forall z \in \mathbb{C}$. From that Prandtl concluded that there is a jump in velocity potential from $x \to 0_-$ to $x \to 0_+$ by

$$\Delta\Phi = \mathbf{2}c \cdot arc\ cos(exp(\pi x/d)) . \tag{A.63}$$

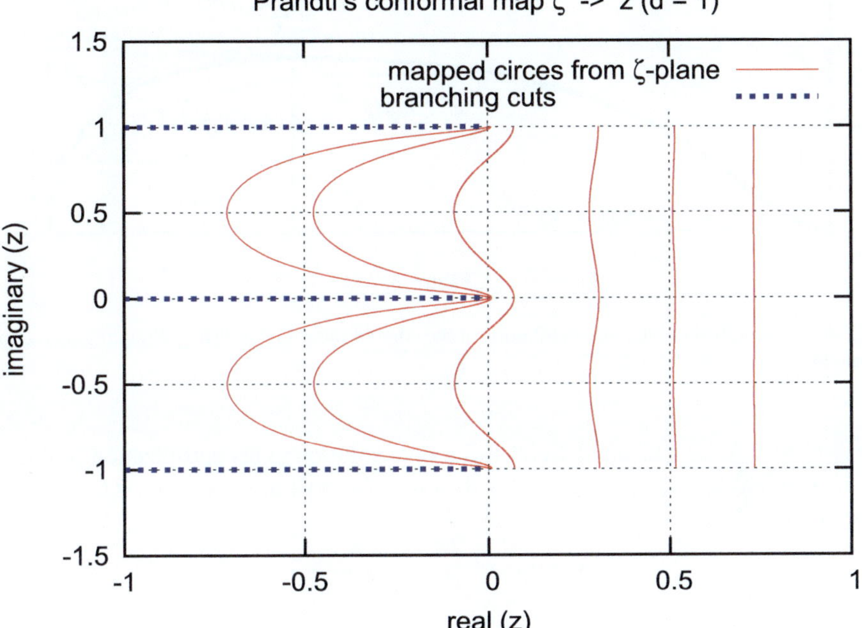

Fig. A.21 Potential flow around a stack of plates induced by a circulatory flow and a conformal mapping

Prandtl further equates the jump in potential to the change in circulation which may be justified because (here Δ means a small finite difference, not the Laplacian)

$$\Delta v = \nabla\Phi \cdot dr = (\Phi_+ - \Phi_-) \cdot dr \tag{A.64}$$

and

$$\Delta\Gamma = \Delta \int_C v \cdot dr = \Delta v \cdot dr \ . \tag{A.65}$$

A slightly different derivation is presented in [10]. Now identifying $-x := R_{tip} - r$, r being the distance from the hub $r \approx R_{tip}$ and demanding $\lim_{d\to 0} \Gamma \to 1$, we have to normalize Φ (and F) according to $c = 1/\pi$ to arrive at

$$\Delta\Phi := F = \frac{2}{\pi} \cdot arc\ cos(exp(-\pi(R_{tip} - r)/d)) \ . \tag{A.66}$$

What remains is to determine d, the spacing of the stacks. As the derivation was an appendix of Betz' optimization problem for a ship propeller giving the highest thrust for a prescribed power ($a < 0$ in wind-turbine terms), Prandtl likens d to the distance of the helical vortex sheets which emerge from the actuator disk and which

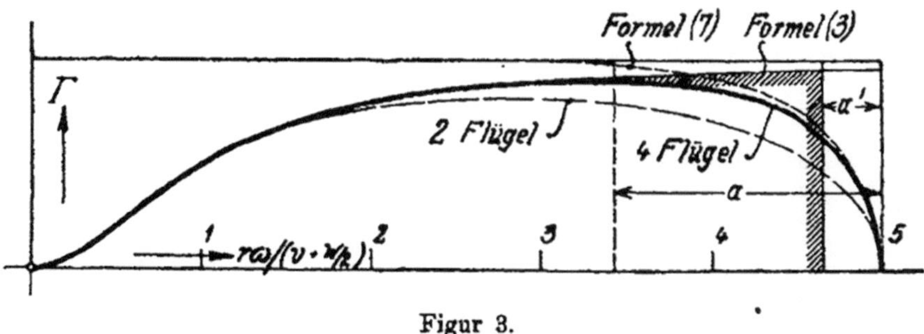

Figur 3.

Fig. A.22 Modified circulation distribution at the edge of the actuator disk by Prandtl's correction from [4]

have to be rigid as a result of Betz' variational solution to his optimization problem. In its original form Prandtl gives for d to zero ($a \approx 0$) order $\lambda = \omega \cdot R/u_\infty$ the tip speed ratio:

$$d_0/R_{tip} = \frac{2\pi}{B\lambda} \cdot \frac{\lambda}{\sqrt{1+\lambda^2}} \,, \tag{A.67}$$

which may be improved [10, 27] by using the flow angle at the tip

$$\tan(\varphi) = \frac{1-a}{1+a'} \cdot \frac{1}{\lambda} \tag{A.68}$$

(a and a' are the usual induction factors from wind-turbine momentum theory in axial and tangential direction) to

$$d/R_{tip} = \frac{2\pi}{B} \cdot sin(\varphi) \,. \tag{A.69}$$

Putting everything together we finally have

$$F = \frac{2}{\pi} \cdot \arccos \exp\left(-\frac{B(R_{tip} - r)}{2R_{Tip} \cdot sin(\varphi)}\right) \,. \tag{A.70}$$

In Fig. A.22 we show the original picture from [4] (Flügel = blade, Formel = equation).

Discussion and Extension to BEM

Obviously there is a lot of arbitrariness in the derivation:

- the conformal map Eq. (A.58) is by no means justified by fluid mechanics of propellers.
- the dependency on the number of blades is an assumption as well.[3]

Glauert [10] gives a summary of Prandtl's and Goldstein's derivations suggesting the improvement of Eq. (A.68) for use of the tip correction (better: finite number of blades) in the Blade Element Momentum (BEM) method.

He introduced (without any detailed reasoning) a *first order correction*

$$\frac{a}{1+a} = \frac{\sigma c_{\tan} \cdot F}{4 \cdot sin^2\varphi} \tag{A.71}$$

$$\frac{a'}{1-a'} = \frac{\sigma c_{nor} \cdot F}{4 \cdot sin\varphi \cdot cos\varphi}. \tag{A.72}$$

In the later classical reviews of de Vries (5) [8] and Wilson, Lissaman and Walker (5) [35], ad hoc *higher corrections* are introduced, meaning that some a, a' are replaced by $a \cdot F, a \cdot F'$.

A thorough discussion of the physical implication (and inconsistencies) was given in [27] with the result that a new model is proposed:

$$F_{new} = \frac{2}{\pi} \cdot \arccos\left[exp\left(-\mathbf{g}\frac{B(R_{tip} - r)}{2R_{tip}sin(\varphi_{r_{tip}})}\right)\right] \tag{A.73}$$

$$g = exp(-c_1(B\lambda - c_2). \tag{A.74}$$

Application to the US NREL Phase VI (around 2000) and Swedish-Chinese FFA-CARDC (around 1990) measurements showed much better predictions of the loading in the tip region.

Comparison to Other Approaches

Sidney Goldstein, at that time a Ph.D. student of Prandtl, solved the problem of finding a (potential-theoretic) flow field for a finite number of blades but for lightly loaded propellers only. The accurate numerical evaluation is regarded to be complicated and was under discussion until recently; see [21]. Figure A.23 gives a comparison for B = 3 and λ (TSR) = 6.

---

[3] In particular it is hard to imagine how a one-bladed rotor is pancaked to a whole disk homogeneously.

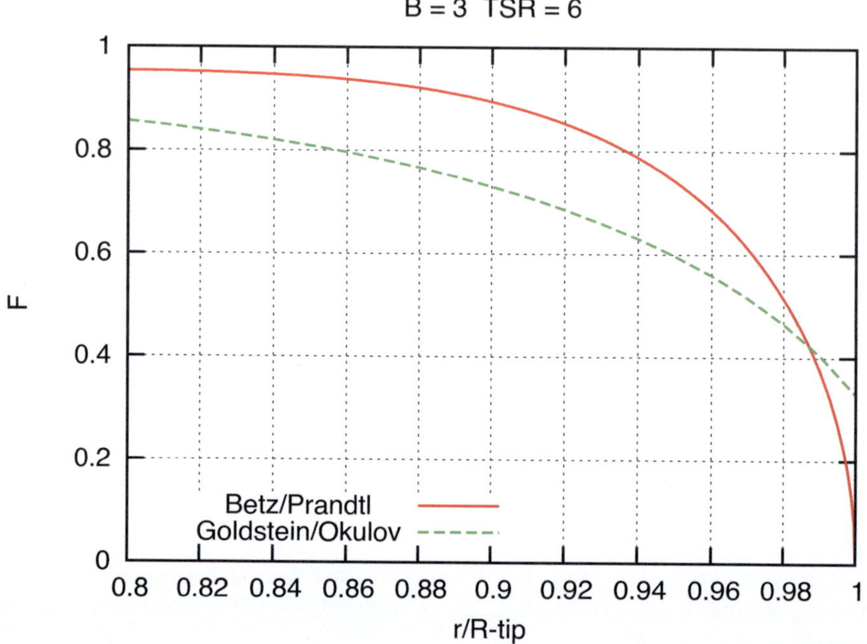

Fig. A.23 Comparison of Prandtl's and Goldstein's optimal circulation distribution

With the emergence of CFD, Hansen and Johansen [12] investigated the flow close to wind-turbine tips using this method. They found strong dependencies on the specific tip shape and confirmed some of the results from [27].

Summary

Prandtl's tip-correction model applies to the drop in circulation at the tip of a rotor and is now almost 100 years in use. Its derivation is remarkably simple and relies only on simple but ad hoc conformal mapping. Its use in BEM codes is essential.

It has to be noted, however, that any generalized use of the original Eq. (A.70) in BEM is ad hoc as well. Semi-analytical vortex methods as in [6] seem to guide the way for improved, well-founded and *simple* tip-correction models.

5.7 BEM model of ideal wind-turbine (Fig. A.24).

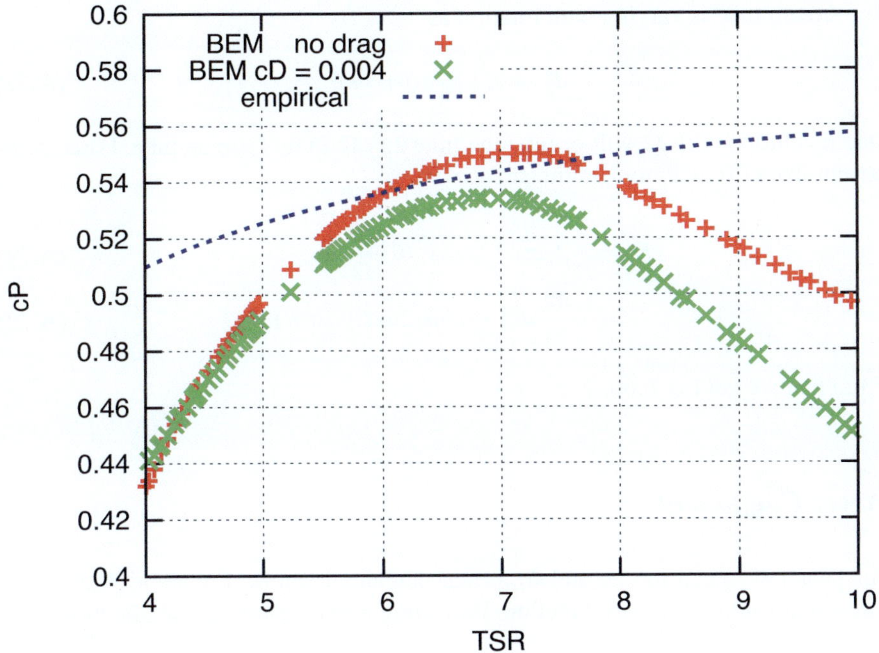

Fig. A.24 Estimation of performance of an ideal rotor (B = 3, $\lambda_{des} = 7$, $c_{L,des} = 1.0$)

A.6   Solutions for Problems of Sect. 6.9

6.1 Wilson's [35] aerodynamic model for a Darrieus rotor.

Averaged Induction

Using Kutta-Joukovsky we start with

$$\Gamma = \pi c W \cdot sin(\alpha). \tag{A.75}$$

Looking at the velocity triangle of Fig. 6.25 (right side), the projected forces (per unit height) from lift only are

$$\mathbf{F} = \rho \pi c \cdot \begin{pmatrix} -v_a v_t cos^2(\theta) \\ v_a cos(\theta) + v_a v_t cos(\theta) sin(\theta) \end{pmatrix}. \tag{A.76}$$

The stream tube is varying with rotation as

$$dy = R \cdot |sin(\theta)| \cdot d\theta \ . \tag{A.77}$$

One revolution with Ω will occupy the time $2\pi/\Omega$ in this stream tube. Force component from lift and momentum balance are

$$F_x = -2\rho\pi c v_a v_t cos^2(\theta)\frac{d\theta}{\Omega} = \tag{A.78}$$

$$= \rho\frac{2\pi}{\Omega}a(1-a)u_1 2u_1 R|cos(\theta)|d\theta. \tag{A.79}$$

From that we get Eq. 6.69.

Power Coefficient

For power we need torque and from that tangential forces. Equation (6.69) gives force in x(= wind = thrust) direction. Projecting into circumferential direction

$$Q = \rho\pi c R u_1^2(1-a)^2 \cdot sin^2(\theta) \ . \tag{A.80}$$

Inserting Eq. 6.69 and averaging over one revolution, one gets Eqs. 6.70–6.72.

Maximum Power

Finding the maximum is just simply derivating Eq. 6.70 to x. The result should be $c_{P,max} = 0.554$ at $x_{max} = 0.802$.

Thrust

We then insert Eq. 6.69 into Eq. A.79 to get

$$c_T(x) = \frac{x}{6} \cdot (3\pi - 4x) \ . \tag{A.81}$$

Discussion

$c_{P,max}$ here is smaller than Betz' limit and from double actuator disk, Problem 5.2. Nevertheless one important point is to have an indication of TSR to be chosen. For maximum power output:

$$\lambda_{opt} = 0.8/\sigma \ . \tag{A.82}$$

6.2 Solution for the problem: Holme's [14] aerodynamical model for a Darrieus turbine. Iterative derivation of the series is not simple. Readers should use the attached FORTRAN code to run it. Figures A.25 and A.26 show results for power and thrust. Notice how close the results are! Remember that both models break down for $x \lesssim 1$.

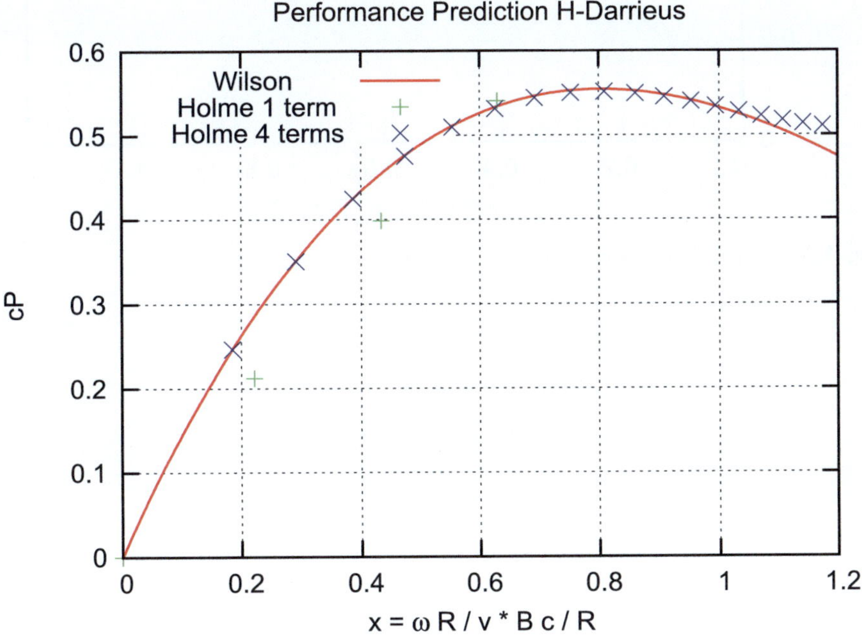

Fig. A.25 Graphical comparison for power prediction

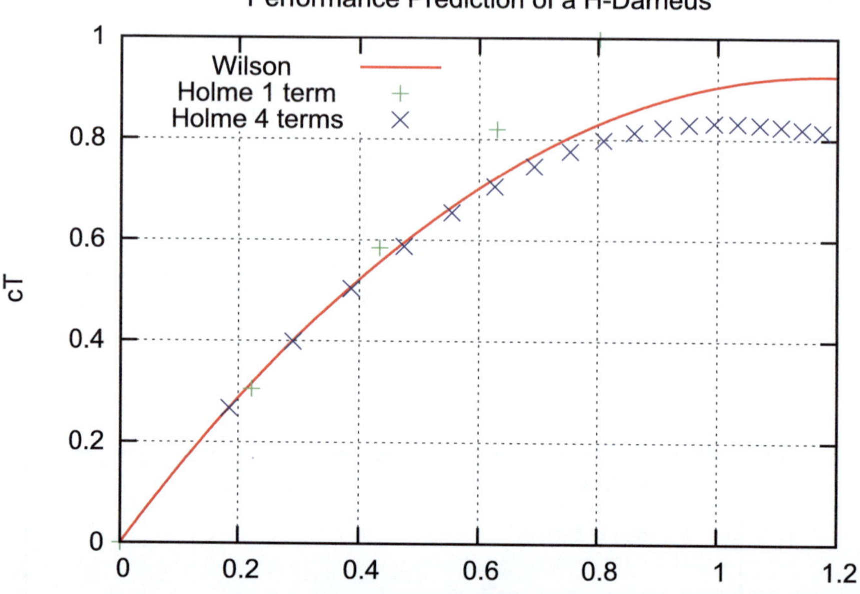

Fig. A.26 Graphical comparison for thrust prediction

FORTRAN source code of Holme's linear approach

| Jul 17, 13 8:30 | holme.f | Seite 1/5 |

```
c
c       nach holme 1976
c
c       redux aps  2005 ... 2013
c
        program holme
c
c       nmax = 8
c
        parameter(nmax=8)
        real mat(nmax,nmax),   b(nmax)
        real aa(0:nmax,nmax),  bb(0:nmax,nmax)
        real la, mu, sig, ko, lamax, lamin
c
        pi = 4.*atan(1.)
        gra = 180./pi
        sig = 0.05
        write(*,101)'sig=',4.*sig
c
        open(unit=10,file='hol.dat',form='formatted',status='unknown')
c
        nl    = 20
        lamin = 0.05
        lamax = 0.7
        dl = (lamax-lamin)/nl
c
        write(10,105)'lam','cp','cx','cy','cmdot','la*sig','sig*mu/0.42'
c
        do ll=0,nl
        la  = lamin+ll*dl
c
        write(*,*)'*********************************'
        write(*,101)'la=',la
        ga  = 1.+ la**2 + sig**2
c
        do i=1,nmax
          do j =1,nmax
            mat(i,j) = 0.0
            aa (i,j) = 0.0
            bb (i,j) = 0.0
          end do
          b(i)    = 0.0
        end do
c
c       hier die aa und bb beseten
c
c       anzahl der gleichungen
c
        ib   = nmax/2
        iend = ib/2
        write(*,*)'nmax ib iend ',nmax, ib, iend
c
c       nur rechte seite der ersten gleichung = 1
c
        b(1) = 1.0
c
c       definiton of Holme's A
c
        do n=0,ib
          do m=1,iend
            ind = 2*m-1
            ko  = float(ind*(-1)**m)/float((n**2-(ind**2)))
```

Mittwoch Juli 17, 2013   holme.f

| Jul 17, 13 8:30 | holme.f | Seite 2/5 |

```
            aa(n,m) = (-1.)**(n/2)*4.*ko*la/pi
          end do
        end do
c
c       B
c
        do n=0,ib
          do m=1,iend
            ind     = 2*m
            ko      = float(ind*(-1)**m)/float((n**2-(ind**2)))
            bb(n,m) = (-1)**((n-1)/2)*ko*4.*la/pi
          end do
        end do
c
c       die ersten 3 gleichungen (7.6)
c-------------------------------------------------------------
c       erste reihe
c
        do n=0,2,2
          do m=1,iend
            mat(1,m) =  mat(1,m) + aa(n,m)
          end do
        end do
        mat(1,1)    = ga + mat(1,1)
        mat(1,ib+2) = la*sig
c
c       zweite reihe
c
        mat(2,1)    =  la*sig
        mat(2,3)    = -la*sig
c
        do n=1,3,2
          do m=1,iend
            mat(2,ib+m) =  mat(2,ib+m) + bb(n,m)
          end do
        end do
c
        mat(2,ib+2) =  ga + mat(2,ib+2)
c
c       dritte reihe
c
        mat(3,ib+2) = -la*sig
        mat(3,ib+4) =  la*sig
c
        do n=2,4,2
          do m=1,iend
            mat(3,m) =  mat(3,m) +  aa(n,m)
          end do
        end do
c
        mat(3,3)    =  ga +  mat(3,3)
c
c       vierte reihe
c
        mat(4,3)    =  la*sig
        mat(4,5)    = -la*sig
c
        do n=3,5,2
          do m=1,iend
            mat(4,ib+m) =  mat(4,ib+m) + bb(n,m)
          end do
        end do
```

holme.f   1/1

```fortran
c
        mat(4,ib+4)  =  ga + mat(4,ib+4)
c
c--------------------------------------------------
c      reihe > nmax/2 die b's nach holme (7.7)
c
c row=ib+1
        mat(ib+1,1)     = sig
        mat(ib+1,ib+1)  = -1.
c
c row=ib+2
        mat(ib+2,1)     = la
        mat(ib+2,2)     = 1.
        mat(ib+2,ib+2)  = sig
c
c row=ib+3
        mat(ib+3,3)     = -sig
        mat(ib+3,ib+2)  = la
        mat(ib+3,ib+3)  = 1
c
c row=ib+4
        mat(ib+4,3)     = la
        mat(ib+4,4)     = 1.
        mat(ib+4,ib+4)  = sig
c
        call gauss(mat,b,nmax)
c
c-------------- debug
c
c      b(1) = 1./(1+la*+2*sig**2+16.*la/(3.*pi))
c      do j=2,ib
c        b(j) = -la*b(j-1)
c      end do
c      do j=ib+1,nmax
c        b(j) = 0.0
c      end do
c
c-------------- debug --------------------------
c
c      write(*,100)((mat(i,j),  j=1,nmax),i=1,nmax)
        write(*,*)'a:'
        write(*,100)( b(i), i=1,nmax/2)
        write(*,*)'b:'
        write(*,100)( b(i), i=1+nmax/2,nmax)
        write(*,*)
c
        teta = -0.75*pi
        tete =  0.75*pi
        nt = 100
        dt   = (tete-teta)/nt
c      write(*,103)'theta ','psi'
        psimax = -1.e3
        psimin =  1.e3
        do i=0,nt
          tet = teta+i*dt
          psi = 0.
          do j=1,ib
            psi = psi + (b(j)*sin(j*tet))/j
            js  = ib + j
            psi = psi - (b(js)*cos(j*tet))/j
          end do
          if(psi.gt.psimax)psimax=psi
```



```fortran
          if(psi.lt.psimin)psimin=psi
c        write(*,102)gra*tet,psi
        end do
        write(*,101)'psimin= ',psimin
        write(*,101)'psimax= ',psimax
c
        cmdot = 0.5*(psimax-psimin)
        write(*,101)'cmdot= ',cmdot
        mu = cmdot*la/sig
        write(*,101)'(TSR)mu= ',mu
        write(*,101)'sig*mu= ',4.*sig*mu
        write(*,*)
c
        cp = 0.0
        do i=1,nmax
          cp = cp + b(i)**2
        end do
        cp = 2.*pi*mu*sig*cp
        write(*,101)'cp = ',cp
c
        cx = 2.*pi*mu*sig*b(1)
        write(*,101)'cx = ',cx
c
        cy = 0.0
        do n=1,ib
          cy = cy + b(n)**2
          j=ib+n
          cy = cy + b(j)**2
        end do
c
        cy = 2.*pi*sig*(mu*b(ib+1)-la*cy)
        write(*,101)'cy = ',cy
c
        so =4.*sig*mu
        write(10,104)mu,cp,cx,cy,cmdot,so,so/0.42
c
        end do
c
        close(unit=10)
c
100     format(6f12.6)
101     format(a20,f10.4)
102     format(2f12.4)
103     format(2a12)
104     format(7f12.4)
105     format(7a12)
c
        end
c----------------------------------------------------------------------
c----------------------------------------------------------------------
        subroutine gauss(a,b,n)
c
        integer n
        real a(n,n), b(n)
        integer i,icol,irow,j,k,l,ll
        integer ipiv(n),indc(n),indr(n)
c
        real big, dum, pivinv
        do j=1,n
          ipiv(j) = 0
        end do
c
```

Gedruckt von CFD

Jul 17, 13 8:30 **holme.f** Seite 5/5

```
        do i=1,n
           big = 0.
           do j=1,n
              if(ipiv(j).ne.1)then
                 do k = 1,n
                    if(ipiv(k).eq.0)then
                       if(abs(a(j,k)).ge.big)then
                          big=abs(a(j,k))
                          irow=j
                          icol=k
                       endif
                    else if(ipiv(k).gt.1)then
                       write (*,*) 'singular matrix 1'
                    end if
                 end do
              end if
           end do
           ipiv(icol)=ipiv(icol)+1
           if(irow.ne.icol) then
              do l=1,n
                 dum = a(irow,l)
                 a(irow,l)=a(icol,l)
                 a(icol,l)=dum
              end do
              dum    = b(irow)
              b(irow)= b(icol)
              b(icol)= dum
          end if
          indr(i)=irow
          indc(i)=icol
          if (a(icol,icol).eq.0.) write(*,*) 'singular matrix 2'
          pivinv=1./a(icol,icol)
         a(icol,icol)=1.
         do l=1,n
            a(icol,l)=a(icol,l)*pivinv
         end do
         b(icol)=b(icol)*pivinv

         do ll=1,n
            if(ll.ne.icol)then
               dum=a(ll,icol)
               a(ll,icol)= 0.
               do l=1,n
                  a(ll,l)=a(ll,l)-a(icol,l)*dum
               end do
               b(ll)=b(ll)-b(icol)*dum
            end if
         end do
      end do
      do l=n,1,-1
         if(indr(l).ne.indc(l))then
            do k=1,n
               dum=a(k,indr(l))
               a(k,indr(l))=a(k,indc(l))
               a(k,indc(l))=dum
            end do
         end if
      end do
      return
      end
```

Mittwoch Juli 17, 2013 holme.f 1/1

6.3 Scully's vortex

$\boldsymbol{\omega} = \nabla \times \mathbf{v}$ in polar coordinates (r, ϕ, z) is

$$
\begin{pmatrix}
\frac{1}{r}\frac{\partial u_z}{\partial \phi} - \frac{\partial u_\phi}{\partial z} \\
\frac{\partial u_\phi}{\partial z} - \frac{\partial u_z}{\partial u_\phi} \\
\frac{1}{r}\left(\frac{\partial r u_\phi}{\partial r} - \frac{\partial u_\phi}{\partial \phi} \right)
\end{pmatrix} . \tag{A.83}
$$

The only component left is

$$
\omega_z = 2 \left(1 + r^{2n} \right)^{-\left(\frac{n+1}{n} \right)} ; . \tag{A.84}
$$

As one may easily see, this is a strictly monotonic decreasing function for $r > 0$.

6.4 Hill's spherical vortex

Readers are urged to consult classical textbooks like [25] or (3) [19]. Nevertheless much work has to be added. It is interesting to notice that this example supports Helmholtz' law that vorticity is convected by the outer flow.

As we have only $\omega_\phi = A \cdot r$ and

$$
u_r = -\frac{1}{r}\frac{\partial \Psi}{\partial z} \tag{A.85}
$$

$$
u_z = \frac{1}{r}\frac{\partial \Psi}{\partial r}, \tag{A.86}
$$

we may integrate Eq. (6.80) via

$$
H_r = \int -u_z \omega_\phi dr \text{ and} \tag{A.87}
$$

$$
H_z = \int u_r \omega_\phi dz . \tag{A.88}
$$

6.5 Solution for Problem: FORTRAN source code of Vortex patch

Solutions


```fortran
c-------------------------------------------------------------
c         (c) Schaffarczyk, UAS Kiel, Germany          +
c-------------------------------------------------------------
c
c         first version 2006
c         vortex patch method for h-darrieus engines
c         after strickland et al trans asme 1979 79 wa/fe6 pp 1 ff
c
c         (c) march 2010
c             all data real*8 (= double preciosion)
c             to improver numerical accuracy
c
c         nbl    = number of blades > 0
c         nvv    = number of vortex patches
c         nrevol = number or revolutions
c
c         nb l number of blades
c
      parameter (nbl   =     3)
      parameter (nvv   = 10000)
c
c         number of revolutions
c
      parameter (nrevol=     7)
c
      integer nx, ny
c
      real*8 circblnew(nbl), circblold(nbl)
      real*8 pb(nbl), blcoor(nbl,2)
      real*8 blvelu(nbl,3), blvel(nbl,3), vblind(nbl,3)

      real*8 vortgam(nbl*(nvv+1)), vortloc(nbl*(nvv+1),2)
      real*8 vortvel(nbl*(nvv+1),2), vortvelold(nbl*(nvv+1),2)
c
      real*8 dr(2), vvind(2), wind(2), rv(2), vec(2), tv(2)
      real*8 t, tx, ty, deg, nox, noy, pi, stp,dt
      real*8 aoa, cl, cd, vx, vy, v, Po, aw, uin
      real*8 ft, ftb, ct, fn, fnb, cn, dsq
c
      real*8 lam, sig, Liftf, Dragf, om, phi, dphit, gasch
      real*8 ak, gk, aku, gku, d2, lstp, vovel
c
c         ***** debug flags ******************
c
      logical freewake, back onblade, wilson
      freewake      = .true.
      backonblade   = .true.
      wilson        = .false.
c
c         ***** debug
c
c         set up initial position of blades
c
      pi  = 4.d0*datan(1.d0)
      deg = 1.8d2/pi
c
      dphibl = 2.d0*pi/nbl
c
c-------------------------------------------------------------
c         H = 1 don't change
c
      H        = 1.0d0
```



```fortran
c-------------------------------------------------------------
c
c***************** input DATA **********************
c
c         r = radius of rotor
c         uin = inflow velocities
c         units are metric
c
      R        = 1.0d0
      uin      = 1.0d0
c
c         sig = solidity = B chord/Radius
c         lam(bda) = tip speed ratio
c
      sig      = 0.30d0
      lam      = 4.0d0
c
c****************************************************
c
      ch       = R*sig/nbl
      om       = lam*uin/R
c
      wind(1) = uin
      wind(2) = 0.0d0
      rho      = 1.225d0
      stp      = 0.5d0*rho*uin**2
      area     = 2.d0*R*H
c
c         time for one revolution
c
      t        = 2.d0*pi/om
c
c         tanf = time from which on transients are gone
c         here = assume 3 revolutions
c
      tanf     = (nrevol-3.d0)*t
      tend     = nrevol*t

      tint     = tend-tanf
      tfre     = nrevol*t
c
c         4 deg steps -> 90
c
      dt       = t/90.d0
      ntime    = tend/dt
      write(*,*)'****************'
      write(*,'(a20,2f6.2,e10.3)'),'ta te dt',tanf,tend,dt
      write(*,*)'****************'
c
      dphit = om*dt
c
c         actual number of vortexpatches
c
      nvort = 0
      nt    = 0
c
      do i=1,nbl*(ntime+1)
         vortgam(i)=0.0
         do k=1,2
            vortloc(i,k)=0.0
         end do
      end do
```

Jul 26, 13 16:15   **VortexPatchDarrieus.f**   Seite 3/12

```fortran
c
      write(*,'(a12,i8)')'b    =',nbl
      write(*,'(a12,f8.3)')'la*sig=',lam*sig
c     write(*,'(a12,f8.3)')'la      =',lam
c
      open(unit=10,file='out',form='formatted',status='unknown')
      open(unit=11,file='for.1',form='formatted',status='unknown')
      open(unit=12,file='vec.1',form='formatted',status='unknown')
      open(unit=13,file='cir',form='formatted',status='unknown')
c
c-----------------------------------------------------------------------
c
c     initial values t = 0
c-----------------------------------------------------------------------
c-----------------------------------------------------------------------
c
c     azimuthal angle(s) of blade(s)
c     phi = 0 along x-axis
c
      do i=1,nbl
         pb(i)=   (i-1)*dphibl
      end do
c
c     cartesian coord (x = 1, y = 2)
c
      do i=1,nbl
         blcoor(i,1) = r*dcos(pb(i))
         blcoor(i,2) = r*dsin(pb(i))
      end do
c
c     wind velocites relative to blades (x = 1, y = 2, absolute value = 3)
c
      do i=1,nbl
         blvelu(i,1) = uin +om*r*dsin(pb(i))
         blvelu(i,2) =      -om*r*dcos(pb(i))
         blvelu(i,3) = dsqrt(blvelu(i,1)**2+blvelu(i,2)**2)
      end do
c
c     angle of attack -> lift -> circulation on blade -> forces
c
      do i=1,nbl
         phi = pb(i)
         vx  = blvelu(i,1)
         vy  = blvelu(i,2)
         v   = blvelu(i,3)
c
c-----------------------------------------------------------------------
c
         tx  =  -dsin(phi)
         ty  =   dcos(phi)
         nox =   dcos(phi)
         noy =   dsin(phi)

         gk  =   nox*vx+noy*vy
         ak  = -(tx*vx+ty*vy)
         aoa = datan2(gk,ak)
c
c-----------------------------------------------------------------------
c
         call lift(aoa,cl)
         circblold(i) = 0.5*cl*ch*blvelu(i,3)
      end do
```

Freitag Juli 26, 2013   VortexPatchDarrieus.f

Jul 26, 13 16:15   **VortexPatchDarrieus.f**   Seite 4/12

```fortran
c
c     shedded circulation = 0
c
      do i=1,nbl
         circblold(i) = 0.0
         circblnew(i) = 0.0
      end do
c
c-----------------------------------------------------------------------
c-----------------------------------------------------------------------
c     time stepping t = n*dt, n > 0
c-----------------------------------------------------------------------
c-----------------------------------------------------------------------
c
      Pbar    = 0.0d0
      Fxbar   = 0.0d0
      Fybar   = 0.0d0
c
      write(11,208)'Phi(1)','aoau','aoa','a','cN','cX','cT','cP','10 cl'
c
c     time step loop
c
      write(*,*)'ntime =', ntime
      do while (nt.le.ntime)
c
         if(nt/100*100.eq.nt)write(*,*)'nt =',nt
c
         nt   = nt+1
         tact = nt*dt
c
         do i=1,nbl*(ntime+1)
            do k=1,2
               vortvel(i,k)    = 0.0
               vortvelold(i,k) = 0.0
            end do
         end do
c
c     new azimuth of blade(s)
c
         do i=1,nbl
            pb(i) = pb(i)+dphit
         end do
c
c****    debug output on display ***********************************
c
         if(nt/10*10.eq.nt) then
            phiout=deg*pb(1)
            write(*,201)'t= ',tact,' phi: ',phiout,' nv:',nvort
         endif
c
c*****************************************************************
c
c     new cartesian coordinates
c
         do i=1,nbl
            blcoor(i,1) = r*dcos(pb(i))
            blcoor(i,2) = r*dsin(pb(i))
         end do
c
c     new driven velocites (x = 1, y = 2, absolute value = 3) relative to b
lades
c
```

VortexPatchDarrieus.f   1/1

Jul 26, 13 16:15   **VortexPatchDarrieus.f**   Seite 5/12

```fortran
      aw = 0.d0
      if (wilson)then
         aw = 0.5d0*sig*lam*dabs(dcos(phi))
         if(aw.gt..5d0)aw=.5d0
         a  = aw
      endif
c
      do i=1,nbl
         blvelu(i,1) = uin +om*r*dsin(pb(i))
         blvelu(i,2) =      -om*r*dcos(pb(i))
         blvelu(i,3) = dsqrt(blvelu(i,1)**2+blvelu(i,2)**2)
      end do
c
      do i=1,nbl
         blvel(i,1) =  blvelu(i,1)-aw*uin
         blvel(i,2) =  blvelu(i,2)
      end do
c
c-------------------------------------------------------------------
--------------
c      correct blade velocity from OLD vortex system (n-1)*dt
c      - if there is any -
c-------------------------------------------------------------------
--------------
c
      if (backonblade)then
c
      do i=1,nbl
         do k=1,2
            vblind(i,k) = 0.0d0
         end do
      end do
c
      if(nvort.ge.nbl)then
         do i=1,nbl
            do j=1,nvort
               dsq = 0.0d0
               do k=1,2
                  dr(k) = blcoor(i,k)-vortloc(j,k)
                  dsq   = dsq+dr(k)**2
               end do
c
               core = 0.5d0*ch
               if (dsqrt(dsq).gt.core)then
            vblind(i,1)=vblind(i,1)-vortgam(j)*dr(2)/(2.d0*pi*dsq)
            vblind(i,2)=vblind(i,2)+vortgam(j)*dr(1)/(2.d0*pi*dsq)
               end if
            end do
         end do
      end if
c     endif (backonblade)
c
      do i=1,nbl
         vblind(i,3) = dsqrt(vblind(i,1)**2+vblind(i,2)**2)
      end do
      endif
c
      do i=1,nbl
         do k=1,2
            blvel(i,k) = blvel(i,k) + vblind(i,k)
         end do
      end do
```


Jul 26, 13 16:15   **VortexPatchDarrieus.f**   Seite 6/12

```fortran
c
c      calculate absolute value of blade velocity
c
      do i=1,nbl
         blvel(i,3) = dsqrt((blvel(i,1))**2+(blvel(i,2))**2)
      end do
c
c-------------------------------------------------------------------
---------
c
c      calculate force-data and new shedded vorticity
c
c      "U" meaning uncorrected without vortex induces velocities
c
c-------------------------------------------------------------------
---------
c
      Fn = 0.0d0
      Ft = 0.0d0
      Fx = 0.0d0
      Fy = 0.0d0
      Po = 0.0d0
      a  = 0.0d0
c
c      BLADE LOOP
cbbbbbbbbbbbbbbbbbbbbbbbbbbbbbbbbbbbbbbbbbbbbbbbbbbbbbbbbbbbbbbbbbbbbbbbbbbbbb
      do i=1,nbl
         phi = pb(i)
         vx = blvel(i,1)
         vy = blvel(i,2)
         v  = blvel(i,3)
c
         vxu= blvelu(i,1)
         vyu= blvelu(i,2)
         vu = blvelu(i,3)
         ab = vblind(i,1)/uin
c+++++++++++++++++++++++++++++++++++++++++++++++++++++++++++++++
         nox =  dcos(phi)
         noy =  dsin(phi)
c
         tx =  -dsin(phi)
         ty =   dcos(phi)
c
c      opposite leg (in german: Gegenkathete) = project to  normal(radial
) direction)
c
         gk = nox*vx+noy*vy
c
c      !!! think carefully of signs !!!!
c
         ak   = -(tx*vx+ty*vy)
c
         aoa  = datan2(gk,ak)
c
         gku  =   nox*vxu+noy*vyu
         aku  = -(tx*vxu+ty*vyu)
         aoau = datan2(gku,aku)
c
c+++++++++++++++++++++++++++++++++++++++++++++++++++++++++++++++
c*************** debug aoa = angel of attack *******************
c
c      nach gasch 3. aufl. pp 350
```

Jul 26, 13 16:15   **VortexPatchDarrieus.f**   Seite 7/12

```fortran
c
c           aoa  = dasin(uin*dcos(phi)/(om*r))
c           aoau = dasin(uin*dcos(phi)/(om*r))
c
c*******************************************************************
c
            call lift(aoa,cl)
            call drag(aoa,cd)
            lstp = 0.5d0*rho*v**2
            Liftf    =        cl*lstp*ch*h
            Dragf    =        cd*lstp*ch*h
c
c----------------------------------------------------------------
c           new vorticity around blade
c
            circblnew(i) = cl*ch*v/2.d0
c
c----------------------------------------------------------------
c
c           force data
c
            Ftb  =     Liftf*dsin(aoa)-Dragf*dcos(aoa)
            Fnb  =     Liftf*dcos(aoa)+Dragf*dsin(aoa)
c
            Fxb  =     Fnb*dcos(phi)-Ftb*dsin(phi)
            Fyb  =     Fnb*dsin(phi)+Ftb*dcos(phi)
c
c           write(*,'(a15,3f8.2)')'ftb fnb fxb',ftb,fnb,fxb
c
            Ft   =     Ft + Ftb
            Fn   =     Fn + Fnb
            Fx   =     Fx + Fxb
            Fy   =     Fy + Fyb
c
            a    =     a  + ab/nbl
c
            Po   =     Po + Ftb*R*om
         end do
c
c        END BLADE LOOP
c
cbbbbbbbbbbbbbbbbbbbbbbbbbbbbbbbbbbbbbbbbbbbbbbbbbbbbbbbbbbbbbbbbbbbbbb
c
c           write(*,'(a15,3f8.2)')'ft fn fx',ft,fn,fx
c
            if (tact.ge.tanf)then
            Pbar     = Pbar  + Po*dt
            Fxbar    = Fxbar + Fx*dt
            Fybar    = Fybar + Fy*dt
            endif
c
            cn  = Fn/(stp*area)
            cx  = Fx/(stp*area)
            cy  = Fy/(stp*area)
            ct  = Ft/(stp*area)
            cpp = Po/(uin*stp*area)
            r1  = deg*aoau
            r2  = deg*aoa
c
c           write(*,'(a15,3f8.2)')'ct cn cx',ct,cn,cx
c
c           azimuthal printout
```

Freitag Juli 26, 2013   VortexPatchDarrieus.f   1/1

Jul 26, 13 16:15   **VortexPatchDarrieus.f**   Seite 8/12

```fortran
            if (wilson)a  = aw
c
            write(11,206)deg*pb(1),r1,r2,a,cn,cx,ct,cpp,1.d1*cl
c
c----------------------------------------------------------------
c           CREATE shedded vorticity from each blade and its initital location
c----------------------------------------------------------------
c
            do i = 1, nbl
c
c              here we have creation of a new vortex
c
               nvort = nvort + 1
c
c                sheded vorticity is -dGamma
c
               vortgam(nvort)= - circblold(i) + circblnew(i)
               circblold(i)  = circblnew(i)
               tv(1) = -0.5d0*ch*dsin(phi)
               tv(2) =  0.5d0*ch*dcos(phi)
               do k=1,2
                  vortloc(nvort,k)=blcoor(i,k)-tv(k)
               end do
c              write(*,203)'CR gam ',vortgam(nvort),' blnr: ',i
            end do
c
c        set initial velocity of new (last nbl) vortices to induced blade velo
city
c
            nvold = nvort-nbl
c
            do i=1,nbl
               newv = nvold+i
               do kc=1,2
                  vortvel(newv,kc) = vblind(i,kc) +wind(kc)
               end do
               vovel = dsqrt(vortvel(newv,1)**2+vortvel(newv,2)**2)
               write(13,204)newv,vortgam(newv),vovel
            end do
c
c----------------------------------------------------------------
c
c           1) store old values
c
            do i=1,nvold
               do k = 1,2
                  vortvelold(i,k)=vortvel(i,k)
                  vortvel(i,k)   = 0.0d0
               end do
            end do
            do i=nvold+1,nbl*(ntime+1)
               do k = 1,2
                  vortvelold(i,k)=0.0d0
               end do
            end do
c
c           2) now calculate influence of bound vorticity
c
            do jref=1,nvort
```

Jul 26, 13 16:15   **VortexPatchDarrieus.f**   Seite 9/12

```fortran
c
c            a) influence of bound vortices on vortex patches
c
      if(jref.lt.nvold)then
         do ib =1,nbl
            d2 = 0.0d0
            do k=1,2
               dr(k)= vortloc(jref,k)-blcoor(ib,k)
               d2      = d2+dr(k)**2
            end do
c
c            cut-off procedure in case vortices are too close
c
            core  = ch
            vindu = circblnew(ib)
            d       = dsqrt(d2)
            if(d.lt.core)then
            d2 = core**2
               do k4=1,2
                  dr(k4)=core*dr(k4)/d
               end do
            end if
c
c            use Biot's and Savart's law for "induction"
c
            vvind(1) = -vindu*dr(2)/(2.*pi*d2)
            vvind(2) =  vindu*dr(1)/(2.*pi*d2)
c
            do k=1,2
               vortvel(jref,k)=vortvel(jref,k) + vvind(k)
            end do
         end do
      end if
c
c            b) influence of other vortices
c
      if (freewake) then
c
         do indu=1,nvort
c
         assume NO influence of vortex on itself
c
            if (indu.eq.jref)goto 100
            d2 = 0.0
            do k=1,2
               dr(k)= vortloc(jref,k)-vortloc(indu,k)
               d2 = d2+dr(k)**2
            end do
c
c            cut-off procedure in case vortices are too close
c
            core  = ch
            vindu = vortgam(indu)
            d       = dsqrt(d2)
            if(d.lt.core)then
               d2 = core**2
               do kk=1,2
                  dr(kk)=core*dr(kk)/d
               end do
            end if
c
            vvind(1) = -vindu*dr(2)/(2.*pi*d2)
```

Freitag Juli 26, 2013   VortexPatchDarrieus.f   1/1

Jul 26, 13 16:15   **VortexPatchDarrieus.f**   Seite 10/12

```fortran
            vvind(2) =  vindu*dr(1)/(2.*pi*d2)
            do k=1,2
               vortvel(jref,k)=vortvel(jref,k) + vvind(k)
            end do
100         continue
         end do
      end if
c
c            jref loop
c
         end do
c
c-----------------------------------------------------------------------
c
c            adjust all locations with induced velocities + wind (eq  (8)
c
c            use "open" differencing schema of strickland loc. cit. eq (8)
c            use "freezed" wake if t > tfre
c
            if (tact.ge.tfre)then
               w1 = 0.0d0
               w2 = 0.0d0
            else
               w1 =  1.5d0
               w2 = -0.5d0
            end if
c
c            konvektions geschw der wirbel bestimmen !!
c
            do i=1,nvort
               do k=1,2
                  vex = w1*vortvel(i,k)+w2*vortvelold(i,k)+wind(k)
                  vortloc(i,k) = vortloc(i,k)+vex*dt
               end do
            end do
c
c-----------------------------------------------------------------------
c
c            now print out vortex locations
c
c-----------------------------------------------------------------------
c
            if(nt.eq.ntime)then
               do kv=1,nbl
                  do i=kv,nvort,nbl
                     write(10,200)i,(vortloc(i,ko), ko=1,2),vortgam(i)
                  end do
                  write(10,200)
               end do
            end if
c
c-----------------------------------------------------------------------
c
c      END time stepping LOOP
c
      end do
c
c      average over integration time
c
      cx   = Fxbar/(   stp*area*tint)
      cy   = Fybar/(   stp*area*tint)
      cp   = Pbar/(uin*stp*area*tint)
c
```

```
      write(*,211)'Ti avrg: lam sig',lam,sig
      write(*,207)'cx cy',cx, cy
      write(*,207)'cn ct',cn, ct
      write(*,207)'cp=',cp
c
c-----------------------------------------------------------------
---
c
c     print out velocity-vector field

      nx =  30
      ny =  50
      xa = -4.d0
      xe = 18.d0
      ya = -5.d0
      ye =  5.d0
      vscal = 0.6d0
c
      dx = (xe-xa)/nx
      dy = (ye-ya)/ny
c
      do ix = 1,nx
         rv(1) = xa+ix*dx
         do iy = 1,ny
            rv(2) = ya+iy*dy
            do k=1,2
               vvind(k)=0.0
            end do
            do j=1,nvort
               dsq = 0.0
               do k=1,2
                  dr(k) = rv(k)-vortloc(j,k)
                  dsq   = dsq+dr(k)**2
               end do
               vvind(1) =  vvind(1)-vortgam(j)*dr(2)/(2.*pi*dsq)
               vvind(2) =  vvind(2)+vortgam(j)*dr(1)/(2.*pi*dsq)
            end do
c
c     hier einfluss der nbl bound vortices auf das v-feld
c
            do ib =1,nbl
               d2 = 0.0
               do k=1,2
                  dr(k)= rv(k)-blcoor(ib,k)
                  d2   = d2+dr(k)**2
               end do
c
c     cut-off procedure in case vortices are too close
c
               core= ch
               vindu = circblnew(ib)
               vindu = 0.0
               d    = sqrt(d2)
               if(d.lt.core)then
                  d2 = core**2
                  do kk=1,2
                     dr(kk)=core*dr(kk)/d
                  end do
               end if
c
               vvind(1) = vvind(1)-vindu*dr(2)/(2.*pi*d2)
               vvind(2) = vvind(2)+vindu*dr(1)/(2.*pi*d2)
```



```
            end do
c
            do k=1,2
               vec(k) = wind(k)+vvind(k)
               vec(k) = vscal*vec(k)
            end do
            write(12,209)(rv(k), k=1,2),(vec(k),k=1,2)
         end do
      end do
c
      close(unit=10)
      close(unit=11)
      close(unit=12)
      close(unit=13)
c
c----------------------------------------------------------------
----------------
c
200   format(i5,3f12.6)
201   format(a6,f8.2,a6,f6.0,a6,i5)
202   format(a10,3f12.2)
203   format(a10,f12.6,a10,i2)
204   format(i5,2f12.6)
205   format(a10,2f12.6)
206   format(3f8.2,6f12.4)
207   format(a6,2f10.4)
208   format(3a8,6a12)
209   format(4f12.6)
211   format(a16,2f6.2)
c
c--------------------------------------------------- end main program ------
----------------
c
      end
c
c----------------------------------------------------------------
----------------
c
      subroutine lift(al,cl)
      real*8 al, cl, clmax, zwpi
c
      zwpi  = 6.2832d0
      clmax = 1.5
c
      cl = zwpi*al
      if(cl.gt. clmax) cl =  clmax
      if(cl.lt.-clmax) cl = -clmax
c
      return
      end
c----------------------------------------------------------------
----------
c
      subroutine drag(al,cd)
      real*8 al, cd
c
      cd = 0.0d0
c
      return
      end
```

A.7 Solutions for Problems of Sect. 7.10

7.1 Logarithmic profile and k-ϵ model [20]
   Introduce scaling variables:

$$0 - c_\mu \frac{u^{\star 4}}{c_\mu} \frac{\chi y}{u^{\star 3}} \frac{u^{\star 2}}{\chi^2 y^2} + \frac{u^{\star 3}}{\chi y} = 0 \,, \tag{A.89}$$

$$-\frac{c_\epsilon}{c_\mu} \frac{u^{\star 4}}{y^2} - c_1 \frac{u^{\star 4}}{\sqrt{c_\mu}} \chi^2 y^2 + c_2 \sqrt{c_\mu} \frac{u^{\star 4}}{\chi^2 y^2} \,. \tag{A.90}$$

7.2 Decaying turbulence and k-ϵ model [20]
   Our solution—as in Problem 7.1—is from [20]: Let $\tau := 1 + \lambda t$. Inserting it in Eqs. 7.81 and 7.82 gives

$$m = n + 1 = -n + 2m - 1 \,, \tag{A.91}$$

$$c_2 = n + \frac{1}{n} \,. \tag{A.92}$$

With $n = 1.3$ given we get $c_2 = 2.06$

7.3 Critical Falkner-Skan (wedge) flow
   Figure A.27 shows the critical values just before separation $\beta = 0.19884$ (3) [34].

7.4 XFoil for DU-W-300 (Figs. A.28, A.29 and A.30)

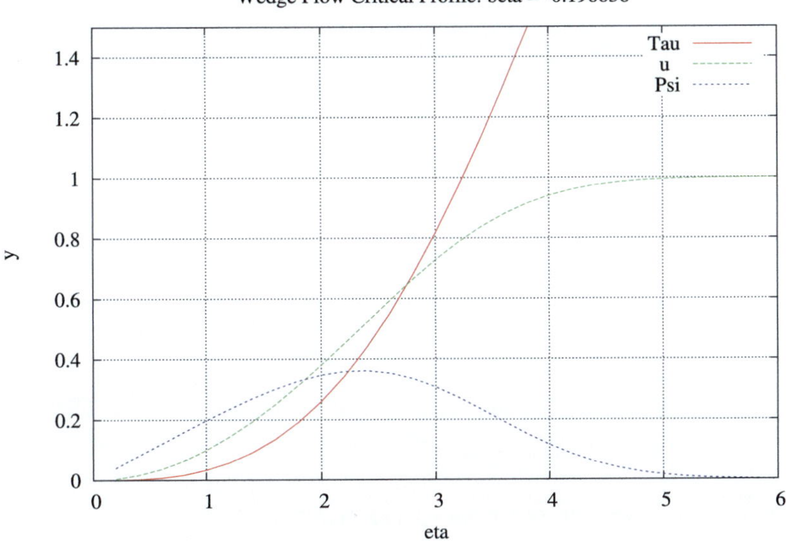

Fig. A.27 Graphical Display of Solution to Problem 7.3

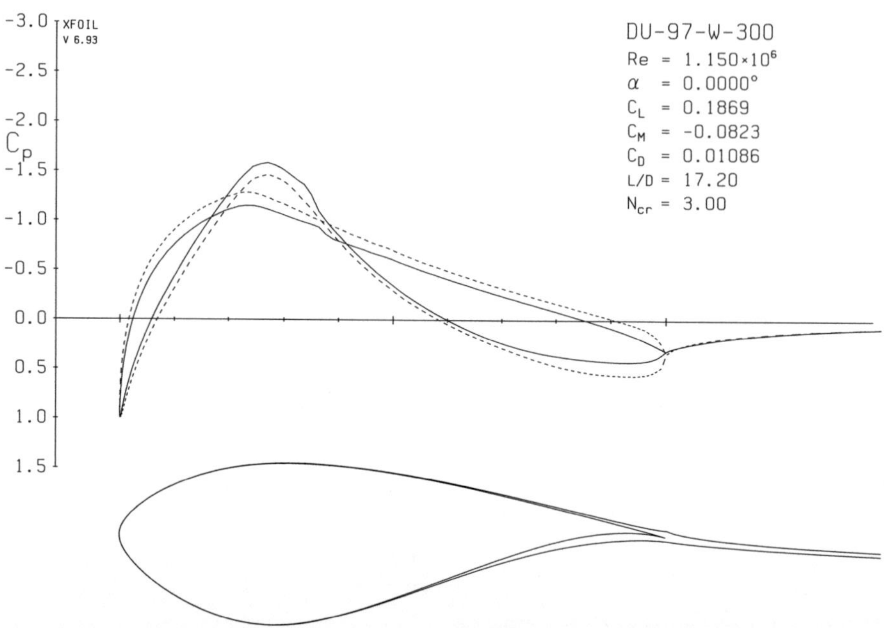

Fig. A.28 XFoil results for DU W 300, Re = 1.5 M, AOA 0 deg

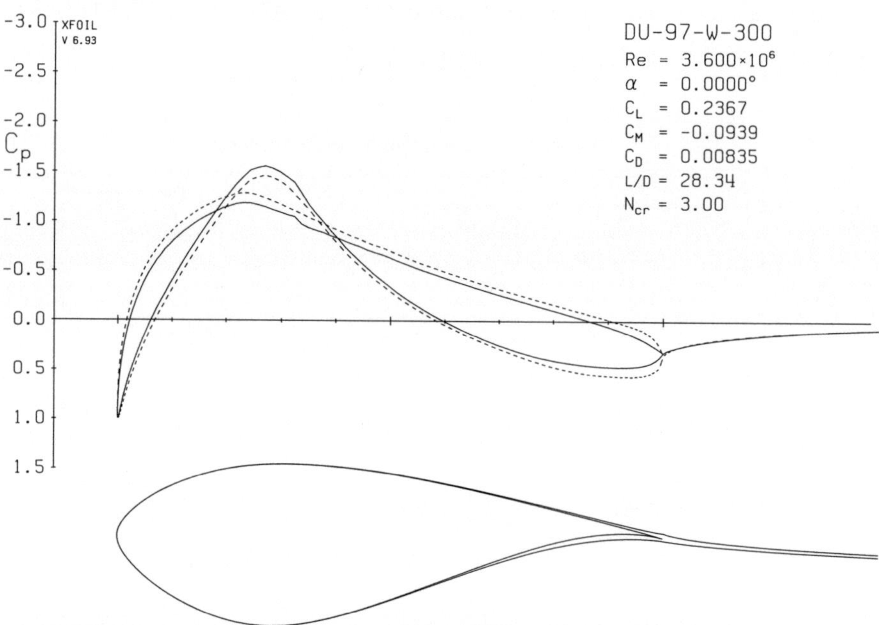

Fig. A.29 XFoil results for DU W 300, Re = 3.6 M, AOA 0 deg

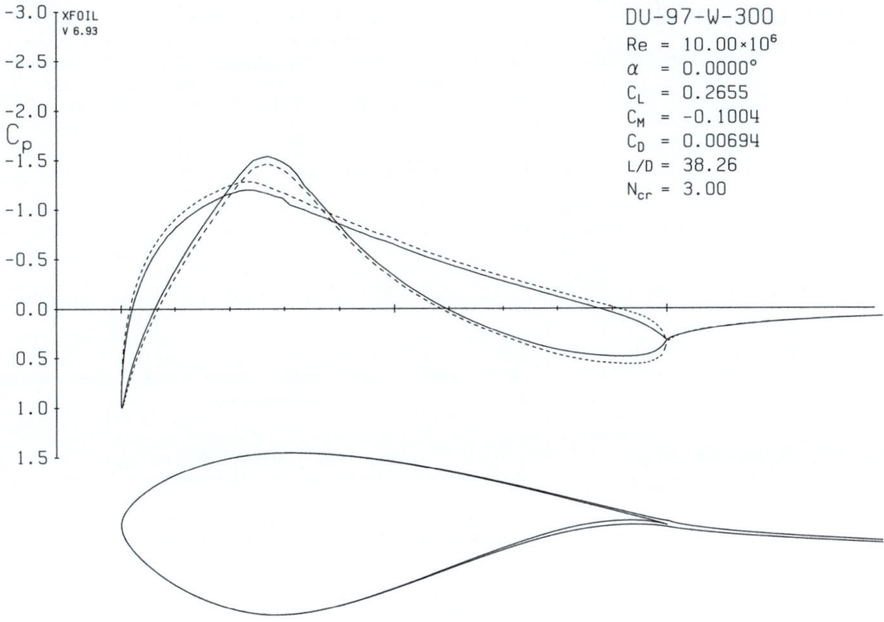

Fig. A.30 XFoil results for DU W 300, Re = 10 M, AOA 0 deg

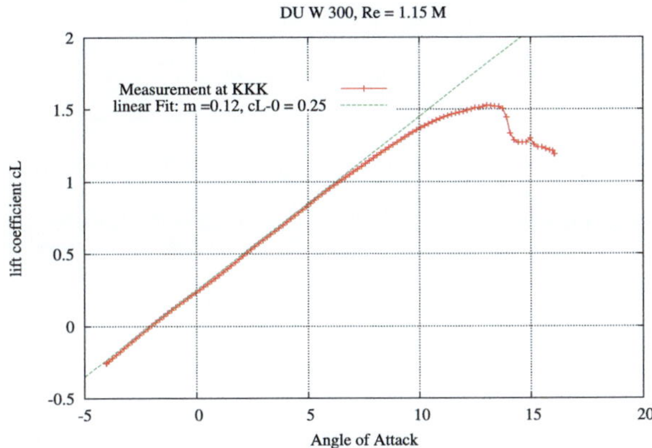

Fig. A.31 Kölner KK, Re = 1.2 M: c_L versus AOA (α)

Measurement at Kölner KK (2002/2003) (Figs. A.31, A.32, A.33, A.34, A.35 and A.36).

7.5 CFD Code (FLOWER) with e^N transition module (Fig. A.37)

Fig. A.32 Kölner KK, Re = 1.2 M: c_L versus c_D

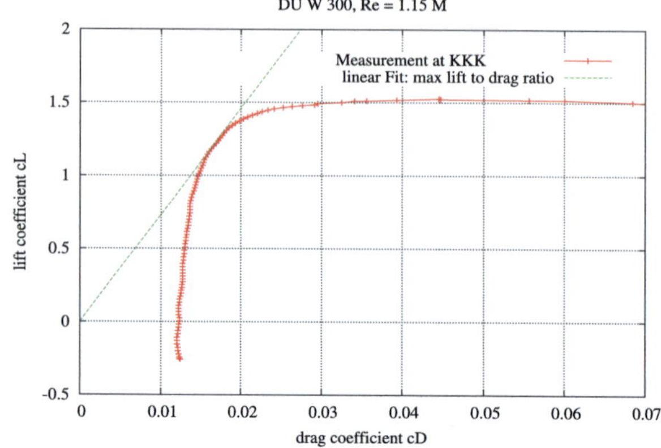

Fig. A.33 Kölner KK, Re = 3.6: c_L versus angle of attack (α)

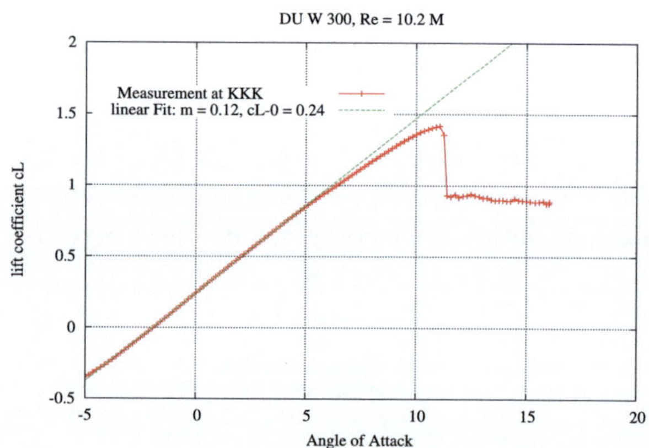

Fig. A.34 Kölner KK, Re = 3.6 M: c_L versus c_D

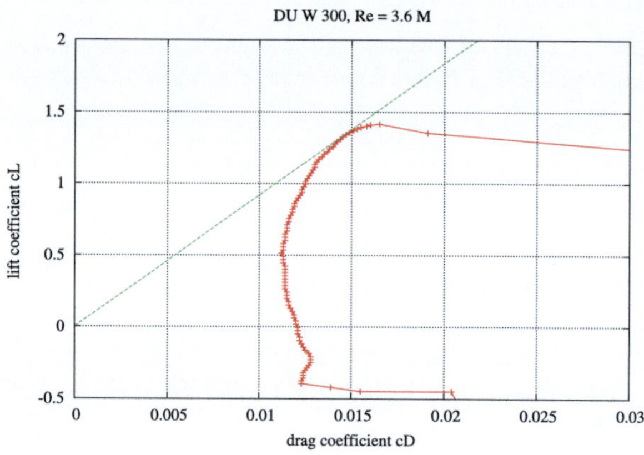

Fig. A.35 Kölner KK, Re = 10 M: c_L versus angle of attack (α)

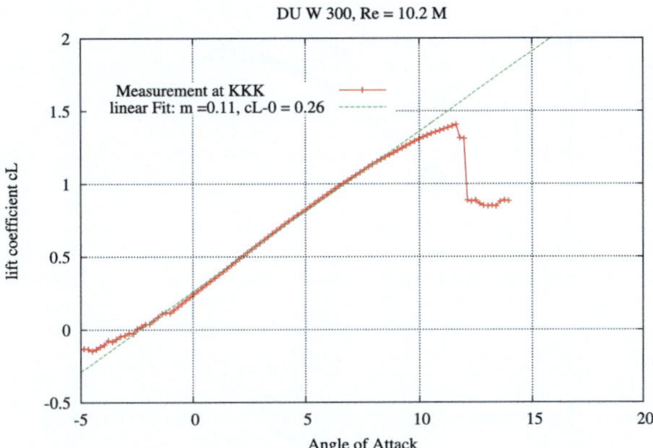

Fig. A.36 Kölner KK, Re = 10 M: c_L versus c_D

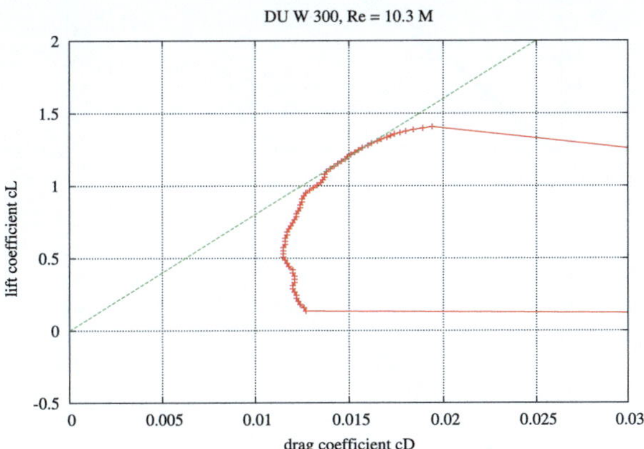

A.8   Solutions for Problems of Sect. 8.9

8.1 Estimation of lift by pressure integration

$$c_{p,i} := \frac{c_{p,i} + c_{p,i+1}}{2} \tag{A.93}$$

$$dx_i := x_{i+1} - x_i \tag{A.94}$$

$$dy_i := y_{i+1} - y_i \tag{A.95}$$

$$c_N = \sum_{i=1}^{N} c_{p,i} \cdot dx_i \tag{A.96}$$

$$c_T = \sum_{i=1}^{N} c_{p,i} \cdot dy_i. \tag{A.97}$$

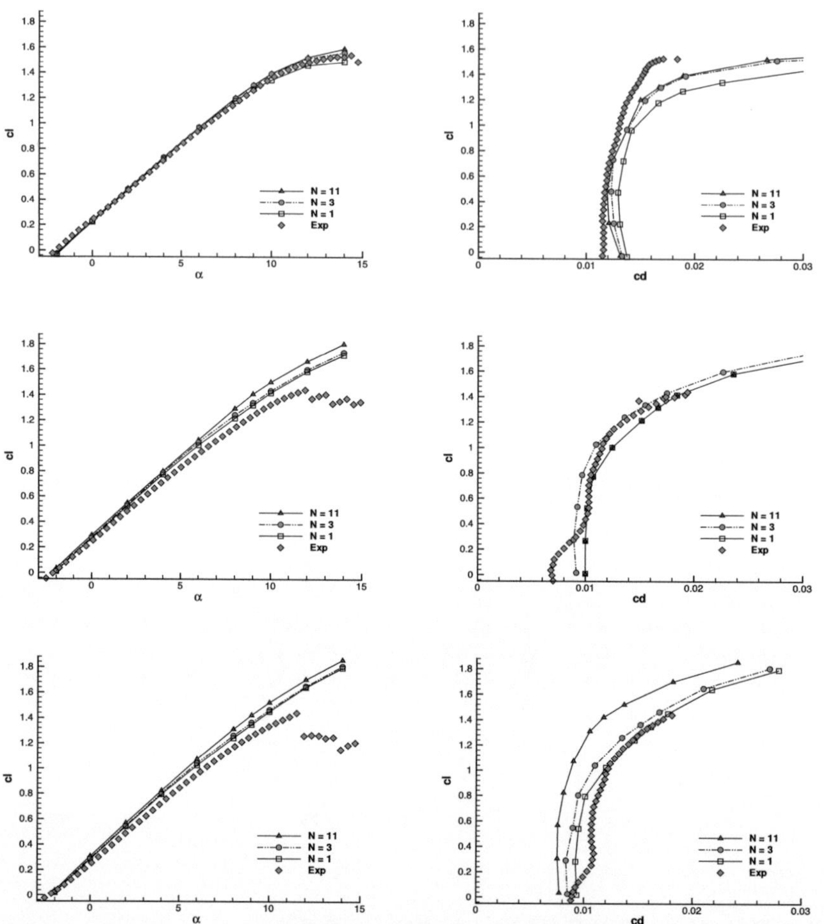

Fig. A.37 DU300-mod: $c_l(\alpha)$ and $c_l(c_d)$ for comparison with measurements at Re $1.2 \cdot 10^6$, $5.8 \cdot 10^6$ and $10.2 \cdot 10^6$

With AOA α known, lift may be calculated via

$$c_L = c_N \cdot cos(\alpha) - c_T \cdot sin(\alpha) \ . \tag{A.98}$$

The results should be $c_L(\alpha = 0) = 0.214$ and $c_L(\alpha = 8) = 1.019$ resp. (Table A.7).

8.2 Estimation of a power curve (Fig. A.38).

8.3 Lift-to-drag ratio for measured data (Fig. A.39).

Table A.7 Explanations of polars, Ma = 0.2, Re = 2.2 M

Number	Description
13	Severe roughness (Karborundum 60) around nose
16	Tripping wire at x/c = 0.15
19	Tripping wire at x/c = 0.30
34	As 13 but with VGs at x/c = 0.35
37	As 13 but with 2% GF
40	Clean
125	Less sever roughness (Karborundum 120) around nose

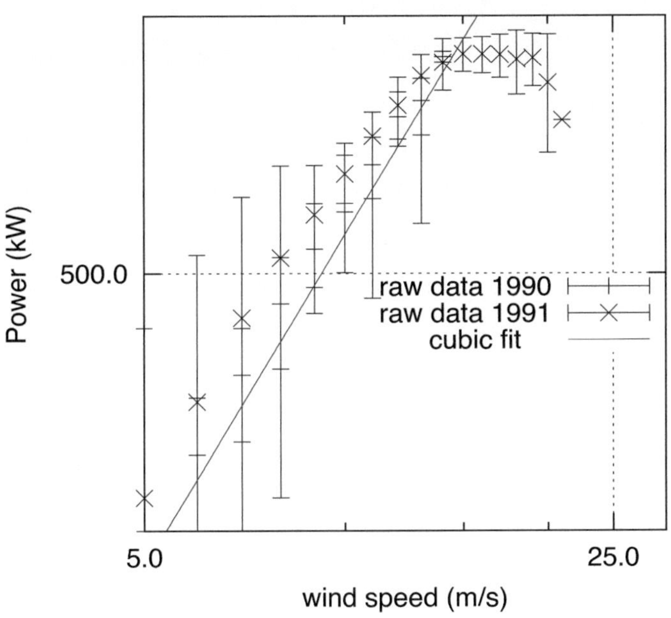

Fig. A.38 Double logarithmic plot

A.9   Solutions for Problems of Sect. 9.6

9.1 Wind Vehicle Part 2: Blade design.

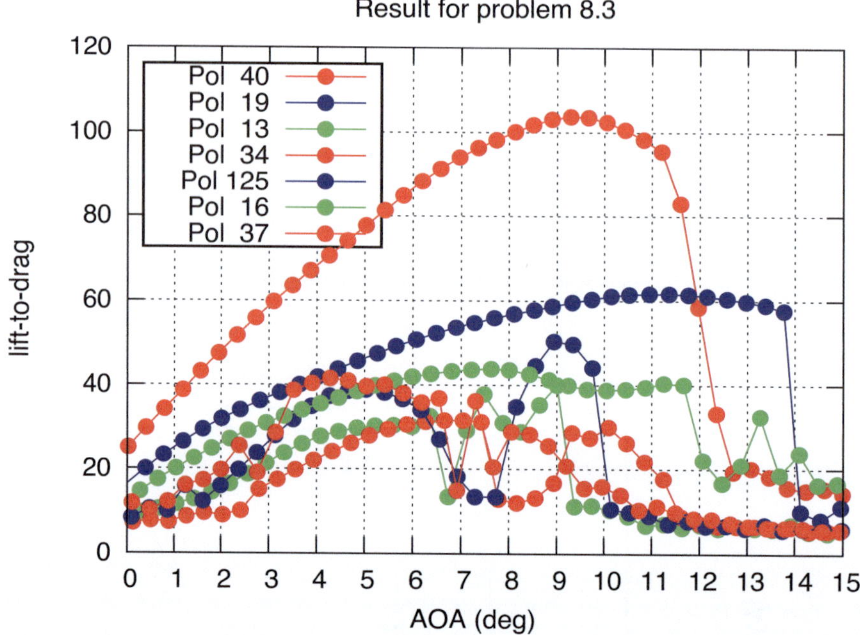

Fig. A.39 Lift-to-drag ratio of well-known DU97-W-300-mod

Summary from Problem 5.5

Maximum power extraction is not meaningful for a wind-driven vehicle. Instead we chose $v_{car}/v_{wind} \to$ max. In addition we use general BEM. As usual we assume all ring-segments to be independent. No radial flow is present and lift is assumed to be the only force acting. Comparing with Fig. A.40 we have

$$dP = \omega r \cdot dL \cdot cos(\varphi) \tag{A.99}$$

with

$$dT = dL \cdot sin(\varphi). \tag{A.100}$$

Now lift is

$$dL = \frac{\rho}{2} w^2 \cdot c_L \cdot c(r) \cdot dr \tag{A.101}$$

Fig. A.40 Blade segments

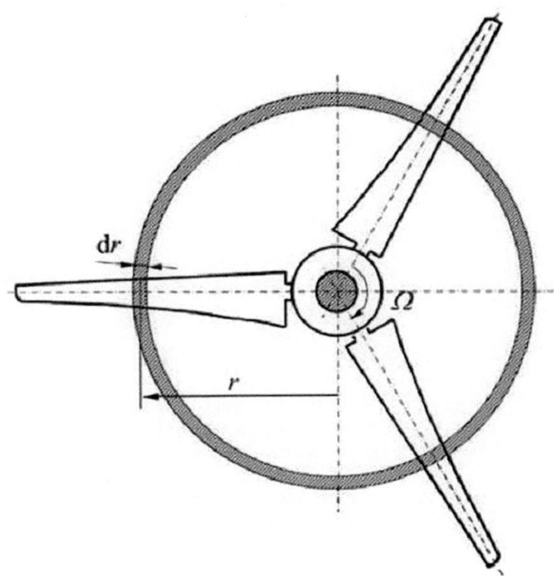

and the total inflow velocity:

$$w^2 = ((1 - a) \cdot v_{wind})^2 + (\omega(1 + a'))^2 \,. \tag{A.102}$$

a and a' again are the usual induction factors (5) [8].

Two ways for optimization have been described:

- Optimize local fluid-angle φ, and choose chord c(r) as usual [17, 32]
- Optimize c(r) and choose AOA so that the profile works at maximum c_L/c_D (5) [9].

Stuttgart [17, 32] gives numbers for comparison. Their lightly loaded (a = 0.2) (2008) rotor had $\lambda = 5$, resulting in $c_P = 0.24$ and $c_T = 0.23$. At $v_{wind} = 8$ m/s, the car's speed was estimated to $v_{car}^{max} = 5$ m/s using a power of $P_{Max} = 250$ W only.

The Danish design (5) [9] may be summarized in equation

$$C_{PROPF,loc} = \eta_P \eta_T \left(1 + \frac{1}{V/V_{wind}}\right) C_{P,loc} - C_{T,loc} \rightarrow \text{max} \,. \tag{A.103}$$

Blade Design Method

We use de Vries (5) [8] and start with

$$C_T = \frac{8}{\lambda^2} \int_0^\lambda (1 - aF)aF \left(1 + \frac{\tan(\varphi)}{L2D}\right) x \, dx, \qquad (A.104)$$

$$C_P = \frac{8}{\lambda^2} \int_0^\lambda (1 - aF)aF \left(\tan(\varphi) - \frac{1}{L2D}\right) x^2 \, dx. \qquad (A.105)$$

Here L2D is the usual lift-to-drag ratio and $x = \lambda\rho$ a local tip speed ratio. For tip correction we use Eq. 5.59 from Chap. 5:

$$F = \frac{2}{\pi} \arccos\left(\exp\left(-B/2 \, (1 - \rho)/(\rho \sin(\varphi))\right), \qquad (A.106)$$

introducing $\rho = r/R$ as the normalized distance to tip.
   Optimization of Eq. (A.103):

$$C_{PROPF,loc} \to \max \qquad (A.107)$$

is performed numerically using a binary search algorithm. The (FORTRAN) source code is shown from "FORTRAN source:Wind Vehicle Design Code" on.
   Remark: Results from this optimization are most easily attained by introducing a dimensionless drag coefficient of the car as shown in Problem 5.5:

$$K := C_D \cdot \frac{A_V}{A_R}. \qquad (A.108)$$

C_D is the bare drag coefficient of the vehicle without a rotor, $D := c_D \cdot \rho/2v^2 \cdot A_V$, A_V and A_R are projected areas of the rotor and the vehicle; see Fig. A.17.
   Inclusion of a constant rolling resistance seems to be problematic. After solving the equations blade geometry (chord- and twist-distribution) may be deduced; see Fig. A.42. For design we used: $B = 3$, $\lambda = 5.5$, design lift $C_L^{des} = 1,0$ and L2D = 80. Diameter of the rotor is 1800 mm and drive-train efficiency is assumed to be $\eta_{Drivetrain} = 0.7$. An interesting question is whether the same results (decrease of loading) may be achieved by simply pitching an ordinary wind turbine blade (Wind Dynamics—3WTR for example). Figure A.19 compares the results from our method with those from Sørensen's [29] approach. It has to be noted that our values should be generally smaller because more effects of losses (C_D tip correction) are included. Figure A.19 shows further which limit K-values may be achieved if unrealistically high c_P are excluded. A further insight is that the reported values from [17] may only be possible if $K \leq 0.005$ which is extremely small (Fig. A.41).

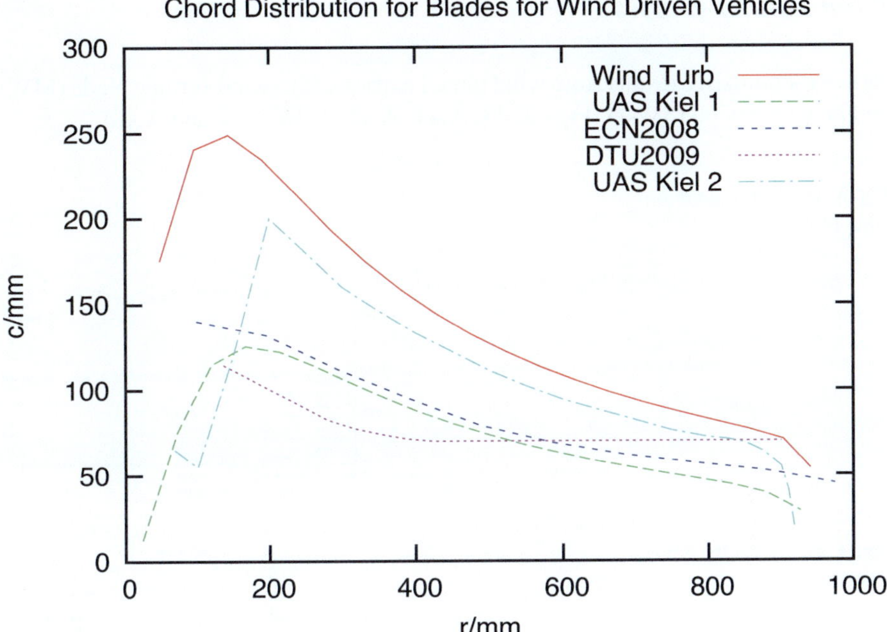

Fig. A.41 Chord distribution of different designs

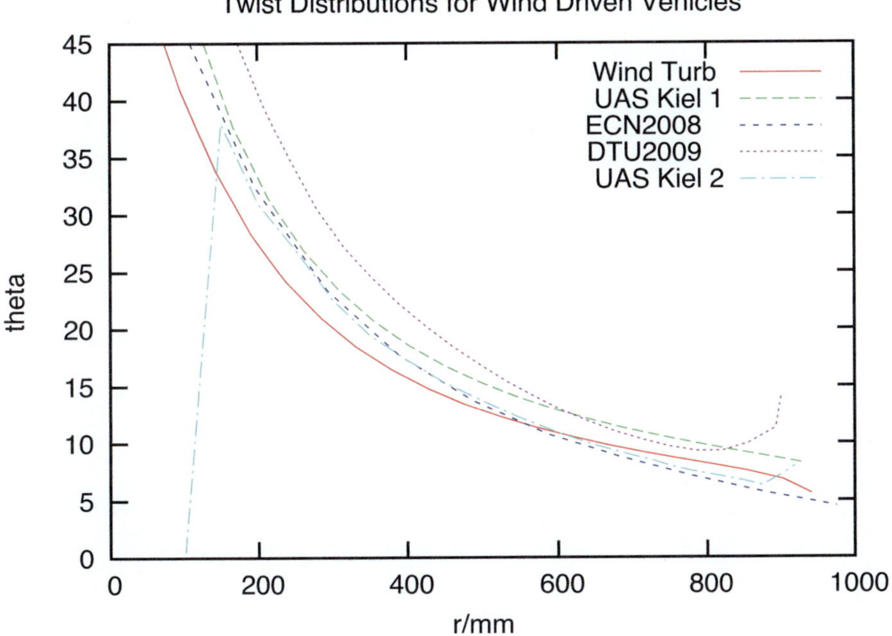

Fig. A.42 Twist distribution of different designs

Wind Tunnel Entries

Here we summarize results from wind tunnel entries with a wind-turbine blade (MW)
and an own design (SW-1) (Figs. A.43, A.44, A.45, A.46, A.47 and A.48).

Fig. A.43 c_P for the multi
wind blades

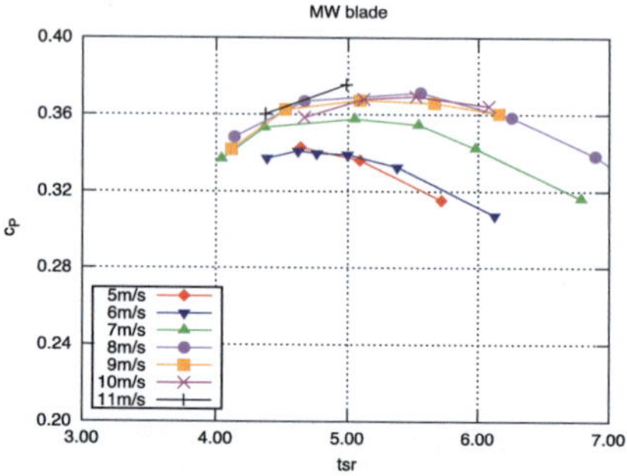

Fig. A.44 c_P for the SW-1
blades

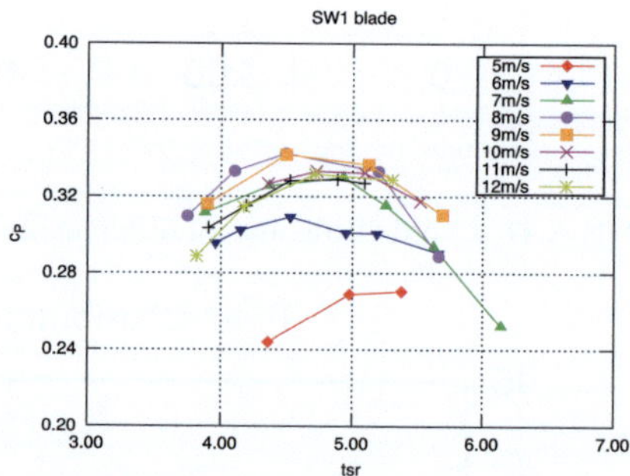

Fig. A.45 c_T for the multi
wind blades

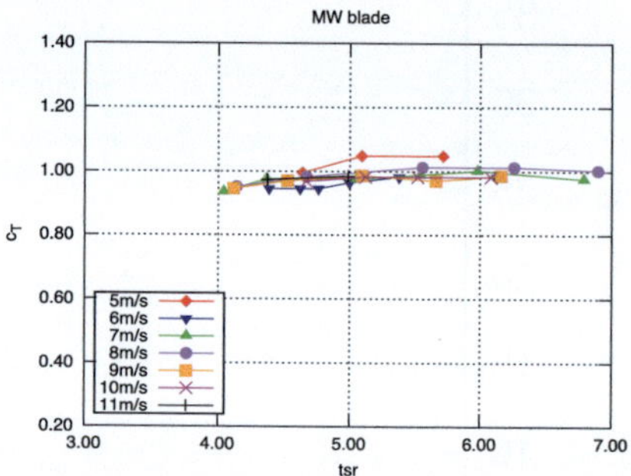

Fig. A.46 c_T for the SW-1 blades

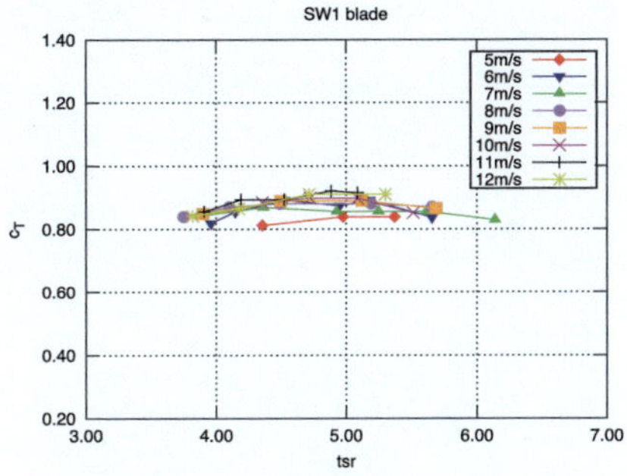

Fig. A.47 c_P/c_T of the multi wind blades

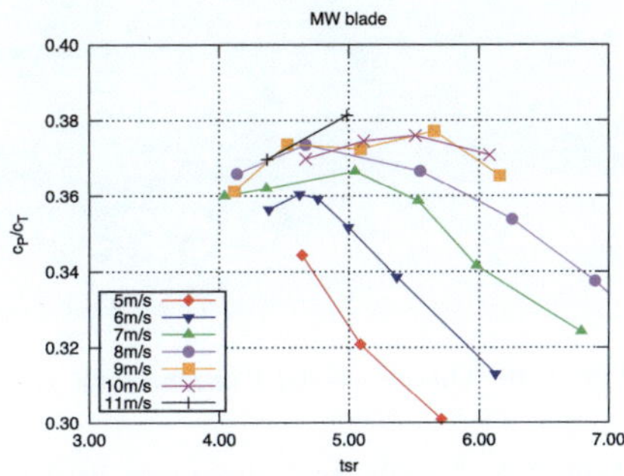

Fig. A.48 c_P/c_T of the SW-1 blades

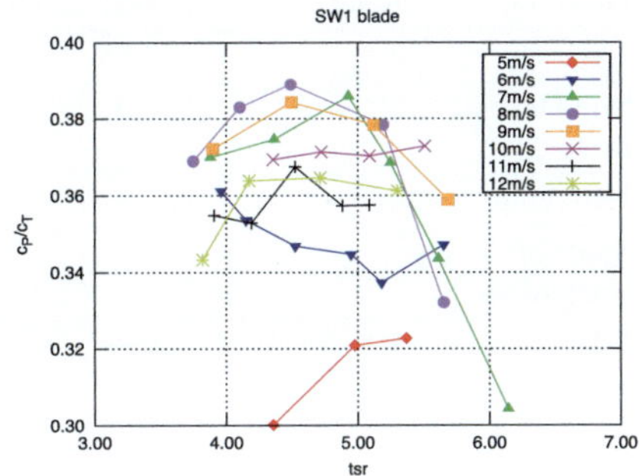

Wind Car Baltic Thunder

Finally we give an impression we see in Fig. A.49 our car from 2019. It received second rank in Table A.8. Figure A.18 already showed a comparison of all results so far.

Fig. A.49 Our car of 2019, second rank

Table A.8 Sample results from race 2019 at Den Helder, The Netherlands

Fahrzeug	Rang	Endurance%	Drag race	Fastest%	Innovation
Chinook	1	105.5	1	114.8	2 Ftab
Wind turbine racer	3	72.9	3	103.3	8
NeoVento-Spirit of amsterdam	4	27.3	4	47.7	4
DTU E	5	20.9	–	40.4	7
Stormy sussex	6	–	–	–	2
REK Yildiz	7	–	–	–	5
Fibershark—Alkmaar	8	–	–	–	6

FORTRAN source: Wind Vehicle Design Code

Sep 05, 13 14:15 — **RA.f** — Seite 1/8

```fortran
c
c-----------------------------------------------
----------
c
c     PROGRAMM opt  (C)  A.P. SCHAFFARCZYK UAS Kiel
c     march 1998, nov 2004, oct 2009
c
c     for CHINOOK, 23. august 2011
c
c     basic equution come from DEVRIES AGARDOgraph 243 (1979), Kap 4.4 ff
c     and
c     Mac Gaunaa et.al. proc. ewec 2009, Eqaution  (36) - attached
c
c-----------------------------------------------
---
c
c     some commends are in GERMAN - remember my discussion afer some beers ....
c
c-----------------------------------------------
----------
c
      COMMON PI,B,LAMBDA,GZ,vratio,etadr
c
      CHARACTER*100 NAMEOUT1,NAMEOUT2, NAMEOUT3
      CHARACTER*100 NAMEIN
      CHARACTER*1   s1, s2
      CHARACTER*6   steuer
c
      REAL PI, LAMBDA, la, B, GZ, etadr
c
C---- NUMERiC constants
c
      PI        =    4. * ATAN(1.)
      EPS       =    1.E-07
      EPSG      =    1.E-05
      MAXI      =  200
      MAXJ      =  100
      thetamax = pi/2.
c
c*************************************************************************
c********** start parameter *********************************************
c
c
C---- to change: re-compile
c
c     number of B(lades), Lambda = Omega R_tip = TSR , G(leit)Z(ahl) = L2D ratio
c     Rtip = Tip-Radius
c     CL-des desing lift
c
      B       =    3.0
      LAMBDA  =    5.5
      GZ      =   70.0
      Rtip    =    0.95
c
c     design lift
c
      CLdes   =    1.00
c
c     aoa(CL-des)
c
      aoa     =  6
c
```

Sep 05, 13 14:15 — **RA.f** — Seite 2/8

```fortran
c     vratio = desing ratio;
c     etadr = drive-train efficieny
c
      vratio   =   0.5
      etadr    =   0.7
cc*********************************************************************************
cc************** end parameter ****************************************************
cc
c
c !!!!!!!!!!!!!!!!!! DON'T CHANGE from here !!!!!!!!!!!!!!!!!!!
c
      write(*,'(2(a10,f6.3))')'vra=',vratio,'eta=',etadr
c
      drag     = 1./gz
c
      intla    = int(10*lambda)
      i1 = intla/10
      i2 = intla-10*i1
      s1 = ACHAR(48+i1)
      s2 = achar(48+i2)
c
      IOOUT21   = 21
      OPEN(UNIT=IOOUT21,FILE='list',FORM='FORMATTED',STATUS='UNKNOWN')
c
      WRITE(21,'(5a12 )')'B','LAMBDA','GZ','Rtip','Design CL'
      WRITE(21,'(5f12.4)') B, LAMBDA, GZ, Rtip,CLdes
c
      Write(21,*)
c
      NAMEOUT1 = 'BTdes.' // s1 //s2
c
      IOOUT1   = 12
      OPEN(UNIT=IOOUT1,FILE=NAMEOUT1,FORM='FORMATTED',STATUS='UNKNOWN')
      WRITE(IOOUT1,'(7a12)')
     + 'x=r/Rtip','xx=x*la','c Cl/Rtip','r/mm','c/mm','Phi/Deg',' F'
      WRITE(*,'(7a12)')
     + 'x=r/Rtip','xx=x*la','c Cl/Rtip','r/mm','c/mm','Phi/Deg',' F'
c
      NAMEOUT2 = 'INDUCTION_FACTORS'
c
      IOOUT2   = 13
      OPEN(UNIT=IOOUT2,FILE=NAMEOUT2,FORM='FORMATTED',STATUS='UNKNOWN')
      WRITE(IOOUT2,'(5a12)')'X','A ax','dphi','dcp',' A phi'
c
      NAMEOUT3 = 'steuer.dat'
c
      IOOUT3   = 14
      OPEN(UNIT=IOOUT3,FILE=NAMEOUT3,FORM='FORMATTED',STATUS='UNKNOWN')
      WRITE(IOOUT3,'(5a10,a5)')'fn','z','pitch','chord','aec','ipr'
c
      aec      = 0.25
      ipr      = 0
      steuer = 'pl.dat'
c
c-----------------------------------------------------------------------
c
      cptot = 0.0
      cttot = 0.0
c
c     blade loop
c
```

Gedruckt von CFD


```
      nx = 20
      dx = 1./nx
c
      DO IX=0,nx-1
c
c     x = r/rtip
c     xx = x * lambda
c
      X    = (0.5+ix)*dx
      XX   = LAMBDA*X
c
c     start from AD
C---  AUSGANGSWERTE NACH ACT DISK SCHAETZEN
c
c     deVries equation  (4.4.18)
c
      THETANULL   = .5*ATAN(1./XX)
      SIGMANULL   = 4.0*(1.-COS(THETANULL))
c
      CCLACT      = 2.*PI*X*SIGMANULL/B
c     write(*,*)'x sigmanull ',x,sigmanull
      DCPALT      = DELTACP(X,SIGMANULL,THETANULL)
      DCTALT      = DELTACT(X,SIGMANULL,THETANULL)
c
c     WRITE(*,*)'from ACTUATOR DISK-THEORY:'
c     WRITE(*,*)'---------------------------'
c     WRITE(*,*)'THETANULL(GRAD): ',180.*THETANULL/PI
c     WRITE(*,*)'CCL              ',CCLACT
c     WRITE(*,*)'DCPALT:          ',DCPALT
c     WRITE(*,*)'DCTALT:          ',DCTALT
c     WRITE(*,*)'---------------------------'
c     WRITE(*,*)
c
      DELTATHETA = .05*THETANULL
c
c     WRITE(*,*)'DELTATHETA(GRAD):',180.*DELTATHETA/PI
c
C-------------------------------------------------------------------
C---  outer (theta) ITERATION
c     maximize (local !) NOT cP but :
c
c     cP-car = eta*(1+ v-wind/v-car)*cP - cT
c
c     WRITE(*,101)'THETA          DCP          DF          '
C101  FORMAT(19X,A42)
c
      J = 0
c
 10   DO WHILE(.TRUE..and.theta.lt.thetamax)
      J = J + 1
c
      IF (J.EQ.1)THEN
         THETA = THETANULL
         FALT  = DELTACPauto(X,SIGMANULL,THETANULL)
      ENDIF
c
C-------------------------------------------------------------------
c     INNer (SIGMA) Iteration to
c     check if constraint G(SIG,THETA) = 0 is fullfilled
c
c     from wilson-lissaman (deVries Eq. 4.4.14)
c     may be some kind of  orthogonalty condition
```



```
c
         I = 0
      DO WHILE(.TRUE.)
      I=I+1
      IF (I.EQ.1)THEN
         SIGMA = SIGMANULL
      ELSE
         SIGMA = SIGMA
     +      - GRX(X,SIGMA,THETA)/DGDSIGMA(X,SIGMA,THETA)
      ENDIF
c
      GRXD = GRX(X,SIGMA,THETA)
c
C---  stop INNER ITERATION:
c
      IF(ABS(GRXD).LT.EPSG.OR.I.GT.MAXI)EXIT
      END DO
      IF(I.GT.MAXI)WRITE(*,*)'WARNing: MAXI reached'
c
C---  END INNER ITERATION
C-------------------------------------------------------------------
c
c
      WRITE(*,*)I,'.INN. ITER.: SIG,TH,GRX ',SIGMA,THETA,GRXD
c
      FNEU  = DELTACPauto(X,SIGMA,THETA)
      DF    = FNEU - FALT
      DFEPS = ABS((FNEU - FALT)/FNEU)
c
c     WRITE(*,100)J,'.AEUSS. ITER: THETA,DCP ',180.*THETA/PI,FNEU,DF
c100  FORMAT(I4,A14,3E14.5)
c
      IF (DF.GT.EPS)THEN
         THETA = THETA + DELTATHETA
         FALT  = FNEU
      ELSE
         DELTATHETA = -0.5*DELTATHETA
         THETA      = THETA + DELTATHETA
         FALT       = FNEU
      ENDIF
c
      IF (DFEPS.LT.EPS.OR.J.GT.MAXJ)EXIT
      END DO
C---  END outer iteration
C-------------------------------------------------------------------
c
c     WRITE(21,*)
c     WRITE(21,*)'---------------------------------------------'
c     WRITE(21,*)'ENDE DER ITERATION: I,J',I,J
c     WRITE(21,*)'ENDE DER ITERATION: '
c     WRITE(21,*)'---------------------------------------------'
c
      CCLOUT = CCL(X,SIGMA,THETA)
      AXOUT  = AAX(X,SIGMA,THETA)
      PHIOUT = APHI(X,SIGMA,THETA)
c
c     WRITE(21,*)'CCL,THETA      ',CCLOUT,180.*THETA/PI
c     WRITE(21,*)'AAX,APHI       ',AXOUT,PHIOUT
C-------------------------------------------------------------------
c
c     compute increment of  cP and  cT bestimmen
```

```
Sep 05, 13 14:15                    RA.f                          Seite 5/8
c
        sth   = sin(theta)
        cth   = cos(theta)
        cth   = cth/sth
        la    = LAMBDA
        ftip  = F(x,theta)
        AA    = AAX(X,SIGMA,THETA)*ftip
c
        dcp   = aa*(1.-aa)*(tan(theta)   -drag)*x*x*dx
        dct   = aa*(1.-aa)*(1.+tan(theta)*drag)  *x*dx
c
        cptot = cptot + dcp
        cttot = cttot + dct
c
C----   output   localen values
c
        WRITE(IOOUT1,'(7F12.4)')X,xx,CCLOUT,1.e3*X*Rtip,
     +        1.e3*CCLOUT*Rtip/CLdes,
     +        180.*THETA/PI,ftip
c
        WRITE(*       ,'(7F12.4)')X,xx,CCLOUT,1.e3*X*Rtip,
     +        1.e3*CCLOUT*Rtip/CLdes,
     +        180.*THETA/PI,ftip
c
        WRITE(IOOUT2,'(5F12.6)')X, AXOUT,PHIOUT, dcp,AA
c
        output for blade desing (CAD) tool
c
        WRITE(IOOUT3,'(a10,F10.4,f10.1,2F10.4,i5)')
     +     steuer,X*Rtip,180*theta/pi,cclout*rtip,aec,ipr
C
C---    END X loop =  1, nx
c
      END DO
c
c     this normalization seems highly unclear !!!
c
      cptot = 8.*la*cptot
      cttot = 8.*cttot
      cpauto = etadr*(1+1./vratio)*cptot - cttot
      aa     = 0.5*(1.-sqrt(1.-cttot))
c
      WRITE(*,*)
      WRITE(*, '(a4,2a8, 4a11 )')'B','LAMBDA','GZ','cP',
     +                     'cT','cP-car/eta','a'
      WRITE(* ,'(i4,2f8.1,5f11.3)') int(B) , LAMBDA , GZ,
     +                     cptot,cttot,cpauto,aa
c---------------------------------------------------------------------
c     WRITE(ioout1, '(a4,a8,    5a9  )')'B','LAMBDA','GZ',
c     +                     'cP','cT','cP-car/K','a'
c     WRITE(ioout1 ,'(i4,2f8.1,5f9.3)') int(B) , LAMBDA , GZ,
c     +                     cptot,cttot,cpauto,aa
c---------------------------------------------------------------------
      CLOSE(IOOUT1)
      CLOSE(IOOUT2)
c
      END
c
C---------------------------------------------------------------------
C---------------------------------------------------------------------
C---------------------------------------------------------------------
c
C--------------------------------------------------
```

```
Sep 05, 13 14:15                    RA.f                          Seite 6/8
      FUNCTION AAX(X,SIG,THETA)
C-------------------------------------------------
C
      COMMON PI,B,LAMBDA,GZ,vratio,etadr
C
      REAL LAMBDA
C
      AAX = SQRT(F(X,THETA)*F(X,THETA)+4.*A(SIG,THETA)*F(X,THETA)*
     1       (1.-F(X,THETA)))
      AAX = (2.*A(SIG,THETA)+F(X,THETA)-AAX)
     1       /(2.*(A(SIG,THETA)+F(X,THETA)**2))
c
c     WRITE(*,*)'debug: AAX= ',AAX
      END FUNCTION AAX
C
C-------------------------------------------------
      FUNCTION APHI(X,SIG,THETA)
C-------------------------------------------------
C
      COMMON PI,B,LAMBDA,GZ,vratio,etadr
C
      REAL LAMBDA
C
      APHI  = SIG/(4.*F(X,THETA)*COS(THETA) - SIG)
C
      IF(APHI.LT.0.)WRITE(21,*)'WARNING: APHI:' ,APHI
      END FUNCTION APHI
C
C-------------------------------------------------
      FUNCTION A(SI,TH)
C-------------------------------------------------
C
      COMMON PI,B,LAMBDA,GZ,vratio,etadr
C
      REAL LAMBDA
C
      A   = SI* COS(TH)/(4.*SIN(TH)*SIN(TH))
C
      IF(A.LT.0)WRITE(21,*)'WARNING: A=',A
      END FUNCTION A
C
C-------------------------------------------------
      FUNCTION F(X,THETA)
C-------------------------------------------------
C
C     Tip-Loss model from L. Prandtl (1919)
c
      COMMON PI,B,LAMBDA,GZ,vratio,etadr
C
      REAL LAMBDA
C
C-------------------------------------------------
      call shen(gc)
      F = gc*.5*B*(1.-X)/(X*SIN(THETA))
C-------------------------------------------------
c
      F = (2./PI) * ACOS(EXP(-F))
c
c     WRITE(*,*)'debug: F= ',F
      END FUNCTION F
C
C-------------------------------------------------
```

Sep 05, 13 14:15   **RA.f**   Seite 7/8

```fortran
      SUBROUTINE shen(gc)
C-------------------------------------------------------------------
C
C     tip loss model from  soerensen, shen et al (2004)
C
      COMMON PI, B, LAMBDA, GZ,vratio,etadr
C
      REAL LAMBDA, gc
C
c     shen
c
      gc = 0.1 + exp(-0.125*(B*lambda-21.))
c
c     kein shen
c
c     gc = 1.
      END
C
C-------------------------------------------------------------------
      FUNCTION CCL(X,SIG,THETA)
C-------------------------------------------------------------------
C
      COMMON PI,B,LAMBDA,GZ,vratio,etadr
C
      REAL LAMBDA
C
      CCL = 2.*PI*X*SIG/B
C
C     WRITE(*,*)'CCL= ',CCL
      END FUNCTION CCL
C
C-------------------------------------------------------------------
      FUNCTION GRX(X,SIG,THETA)
C-------------------------------------------------------------------
C
      COMMON PI,B,LAMBDA,GZ,vratio,etadr
C
      REAL LAMBDA
C
      GRX=LAMBDA * X -
     +      AAX(X,SIG,THETA)*(1.-AAX(X,SIG,THETA)*F(X,THETA))*TAN(THETA)
     +      /(APHI(X,SIG,THETA)*(1.-AAX(X,SIG,THETA)))
C
C     WRITE(21,*)'GRX= ',GRX
C     WRITE(21,*)'L,X,AAX,TH,APHI',LAMBDA,X,AAX(X,SIG,THETA),THETA,
C    +                            APHI(X,SIG,THETA)
      END FUNCTION GRX
C
C-------------------------------------------------------------------
      FUNCTION DELTACPauto(X,S,T)
C-------------------------------------------------------------------
C
      COMMON PI,B,LAMBDA,GZ,vratio,etadr
C
      REAL LAMBDA
C
      DELTACPauto = etadr*(1.+1./vratio)*deltacp(x,s,t)-deltact(x,s,t)
C
      END FUNCTION DELTACPauto
C
C-------------------------------------------------------------------
      FUNCTION DELTACT(X,S,T)
```


Sep 05, 13 14:15   **RA.f**   Seite 8/8

```fortran
C-------------------------------------------------------------------
C
      COMMON PI,B,LAMBDA,GZ,vratio,etadr
C
      REAL LAMBDA
C
      xx = lambda*x
C
      DELTACT = (TAN(T)/GZ + 1.)
     +         *      AAX(X,S,T)*F(X,T)
     +         *(1.-AAX(X,S,T)*F(X,T))/XX
C
      END FUNCTION DELTACT
C
C-------------------------------------------------------------------
      FUNCTION DELTACP(X,S,T)
C-------------------------------------------------------------------
c     formel (6) auf seite 4-17
c
C-------------------------------------------------------------------
C
      COMMON PI,B,LAMBDA,GZ,vratio,etadr
C
      REAL LAMBDA
C
      DELTACP = (-1./GZ+TAN(T))
     +         *      AAX(X,S,T)*F(X,T)
     +         *(1.-AAX(X,S,T)*F(X,T))
C
c     WRITE(*,*)'debug: DELTACP= ',DELTACP
      END FUNCTION DELTACP
C
C-------------------------------------------------------------------
      FUNCTION DGDTHETA(X,S,T)
C-------------------------------------------------------------------
C
      COMMON PI,B,LAMBDA,GZ,vratio,etadr
C
      DT = T/100.
      DGDTHETA   = (GRX(X,S,T+DT)-GRX(X,S,T))/DT
C
C     WRITE(21,*)'DGDTHETA= ',DGDTHETA
      END FUNCTION DGDTHETA
C
C-------------------------------------------------------------------
      FUNCTION DGDSIGMA(X,S,T)
C-------------------------------------------------------------------
C
      COMMON PI,B,LAMBDA,GZ,vratio,etadr
C
      DS = S/100.
      DGDSIGMA   = (GRX(X,S+DS,T)-GRX(X,S,T))/DS
C
C     WRITE(21,*)'DGDSIGMA= ',DGDSIGMA
      END FUNCTION DGDSIGMA
C
C-------------------------------------------------------------------
```

About the Author

The author spent his childhood gazing at the blue sky of Northwestern Germany (*Star-fighters* flashing by) and also watching astonished on black-and-white TV the rocket *Saturn V* launches that brought the first men to the moon in the 1960s.

Interested in mathematics and physics, he attended University Göttingen, well-known for mathematics and physics but also for Ludwig Prandtl's *Aerodynamische Versuchsanstalt*. After receiving his PhD in some kind of Statistical Mechanics, he came into deeper touch with fluid mechanics after his employment in industry. There he learned to apply all his learned theoretical knowledge and skills to solve practical problems. In 1992 he joined *University of Applied Sciences Kiel, Germany* and worked further on computational fluid dynamics and, since 1997, on the aerodynamics of wind turbines.

<table>
<tr><td>Kiel, Germany,
April 2014</td><td align="right">*Alois Peter Schaffarczyk*
∞</td></tr>
</table>

A. P. Schaffarczyk, *Introduction to Wind Turbine Aerodynamics*, Green Energy and Technology, https://doi.org/10.1007/978-3-031-56924-1

Glossary

ABL   Atmospheric boundary layer

AOA   Angle-Of-Attack: angle between chord line (longest line between nose and tail of an aerodynamic profile) and inflow wind

Boundary layer   A concept introduced by Ludwig Prandtl in which the influence of viscosity in high Reynolds number flow is confined in a thin layer close to the wall of a body

GROWIAN   GROsse WInd ANlage = large wind turbine

Hub   Height of main shaft

Ising model   Model system in statistical physics

Kölner KK   Kölner Kryo Kanal = Cryogenic wind tunnel at Cologne, Germany. Here Reynolds number is increased by cooling the tunnel gas (nitrogen) down to $-200°K$

Laminar Flow   A flow type realized if Reynolds number is sufficiently small

Reynolds Number   Local ratio of inertial to viscous forces in fluid flow

Tensor   In most cases represented by a matrix, they are defined mathematically as linear transformations

Turbulence   Synonymous to turbulent flow, a state opposed to laminar flow which is large temporal and spacial fluctuations

Vortex generators   Small upright oriented triangles which shed small vortices into the boundary layer and may prevent early stall

Windmill   A windmill is an early form of a *working machine* which was (is) used for grinding wheat

Wind index   A positive number which correlates the annual average wind speed to a long-term average

Wind turbine   A wind turbine is an engine which converts kinetic energy from the wind, mostly into electricity.

© The Editor(s) (if applicable) and The Author(s), under exclusive license to Springer Nature Switzerland AG 2024

A. P. Schaffarczyk, *Introduction to Wind Turbine Aerodynamics*, Green Energy and Technology, https://doi.org/10.1007/978-3-031-56924-1

References

1. Babinsky H (2003) How do wings work? Phys Edu 8:6
2. Bachelor GK (1967) An introduction to fluid dynamics. Cambridge University Press, Cambridge, UK
3. Beljaars ACM (1987) The influence of sampling and filtering on measured wind gusts. J Atmos Oceanic Technol 4:613–626
4. Betz A (1919) Schraubenpropeller mit geringstem energieverlust. Nachrichten der Königlichen Gesellschaft der Wissenschaften zu Göttingen 193–217
5. Betz A (1964) Konforme Abbildung. Springer, Berlin/Göttingen/Heidelberg
6. Branlard E, Dixon K, Gaunaa M (2012) An improved tip-loss correction based on vortex code results. In: Copenhagen D (ed) Proceedings EWEA 2012
7. Carlin P (1996) Analytic expressions for maximum wind turbine average power in a Rayleigh wind regime. Technical report, NREL/CP-440-21671, Golden, CO, USA
8. de Vries O (1979) Fluid dynamic aspects of wind energy conversion. Technical report AGAR-Dograph, 242, NATO, Neuilly sur Seine, France
9. Gaunåa M, Øye S, Mikkelsen R (2009) Theory and design of flow driven vehicles using rotors for energy conversion. In: Proceedings EWEC, Marseille, France
10. Glauert H (1936) Aerodynamics of propellers. J. Springer, Berlin
11. Gontier H et al (2007) A comparison of fatigue loads of wind turbine resulting from a non-gaussian turbulence model versus standard ones. In: DTU (ed) The science of making torque from wind. Lyngby, Danmark, IoP, p 012070
12. Hansen MOL (2004) Aerodynamics of wind turbines, 2nd edn. earthscan, London, UK
13. Hansen MOL (2008) Aerodynamics of wind turbines, 2nd edn. Earthscan, London, UK
14. Holme O (1976) A contribution to the aerodynamic theory of the vertical-axis wind turbine. In: Proceedings of the international symposium on wind energy systems, pp C - C4-72, UK, pp 4–55
15. Kleinhans D (2008) Stochasitische Modellierung komplexer Systeme. PhD thesis, Dissertation Universität Münster in German
16. Lang S (2000) Complex analysis, 4th edn. Springer, New York
17. Lehmann J, Miller A, Capellaro M, Kühn M (2008) Aerodynamic calculation of the rotor for a wind driven vehicle. In: Bremen G (ed) Proceedings of the DEWEK
18. Loth JL, McCoy H (1983) Optimization of Darrieus turbines with an upwind and downwind momentum model. J. Energy 7(4):313–318

19. Milne-Thomson LM (1996) Theoretical hydrodynamics, 5th edn. Dover Publications, New York, USA

20. Mohammandi B, Pirronneau O (1994) Analysis of the K-Epsilon turbulence model. Wiley, Chichester, Uk

21. Okulov VL, Sørensen JN (2008) Refined Betz limit for rotors with a finite number of blades. Wind Energy 11:415–426

22. Paulsen US (1995) Konceptundersøgelse nordtanks 500/41 strukturelle laster. Technical report Risø-I-936(DA), Risø National Lab., Roskilde, Denmark

23. Philip J (2013) Spatial averaging of velocity measurements in wall-bounded turbulence: single hot wires. Meas 24

24. Renner C (2002) Markowanalysen stochastischer fluktuierender Zeitserien. PhD thesis, Carl von Ossietzky Universität, Oldenburg, Germany

25. Saffman PG (1992) Vortex dynamics. Cambridge University Press, Cambridge, UK

26. Sharpe DJ (2004) A general momentum theory applied to an energy-extracting actuator disk. Wind Energy 7:177–188

27. Shen WZ, Mikkelsen R, Sørensen JN (2005) Tip loss corrections for wind turbine computations. Wind Energy 4:457–475

28. Sørensen JN Aero-mekanisk model for vindmølledrevet køretøj. unpublished report, Copenhagen, Denmark, no year

29. Sørensen JN Aero-mekanisk model for vindmølledrevet køretøj. Unpublished report, Copenhagen, Denmark, no year (in Danish)

30. Sørensen JN, van Kuik GAM (2010) General momentum theory for wind turbines at low tip speed ratios. Wind Energy 43:427–448

31. Spera D (ed) Wind turbine technology, 2nd edn. ASME Press, New York, USA

32. Lehmann J, Kühn M (2009) Mit dem wind gegen den wind - das windfahrzeug inventus ventomobil. Physik in unserer Zeit 4:176–181

33. van Kuik G (2013) On the generation of vorticity in rotor & disc flows. In: Proceedings of the ICOWES2013, Copenhagen, Denmark

34. White FM (2005) Viscous fluid flow, 3rd edn. Mc Graw Hill, New York, USA

35. Wilson RE, Lissaman PBS, Walker SN (1976) Aerodynamic performance of wind turbines. Technical report, Corvallis, Oregon, USA

Index